D0900149

High-Throughput Screening in Heterogeneous Catalysis

Edited by
Alfred Hagemeyer, Peter Strasser, Anthony F. Volpe, Jr.

Further Titles of Interest

K. C. Nicolaou, R. Hanko, W. Hartwig (Eds.)

Handbook of Combiatorial Chemistry
From Theory to Application

2002, ISBN 3-527-30509-2

J. Gasteiger (Ed.)

Handbook of Chemoinformatic (4th Edition)

2003, ISBN 3-527-30680-3

H. U. Blaser, E. Schmidt (Eds.)

Asymmetric Catalysis on Industrial Scale

2003, ISBN 3-527-30631-5

B. Cornils (Ed.)

Catalysis A–Z (2nd Edition)

2003, ISBN 3-527-30373-1

High-Throughput Screening in Heterogeneous Catalysis

Edited by
Alfred Hagemeyer, Peter Strasser,
Anthony F. Volpe, Jr.

WILEY-VCH Verlag GmbH & Co. KGaA

Editors:

Dr. Alfred Hagemeyer
Dr. Peter Strasser
Dr. Anthony F. Volpe, Jr.

Symyx Technologies, Inc.
3100 Central Expressway
Santa Clara, CA 95051
USA

Library of Congress Card No.: applied for
A catalogue record for this book is available from the British Library.

Bibliographic information published by Die Deutsche Bibliothek
Die Deutsche Bibliothek lists this publication in the Deutsche Nationalbibliografie; detailed bibliographic data is available in the Internet at <http://dnb.ddb.de>.

Printed in the Federal Republic of Germany
Printed on acid-free paper

Composition K+V Fotosatz GmbH, Beerfelden
Printing betz-druck GmbH, Darmstadt
Bookbinding Litges & Dopf Buchbinderei GmbH, Heppenheim

ISBN 3-527-30814-8

Foreword

Nature has utilized catalysis by employing enzymes since time immemorial. For example, the enzyme urease is one of the most effective catalysts nature has designed. It is capable to catalyze the decomposition of urea 10^{14} times faster than what is possible in the uncatalyzed elimination reaction. In this highly efficient reaction, nature (bacteria) decomposes the urea produced (in the Krebs Cycle) by mammals to ammonia and carbon dioxide for future uptake by plants, which is one of the requirements for their growth and ultimate ingestion by man; thus completing the life cycle. Nature has had time in its favor to ingeniously design its catalytic systems. Man has had a much shorter time span to discover desired catalysts for the production of useful intermediates and products and thereby influencing the course of modern life.

It was Berzelius, who in 1835 first coined the word catalyst, describing it as a substance capable of accelerating reactions without itself being destroyed in the process. Some fifty years later, in 1891 Oswald developed his theory of catalysis putting it on a thermodynamic basis and proposing the concept of microscopic reversibility. Then in the 1920's Mittasch became one of the first to promote the idea of a rational scientific approach to the design of catalytic materials. This noble goal is some eighty years later still a subject of much debate and want.

While nature had time in its favor, it had to work with only a limited number of elements from which to design its key catalytic functionalities (e.g. nickel in urease), and being limited by temperature constraints, man on the other hand has the whole periodic table at his disposal from which to select key catalytic elements and is not bound by any significant temperature constraints. Thus, even in the absence of any unified catalysis theory, the ingenuity of man and his persistence to succeed, and by utilizing working hypotheses and refining them through experimental feed back, has discovered over the past one hundred years several major industrial processes based on heterogeneous catalysis. Among these are: the IG Farben Haber-Bosch process for the synthesis of ammonia utilizing Fe-Al-oxide based catalysts; the Houdry catalytic cracking process using Si-Al-oxides; the UOP Platforming process using Pt-Al-oxides; the SOHIO acrylonitrile process using Bi-Mo-oxides; and the Mobil aromatics isomerization process using ZSM-5 zeolites, to name but a few.

All of these mentioned commercial processes have had a major impact on the industrial world and hence indirectly also on man's improved lifestyle over the years. Currently and for the foreseeable future, the majority of industrial pro-

High-Throughput Screening in Chemical Catalysis
Edited by A. Hagemeyer, P. Strasser, A. F. Volpe, Jr.

ISBN: 3-527-30814-8

cesses for the manufacture of petrochemical intermediates and organic chemicals are based on heterogeneous catalysis.

The proverbial question remains, how can we accelerate the process of catalyst discovery? Looking back, it is obvious that a great amount of ingenuity and chemical intuition, combined with hard work has gone into the discovery process. It is very unlikely that purely empirical research would have led to the important catalytic process discoveries of the past one hundred years. This is particularly true because many of the tools available to the researcher today were not available to him or her some fifty years ago. For example, in the 1950's GC and NMR were not routinely available and product analysis became one of the major bottlenecks in catalytic research. For these reasons it was common then to experiment with 100 g catalyst charges in order to collect sufficient material for product analysis, which consisted often of making appropriate organic derivatives so that they could be analyzed and quantized by chemical and IR analyses. Indeed a tedious and time consuming process that often only allowed one experiment to be performed per week! By the 1960's the analytical methods had improved to the point that many exploratory experiments could be performed on 5 g samples, which decreased to milligram samples by the 1970's and thereafter. Nonetheless, most of the exploratory catalytic research was performed in 1 to 5 g microreactors. Some two or three tube microreactors were used in industrial laboratories by the 1980's, but those were rather rare and single tube microreactors much more common.

Now with the advent of high throughput screening techniques, also known as combinatorial methodologies, pioneered by Symyx Technologies and the subject of this book, the catalyst researcher is provided with an entirely new tool which allows for experimentation to be accelerated by a factor of 10^2 to 10^4. This indeed is remarkable as is amply described and dissected in the different contributions of this book, for it is now possible to combine chemical know-how, experience, intuition and fast experimentation all at once.

It is believed that modern catalytic research has been altered by the advent of combinatorial methodology permanently and that it will behoove the researcher to employ these methods to enhance the process of discovery, thus shortening significantly the time from inception, confirmation and ultimate commercialization of promising catalytic systems. Combinatorial methodology is a tool that all catalyst researchers should avail themselves of in order to enhance their discovery process. It is a tool and in itself not exclusively sufficient for the discovery process, as several contributors to this book attest. Combinatorial methodology needs to be combined with sound chemical knowledge, structural and surface analyses, well thought out working hypotheses, experience, intuition and theory, to achieve optimum results.

Robert K. Grasselli
Center for Catalytic Science and Technology
University of Delaware, USA, and
Institute of Physical Chemistry
University of Munich, Germany

Preface

Rising economic demands for higher efficiency and productivity in Research and Development in the chemical and refining industries have led to the implementation of high throughput, or combinatorial, methods in heterogeneous catalysis. The key drivers have included the desire to reduce the time-to-market for new and optimized catalysts and processes, increased probability of success and better intellectual property protection from the ability to perform many more experiments than in the past, shorter/more projects possible per unit time, and the increased organizational efficiency resulting from improved data storage, access, analysis, and sharing. The number of experiments that can now be screened can be orders of magnitude higher than using traditional methods.

The combinatorial process in catalysis allows the exploration of large and diverse compositional and parameter spaces by utilizing integrated workflows that include software-assisted design of diverse, high-density assemblies, or arrays, of potential catalytic materials (known as "libraries"), and high-throughput synthesis, characterization, and screening techniques that are characterized by the use of robotics and advanced software. The integrated synthesis and screening of a plurality of catalysts in library format has been recognized as an essential factor. Equipment miniaturization and integrated data management systems are also key aspects of successful workflows. The development and implementation of these methods requires the involvement of unconventional engineering and software resources not commonly available at chemical, refining and petrochemical companies where heterogeneous and homogeneous catalysis is practiced.

This book will describe the current state of the art synthesis and screening techniques for high throughput experimentation in chemical catalysis with a focus on technology developed over the last 2–3 years. It will provide an up-to-date overview of the current status and advances that have been made in this rapidly growing field in both academia and industry. The targeted readership is the advanced-level student, the catalytic or solid-state chemist in industrial and academic R&D and engineers specializing in reactor technology, detection schemes and automation.

It has been a great pleasure and distinction for us to assemble a diverse group of distinguished international authors from both academia and industry, each contributing the most up-to-date results and status in their application of high

High-Throughput Screening in Chemical Catalysis
Edited by A. Hagemeyer, P. Strasser, A. F. Volpe, Jr.

ISBN: 3-527-30814-8

throughput methodologies. The book covers reactor technology and integrated synthesis and screening workflows, experiment design and search strategies, detection schemes, and applications to liquid and gas phase heterogeneous catalysis, fuel cell electrocatalysis and homogeneous catalysis. Diverse catalyst systems, such as mixed metal oxides, supported metals, microporous systems/zeolites, mesoporous sieves, as well as diverse chemical transformations (oxidations, dehydrogenations, C1 chemistry, emissions control, petrochemical transformations, hydrogenations, fuel processors) are discussed. We hope that this book will clearly demonstrate the applicability, utility and advantage of combinatorial and high throughput methodologies in chemical catalysis and that the reader will benefit from this snapshot of a rapidly developing field in applied materials science.

We would like to thank all the authors for their contributions. We are thankful to Silvia Lee for her help in putting the book together.

Santa Clara, May 2004

Alfred Hagemeyer
Peter Strasser
Anthony F. Volpe, Jr.

Contents

High-Throughput Screening in Chemical Catalysis
Edited by A. Hagemeyer, P. Strasser, A.F. Volpe, Jr.

ISBN: 3-527-30814-8

List of Contributors

Editors

Alfred Hagemeyer
Symyx Technologies, Inc.
3100 Central Expressway
Santa Clara, CA 95051
USA

Peter Strasser
Symyx Technologies, Inc.
3100 Central Expressway
Santa Clara, CA 95051
USA

Anthony F. Volpe, Jr.
Symyx Technologies, Inc.
3100 Central Expressway
Santa Clara, CA 95051
USA

Contributors

Sam Bergh
Symyx Technologies, Inc.
3100 Central Expressway
Santa Clara, CA 95051
USA

Keith Cendak
Symyx Technologies, Inc.
3100 Central Expressway
Santa Clara, CA 95051
USA

Konstantinos Chondroudis
Symyx Technologies, Inc.
3100 Central Expressway
Santa Clara, CA 95051
USA

Avelino Corma
Instituto de Tecnología Química
Universidad Politecnica de Valencia
(UPV-CSIC)
Av. Los Naranjos s/n
40619 Valencia
Spain

Alexander Cross
hte Aktiengesellschaft
Kurpfalzring 104
69123 Heidelberg
Germany

Dirk Demuth
hte Aktiengesellschaft
Kurpfalzring 104
69123 Heidelberg
Germany

Martin Devenney
Symyx Technologies, Inc.
3100 Central Expressway
Santa Clara, CA 95051
USA

High-Throughput Screening in Chemical Catalysis
Edited by A. Hagemeyer, P. Strasser, A. F. Volpe, Jr.

ISBN: 3-527-30814-8

Klaus Drese
Institut für Mikrotechnik
Mainz GmbH
Carl-Zeiss-Strasse 18–20
55129 Mainz
Germany

Qun Fan
Symyx Technologies, Inc.
3100 Central Expressway
Santa Clara, CA 95051
USA

David Farrusseng
Institute de Recherches
sur la Catalyse – CNRS
2, Av. Albert Einstein
69626 Villeurbanne Cedex
France

Olga Gerlach
hte Aktiengesellschaft
Kurpfalzring 104
69123 Heidelberg
Germany

Daniel Giaquinta
Symyx Technologies, Inc.
3100 Central Expressway
Santa Clara, CA 95051
USA

Sasha Gorer
Symyx Technologies, Inc.
3100 Central Expressway
Santa Clara, CA 95051
USA

Alfred Haas
hte Aktiengesellschaft
Kurpfalzring 104
69123 Heidelberg
Germany

Martin Holeňa
Institut für Angewandte Chemie
Berlin-Adlershof
Richard-Willstätter-Strasse 12
12489 Berlin
Germany

Guido Kirsten
Lehrstuhl für Technische Chemie
Universität des Saarlandes
Im Stadtwald
66125 Saarbrücken-Dudweiler
Germany

Jens Klein
hte Aktiengesellschaft
Kurpfalzring 104
69123 Heidelberg
Germany

Tetsuhiko Kobayashi
National Institute of Advanced
Industrial Science and
Technology (AIST)
Research Institute for
Ubiquitous Energy Devices
1-8-31 Midorigaoka, Ikeda
Osaka 563-8577
Japan

Wilhelm F. Maier
Lehrstuhl für Technische Chemie
Universität des Saarlandes
Im Stadtwald
66125 Saarbrücken-Dudweiler
Germany

Thomas Maschmeyer
The University of Sydney
School of Chemistry, F11
Sydney, NSW, 2006
Australia

Claude Mirodatos
Institute de Recherches
sur la Catalyse – CNRS
2, Av. Albert Einstein
69626 Villeurbanne Cedex
France

Andreas Müller
Institut für Mikrotechnik
Mainz GmbH
Chemical Process Technology
Carl-Zeiss-Strasse 18–20
55129 Mainz
Germany

Vince Murphy
Symyx Technologies, Inc.
3100 Central Expressway
Santa Clara, CA 95051
USA

John M. Newsam
fqubed
6330 Nancy Ridge Drive
Suite 107
San Diego, CA 92121
USA

Paolo P. Pescarmona
Technische Universiteit Delft
Laboratory of Applied Organic
Chemistry and Catalysis
DelftChemTech
Julianalaan 136
2628 BL Delft
The Netherlands

Stephan A. Schunk
hte Aktiengesellschaft
Kurpfalzring 104
69123 Heidelberg
Germany

José M. Serra
Instituto de Tecnología Química
Universidad Politecnica de Valencia
(UPV-CSIC)
Av. Los Naranjos s/n
40619 Valencia
Spain

Andreas Sundermann
hte Aktiengesellschaft
MPJ für Microstrukturphysik
Weinberg 2
06120 Halle
Germany

Wolfram Stichert
hte Aktiengesellschaft
Kurpfalzring 104
69123 Heidelberg
Germany

Wolfgang Strehlau
hte Aktiengesellschaft
Kurpfalzring 104
69123 Heidelberg
Germany

Howard W. Turner
Symyx Technologies, Inc.
3100 Central Expressway
Santa Clara, CA 95051
USA

Jan C. van der Waal
Avantium Technologies B.V.
Zekeringstraat 29
1014 BV, P.O. Box 2915
1000 CX Amsterdam
The Netherlands

Leon G. A. van de Water
Universiteit Utrecht
Department of Inorganic
Chemistry and Catalysis
Sorbonnelaan 16
3584 CA Utrecht
The Netherlands

Uwe Vietze
hte Aktiengesellschaft
Kurpfalzring 104
69123 Heidelberg
Germany

W. Henry Weinberg
Symyx Technologies, Inc.
3100 Central Expressway
Santa Clara, CA 95051
USA

Yusuke Yamada
National Institute of Advanced
Industrial Science and
Technology (AIST)
Research Institute for
Ubiquitous Energy Devices
1-8-31 Midorigaoka, Ikeda
Osaka 563-8577
Japan

Torsten Zech
hte Aktiengesellschaft
Kurpfalzring 104
69123 Heidelberg
Germany

1
Impact of High-Throughput Screening Technologies on Chemical Catalysis

W. Henry Weinberg and Howard W. Turner

1.1
Introduction

Traditional methods of catalyst synthesis and testing are slow and inefficient. Whether one considers either homogeneous or heterogeneous catalyst systems, the root causes for this problem are the same: the ability to predict theoretically the optimal catalyst structure, composition, and synthesis conditions is poor to non-existent, and the catalyst and catalyst formulations are prepared and tested one at a time in a manually intensive fashion. While chemical principles and knowledge of the literature guide the chemist in designing targets and experiments, even in the most well understood areas of catalysis the parameter space that one needs to explore is huge. The result is that the chemist using traditional methods must navigate a complex and unpredictable diversity space with a limited data set to make discoveries, a situation that is perhaps acceptable for the optimization of known systems where the synthesis–structure–property surface is smooth, but unacceptably inefficient for the optimization of systems where this surface is jagged or for the discovery of unprecedented catalytic systems. The reliance on traditional methods of catalyst research leads to a bottleneck in the supply of fundamentally new classes of catalytic materials (which we refer to as "hits") and enormous competition in industrial and academic laboratories during the optimization of the precious few new systems that are discovered and published. Given the inefficiency of traditional optimization of new homogeneous and heterogeneous catalytic systems, the competition in recent decades is not surprising, such as for metallocene olefin polymerization catalysts and the MoVNb-oxide partial oxidation catalysts discussed below.

Until the publication of Schultz [1] in 1995 describing broadly "combinatorial" or "high-throughput" methods for materials discovery, including homogeneous and heterogeneous catalysts, and the creation of Symyx Technologies, the first company dedicated to developing and applying these methods, efforts to improve the efficiency of R&D were largely limited to enhancing analytical techniques to better understand catalyst structure, computational approaches for the prediction of structure–property relationships, improving the precision of laboratory reactor data and its correlation to the commercial process, and the use of statistically de-

High-Throughput Screening in Chemical Catalysis
Edited by A. Hagemeyer, P. Strasser, A. F. Volpe, Jr.

ISBN: 3-527-30814-8

signed experiments to minimize the number of experiments necessary to advance a target. Miniaturization of reactor screening technologies existed before 1995, but efforts to create highly automated and massively parallel workflows did not. Combinatorial methods were developed in the 1980s to improve the efficiency of drug discovery, and these involve various high-dimensional experimental techniques including the use of "split-pool" synthetic procedures where mixtures of thousands of compounds are created on beads, parallel synthesis to produce a collection of related organic compounds known as a "library", and new property-screening technologies that allow the scientist to sort through large collections of potential leads in an efficient and increasingly precise fashion. Philosophically related to these methods, but differing substantially in application, Symyx developed a hierarchical approach to create entire workflows for the synthesis and screening of homogeneous and heterogeneous catalyst libraries. The hierarchical screening philosophies are similar but the physical embodiments of high-throughput drug discovery and high-throughput catalyst discovery and optimization are utterly different. These catalyst discovery workflows can be considered as assembly lines, which allow one to methodically and efficiently generate arrays of new classes of materials in a specific format (a library) designed to maximize the ease of screening for various catalytic transformations, and, then, upon the discovery of "hits" to optimize them efficiently to create "leads" which become commercial development candidates for the targeted process and product. In the following we refer to the use of these workflows as "high-throughput research and development" (HT-R&D) and define this term to include both the synthesis and property evaluation of the catalyst libraries.

The hierarchical workflow shown in Fig. 1.1 illustrates three distinct phases of research leading to commercialization.

The first phase, known as "primary screening", is designed for broadly and efficiently screening a large and diverse set of *families* of materials that logically *could* perform the desired catalytic transformation. It is during this phase that "hits",

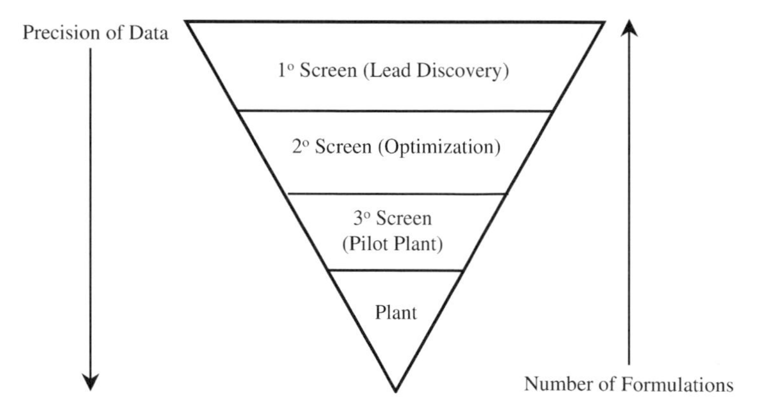

Fig. 1.1 Hierarchical high-throughput R&D workflow.

truly new classes of materials that show promise for a specific catalytic transformation, can be efficiently discovered, and a key bottleneck in the R&D process can be eliminated. Predictably, it is this phase with which the traditional catalytic chemist is the most uncomfortable. All well trained chemists have been taught rigor in the synthesis and characterization of the materials they produce before testing for the desired application. They have also been taught to maximize the quality and precision of the catalytic testing data by using reactors that have a high level of process control and that are as close as possible to the "real" process in which the material will be commercialized. These are all good things and are necessary, but only at a later stage of the hierarchical workflow where optimization and commercial development occur. While we seek to maximize the quality of data and mimic the "real" process conditions during primary screening, the sheer volume of experiments, often thousands per day, makes it extremely difficult to obtain conventional laboratory quality data. Fortunately, this does not pose a problem. In some cases, to get the throughput necessary for a program, we do not screen for the exact property that we seek, but rather screen for an easier and faster observable that represents a necessary but not sufficient property of the targeted material [2]. We trade precision for throughput in a rational way that increases the probability of success and shortens time to commercialization. The key in this enterprise is to create a validated primary screen (i.e., a screen where we have proven that we can "rediscover" state-of-art catalysts and observe known trends). In a primary screen we seek qualitative trends in the data to eliminate families of materials from diverse libraries and to identify hits that have the *potential* to become a lead, i.e., materials that warrant testing in a secondary screen. We design our primary screening technologies to minimize both false negatives and false positives, the latter of which waste time and money in the slower and more precise hit validation phase. To the uninitiated the issue of false negatives is usually the biggest concern; methods of minimizing the risk of "missing a hit" have been well described for homogeneous catalyst workflows [2] and Chapter 3 discusses this issue as it relates to primary screening for heterogeneous catalysts.

Secondary screening plays two roles. First, it validates or eliminates hits generated from the primary screen. As mentioned above, often the primary screen seeks to identify a property that is easy to measure in a high-throughput fashion that is a necessary but not sufficient condition for the hit to be active and/or selective for the targeted chemical transformation. Here, the secondary screen is the first test for the "real" property, and often hits fall out of the program at this stage. In a second role for secondary screening, once a hit, identified by primary screening or identified in the literature, has been validated, secondary screening tools are used to optimize the hit to create a "lead" material. In secondary screening, rigorous catalyst synthesis procedures are important, and, since most optimizations require multiple modifications with small improvements in performance at each successful step, the data quality and precision need to be on a par with a typical laboratory reactor. Therefore, automated laboratory-scale synthetic techniques and highly parallel reactor systems for various processes have been developed. Some of the systems invented at Symyx Technologies are described in Chapter 3. This technology has evolved to the point that the quality of the data obtained during the secondary screening stage allows for the synthesis and screening of hundreds of catalytic materials per week with data quality that is equivalent or superior to that obtained using conventional laboratory technologies.

Optimized leads identified in the secondary screening phase of a HT-R&D program are then taken to the tertiary screening phase to generate commercial development candidates. This usually takes the form of a conventional mini- or pilot-plant.

Finally, the workflow must include advanced experimental design, library design, data management, database, and data mining software. When taken together and executed properly, HT-R&D programs generate vast data sets with concomitant knowledge derived therefrom, shorten time scales for completing a program, increase the probability of success of the program, and increase the strength of an intellectual property portfolio. The talented synthetic chemists should not fear that HT-R&D technologies will render them less important because robots do the chemistry. The opposite is true for several reasons. First, in a HT-R&D environment the number of variables in diversity space that one is able to address is so large that insightful and creative thinking based on sound chemical knowledge and principles is necessary to reduce the number of experiments to a realistic level. This is no different than in traditional laboratories; it is just that one can explore one's concepts more rapidly and more thoroughly. Secondly, since the number of projects one can initiate and conclude in one's career is substantially larger, the chemist will be constantly challenged to create new concepts for new problems. Finally, the increase in R&D efficiency due to HT-R&D technologies will result in a lower unit cost for research and increase the desire for industry to invest in the development of advanced technologies.

To illustrate the power of integrated workflows consisting of primary and secondary screening technologies we briefly describe two examples from our own laboratories.

1.2
Application of HT-R&D Methods in Heterogeneous Catalysis

The discovery of catalytic systems and processes that selectively convert unsaturated hydrocarbons such as ethane and propane into higher value chemicals, such as ethylene, acrylic acid, and acrylonitrile, is a key R&D goal within the chemical and petrochemical industries. The dominant process targeted for these advances involves the use of heterogeneous catalysis in either fixed bed or fluidized bed processes. The chemical challenge is great due to the difficulty of selectively activating saturated hydrocarbons at low temperature, the dominance of inherently unselective free radical pathways at high temperatures, and the fact that the desired products are often more easily oxidized than the saturated hydrocarbon starting material, leading to low selectivity at commercially viable conversions.

In 1978 Union Carbide scientists reported the discovery and optimization of oxidative dehydrogenation catalysts for converting ethane into ethylene that were based on mixed metal oxides consisting of Mo, V, and Nb [3]. This study described a systematic evaluation of the effect of composition (i.e., the ratio of the three metals in the tertiary composition) on performance as measured by space–time yield. The composition $Mo_{0.72}V_{0.26}Nb_{0.02}O_x$ was reported to be optimum.

More than a decade after the publication of the MoVNb catalyst system, scientists at Mitsubishi Chemical reported that modifying this family of mixed metal oxides with Te produced a catalyst for the amoxidation of propane to acrylonitrile [4] and the oxidation of propane to acrylic acid [5]. Modification of the Union Carbide catalyst system with Te was probably not a random choice as it is a known propylene activator [5 b] and the molybdate phase $TeMoO_x$ oxidizes propylene into acrolein and ammoxidizes propylene to acrylonitrile [6], a key intermediate in the commercial production of acrylic acid using Mo-based oxides. Significant efforts to optimize this and related mixed metal oxides continues for the production of both acrylic acid and acrylonitrile, with the main participants being Asahi, Rohm & Hass, BASF, and BP.

In 1998 scientists at Hoechst reported that the addition of Pd to the MoVNb ethane dehydrogenation catalyst enabled the efficient production of acetic acid from ethane [7]. Doping of this known ethane dehydrogenation catalyst with Pd was probably not random, but predicted on the basis of the classical Wacker catalysis.

Fig. 1.2 summarizes the lineage of discoveries based on the MoVNb "hit" published in 1978 and is included to emphasize the importance of the discovery of new starting points in chemical catalysis.

Symyx entered this competition in 1997 in collaboration with Hoechst with the goal of creating and validating primary and secondary synthesis and screening technologies and the use of this workflow to broadly explore mixed metal oxide compositions so as to discover and optimize new "hits". The initial goal was a 10-fold increase in the space–time yield relative to the state-of-the-art MoVNb system for the ethane oxidative dehydrogenation reaction to ethylene.

In the workflow used in this program (Fig. 1.3) *primary* screening is carried out in "wafer" format. The libraries are synthesized from soluble metal precursors using specialized library design software [8] and liquid-dispensing robots in a ter-

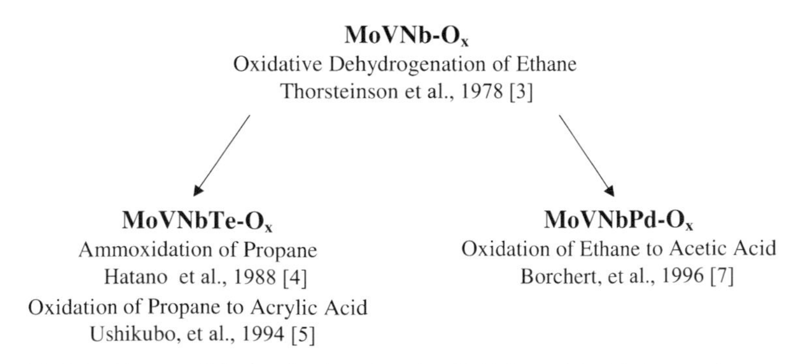

Fig. 1.2 Lineage of discoveries based on the MoVNb partial oxidation catalyst.

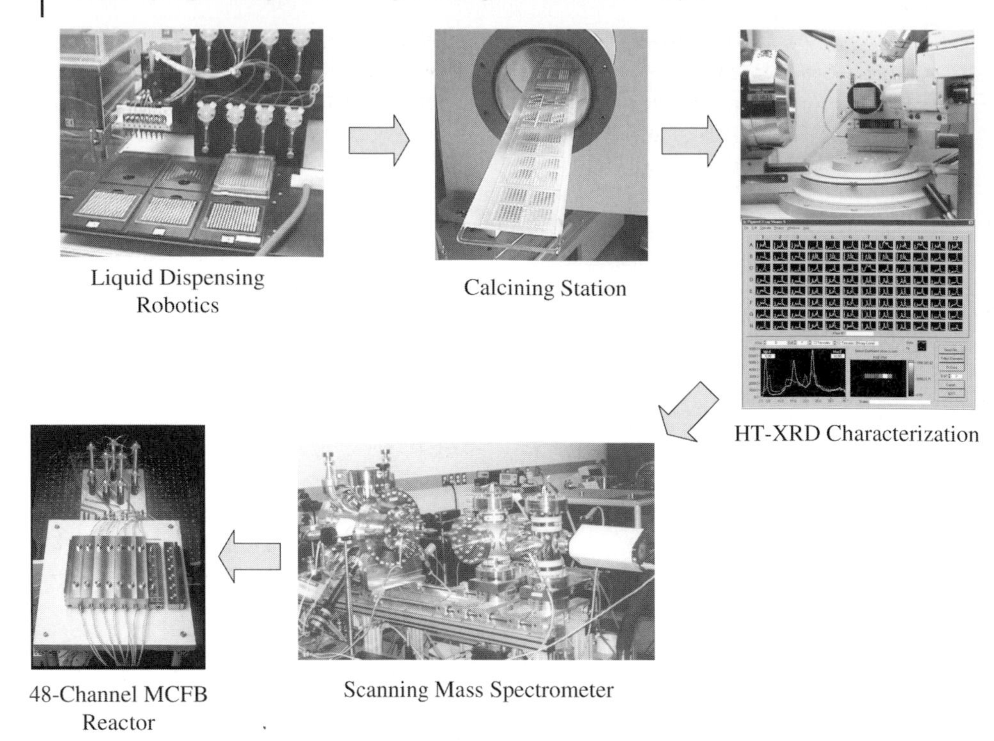

Fig. 1.3 HT-R&D workflow used in the oxidative dehydrogenation of the ethane program at Symyx Technologies.

tiary phase diagram format. The automated process involves creating arrays of mixtures of metal precursors under conditions where the mixtures remain in solution and then depositing 1–2 μl aliquots onto a quartz wafer that has been etched to create an ordered array of micro-wells. Multiple "daughter" wafers are prepared and then calcined under various conditions to minimize the chances of false negatives and/or to help identify optimal processing conditions for bulk catalyst synthesis. Variable space for primary screening includes elemental composition, choice of support (starting material wafers with microgram quantities of powders derived from any commercial or proprietary supports can be produced using similar robotics), choice of metal precursor, and calcining conditions. The wafer-based libraries can be characterized by scanning XRD using commercially available equipment and proprietary methods [9] when appropriate.

The performance of each library member can be screened by several proprietary primary screening technologies [10], including the scanning mass spectrometer (SMS) technology shown in Fig. 1.3. The wafer is placed on a motion control stage capable of positioning a single library element approximately 100 microns below a probe that flows the feed of the starting material over the catalyst surface and removes reaction products to a mass spectrometer and/or other detector technologies. The individual catalyst elements are heated to a preset reaction temperature using an IR laser from the backside of the quartz wafer and an IR camera is used to monitor temperature. The power of the laser is adjusted to control temperature. This is a rapid serial method requiring approximately 2–3 min per sample. The system is fully automated and, after the library is placed on the motion control stage and the experiment initiated, the entire library can be screened unattended.

Several proprietary *secondary* screening technologies for fixed bed processes have been described [11], one of which is depicted in Fig. 1.3. The 48-channel reactor includes a single feed system that supplies reactants to a set of flow restrictors. The flow restrictors divide the flows evenly among the reactors. The back pressure created by the flow restrictors is designed to be large compared with any pressure drop caused by the catalyst bed or downstream plumbing to ensure that an even flow occurs through each reactor. The multichannel fixed bed reactor (MCFB) shown in Fig. 1.3 and used in the ethane oxidative dehydrogenation program is in a 6×8 rectangular array format and can accommodate up to 100 mg of solid catalyst in each reactor.

A more detailed description of heterogeneous catalyst library synthesis (primary and secondary), primary screening, and secondary screening technologies is given in Chapter 3 and references therein.

The first step in this or any new HT-R&D program is to validate the workflow. In the oxidative dehydrogenation of ethane program, primary screening validation was accomplished by "rediscovering" the trends and optimum composition for the MoVNb catalyst system described in 1978 by Union Carbide. Duplicate 66-member libraries (11 members on each diagonal of the phase diagram) were synthesized using soluble Mo, V, and Nb precursors. The libraries were calcined and then characterized by scanning XRD, confirming that the thick film library elements had similar phase composition to bulk samples prepared using traditional methods. The libraries were placed on the SMS motion control stage and screened in a rapid serial mode using a mass spectrometer to quantify CO_x and a laser pump–probe measurement to quantify ethylene. These detectors were calibrated and together allowed the ranking of both activity and selectivity at low conversion.

Fig. 1.4 compares, in topological format, the space–time yield versus composition for the data presented in the 1978 Union Carbide publication and the activity rankings observed in the SMS in an experiment that took less than 4 h, most of which was unattended. The correlation is remarkable. The primary synthesis and screening components of the workflow were thus validated. "Hit" criteria were established that involved ranking the yield of the reaction over the various catalysts (the activity figure of merit multiplied by selectivity). The "hit" criteria performance bar increased as the discovery program evolved and improved systems were discovered.

The 48-channel MCFB reactor and the catalyst synthesis workflow components were similarly validated in experiments where bulk samples were prepared in library format, screened in the array format, and the data compared with known examples. This part of the workflow was used for initial hit validation and to opti-

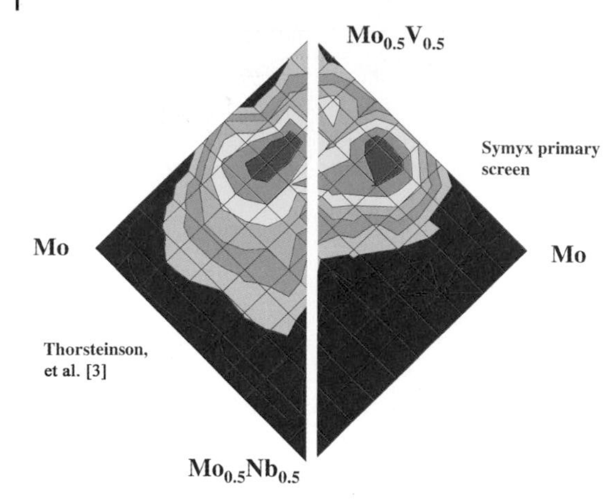

Fig. 1.4 Comparison of STY with the composition as measured by Thorsteinson et al. [3] using traditional methods and by Symyx scientists using wafer-based primary screening technologies where the innermost enclosed areas represent the highest STY.

mize those systems that exhibited acceptable performance under commercially realistic conditions.

The discovery program began by broadly screening ternary mixed-metal oxide compositions. Although many catalyst compositions can be prepared and screened using this workflow, testing of all possible ternary mixed-metal oxide compositions is impractical. The combination of all possible ternary combinations (assuming 70 metal oxides) with 10% gradient steps results in millions of unique compositions. This number increases by many factors as one explores multiple library processing conditions and alternate metal precursor options. Thus, priority decisions based on sound chemical knowledge were made. Since the focus of the research was the low temperature partial oxidation of ethane to ethylene the scope of the search could be reduced by the assumption that the ternary mixed-metal oxides should contain at least two different metal oxides that can be reduced by hydrocarbons and their reduced forms oxidized by molecular oxygen at low temperature. The third metal oxide component was generally designed to act as a matrix or stabilizer and was not required to be redox-active under catalytic conditions. In this way, synergy between different redox-active metals could be explored. With these assumptions the diversity space was narrowed to about 100 000 composition and processing experiments.

Nickel-based systems containing certain other metals such as Ta, Nb and Ce emerged as lead candidates. The best Ni-based catalysts showed activities as measured by SMS that were 50–100× that of the best MoVNb systems. In secondary screening the Ni-based systems distinguished themselves from the Mo-based systems in terms of both space–time yield and the unique and highly desired property of having a flat selectivity versus conversion (i.e., selectivity vs. temperature) relationship relative to the state-of-the-art MoNbV systems. Fig. 1.5 shows a timeline of how the performance characteristics of the best Ni-based systems evolved in terms of activity measured in the primary screen, space–time yield measured in the MCFB secondary screen, and in a ternary screen carried out in pilot plant

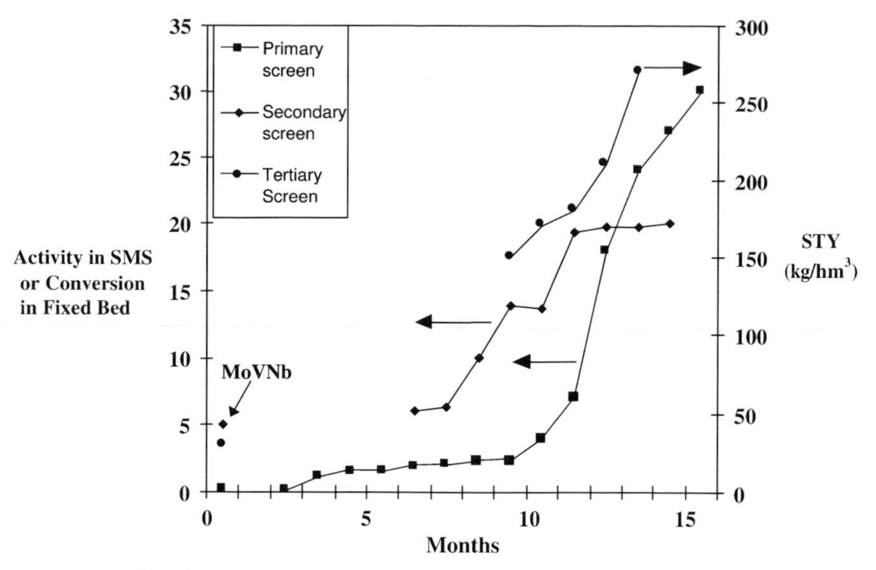

Fig. 1.5 Timeline showing progress in primary, secondary, and tertiary screening experiments.

studies on a 20 g scale. In less than 18 months, a small team of scientists were able to broadly search composition and processing space and develop a new mixed-metal oxide catalyst system with space–time yields ten-fold greater than the previous state-of-the-art system.

In early 1997, Schuurman, et al. reported Ni-based catalysts for the oxidative dehydrogenation of ethane to ethylene [12], but with only a limited ability to explore composition space the results were not compelling, and the real potential for Ni-based systems was missed. Tab. 1.1 shows secondary screening performance data for pure Ni oxide and binary and ternary Ni compositions containing Ta and Nb. The pure Ni catalyst is poor in terms of both conversion (11%) and selectivity (54%). The Ni catalyst containing 12% Ta was essentially the same (12% conversion and 55% selectivity). Increasing the Ta concentration in the binary to 38%

Tab. 1.1 Comparison of performance data of various Ni-based catalysts tested in secondary screening.

Catalyst composition	*Ethane conversion (%)*	*Ethylene selectivity (5)*
Ni	10.5	53.6
$Ni_{0.88}Ta_{0.12}$	12.0	54.9
$Ni_{0.62}Ta_{0.38}$	18.6	84.4
$Ni_{0.89}Nb_{0.11}$	17.7	81.6
$Ni_{0.63}Nb_{0.37}$	19.0	84.7
$Ni_{0.70}Ta_{0.10}Nb_{0.20}$	20.0	85.2
$Ni_{0.62}Ta_{0.10}Nb_{0.28}$	20.5	86.2

$O_2/C_2H_6/N_2 = 0.088/0.42/0.54$ sccm at 300 °C with ca. 50 mg catalysts.

nearly doubled the activity (19% conversion) and increased the selectivity from 55% to 84%. Not observing a positive effect in the first Ta experiment in a conventional program would likely have discouraged the scientist from further exploration and unless the scientist's first experiment was in the region of composition space where the dramatic effect is observed, the catalyst would probably have been missed. This story is documented in numerous patents and publications [13], and this is a classic example of how a validated high-throughput workflow including a primary screening can compress the discovery timeline while leading to the discovery of completely new catalyst systems and starting points for future advances.

1.3 Application of HT-R&D Methods in Homogeneous Catalysis

The second example of the power of integrated workflows consisting of primary and secondary screening is in homogeneous catalysis – specifically single-site catalysts for the production of polyolefins from ethylene, propylene, and other 1-olefins. Approximately 50% of all plastics and elastomers produced today are polyolefins. The dominant process for producing polyolefins involves the use of catalysts at relatively low pressures. This technology began in 1955 with the discovery of Ziegler-Natta (ZN) catalysts derived from mixtures of transition metal salts such as $TiCl_3$ and alkyl aluminum reagents that polymerized ethylene at low pressure [14]. In 1956 the stereoselective polymerization of propylene to produce highly crystalline isotactic polypropylene (i-PP) was reported by Natta [15]. Today, worldwide capacity for the production of i-PP exceeds 65 billion pounds.

Traditional ZN catalysts are typically complex heterogeneous systems, consisting of multiple active sites each of which produces polymers and copolymers with different structure (e.g., tacticity, molecular weight, composition). The result is the production of polymer blends. Controlling blend composition through modification of the heterogeneous catalyst surface was challenging and dominated R&D in this area for decades.

The first well defined and structurally characterized olefin polymerization catalysts were reported in 1982. The two systems, one a bis-cyclopentadienyl lutetium alkyl complex discovered at Dupont [16] and the other a tantalum alkylidene hydride complex discovered at MIT [17], were significant advances in terms of understanding mechanisms, guiding future catalyst design, and demonstrating that it was possible to generate families of well-defined homogeneous catalyst systems. They did not, however, possess commercially viable performance due to their high cost, low activity, and limited comonomer incorporation.

In early 1981 Sinn and Kaminsky reported that mixtures of bis-cyclopentadienylzirconiumdimethyl (Cp_2ZrMe_2, a "metallocene") and methylalumoxane (MAO: a complex and ill-defined mixture of oxo-bridged alkylaluminum oligomers produced by the partial hydrolysis of trimethylaluminum) produced a highly active and long-lived olefin polymerization catalyst [18]. The unique properties of this catalyst system included extremely high activity, long lifetime, and the production of polymers with a narrow molecular weight distribution (MWD) of 2.0, the theoretical MWD for a single active site catalytically producing polymer chains under steady-state process conditions. The complex nature of the MAO cocatalyst made characterization of the structure of the active site difficult so that it was not until 1997 that the mechanism of activation and the structure of the active site were reported. Fig. 1.6 shows the structure of the first metallocene activated by MAO, the structure of the active site as reported by the Exxon Chemical Company in 1997 [19, 20] in patents focusing on new classes of well-defined MAO cocatalyst alternatives, and the mechanism of polymerization.

In 1983 the Exxon Chemical Company reported the discovery that the properties of the metallocene catalyst (e.g., molecular weight, comonomer selectivity, and activity) could be controlled by the choice of metal (Ti, Zr, or Hf) and by substituting and/or bridging the two cyclopentadienyl rings on the catalyst precursor [21].

By the mid-1980s every major polyolefin producer was evaluating metallocene single-site catalysts for possible commercial applications. These efforts involved taking advantage of the ability to tune the properties of the catalyst by modification of the Cp-ligand structure, the development of advanced polyolefin products by taking advantage of the single-site behavior of the catalyst systems that allow, for example, enhanced control of polymer microstructure, MWD, and blend composition, and developing methods of supporting these systems to produce heterogeneous single-site catalysts that could be used in slurry and fluidized bed process technologies. These efforts continue today and polyolefin producers have commercialized a wide variety of high value ethylene and propylene-based products in all of the major commercial processes. The field of single-site catalysis has captured the interest of some of the best scientists in academia and represents one of the

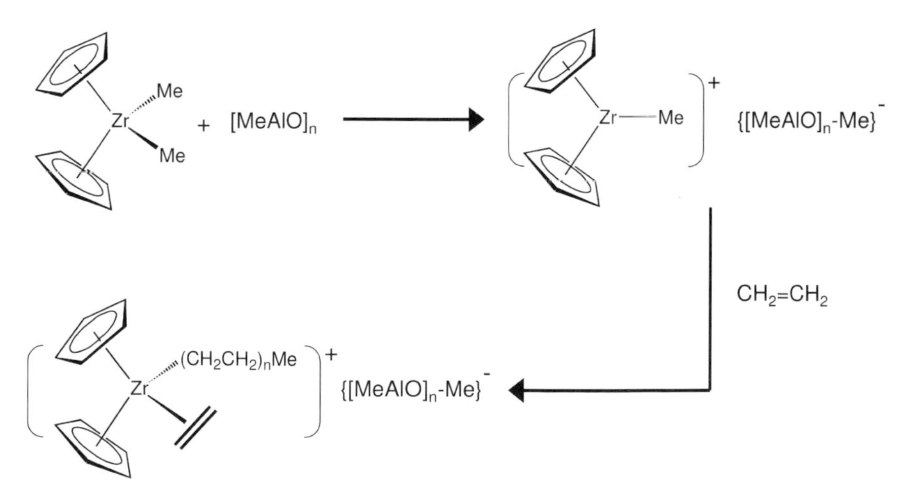

Fig. 1.6 Metallocene activation by MAO, structure of the active site, and ethylene polymerization.

most significant R&D efforts in the chemical industry. The history and evolution of this remarkable story is chronicled in numerous reviews [22].

One subset of this field is the development of metallocene catalysts for the production of isotactic polypropylene (i-PP), a high-melting semicrystalline plastic. The i-PP is stereoregular: the alternating chiral centers on the backbone of the polymer are the same enantiomer (all *R* or all *S*). The high crystallinity of i-PP requires high stereoregularity. In theory, one way to polymerize the prochiral propylene monomer to form i-PP would be the use of a chiral catalyst. Wild and Brintzinger reported the synthesis of chiral metallocenes suitable for this purpose in 1982 [23]. Ewen of the Exxon Chemical Company reproduced the synthesis of one of the chiral systems, the titanium analogue, and reported that i-PP could be produced by the homogeneous chiral titanocene/MAO catalyst in 1984 [24].

The titanium-based system reported by Ewen was unimpressive in terms of performance (low activity and low melting point), but it ignited intense competition within the polypropylene industry to develop a commercially viable catalyst system. The target was to develop a system that could produce high-melting (160–165 °C due to low stereo- and regio-errors), high molecular weight i-PP, with high activity using a single-site catalyst that retained these properties when supported on a commercial carrier such as silica and used under commercially viable conditions in a slurry or gas phase process. Highlights of the evolution of this technology are shown in Fig. 1.7.

Natta and Breslow reported the first metallocene catalyst in the mid-1950s by using Et_2AlCl_2 as the cocatalyst [25]. The system had very low activity and life-

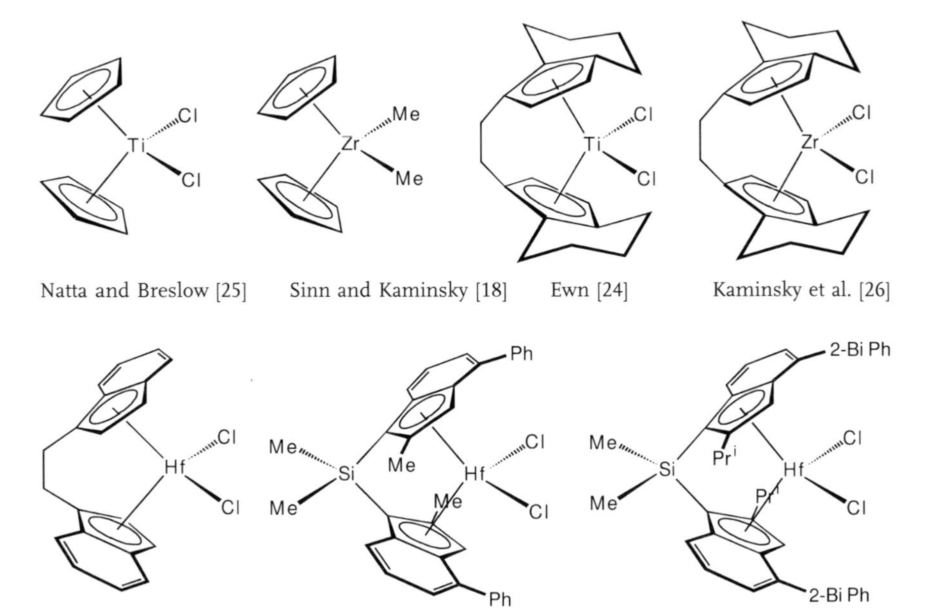

Fig. 1.7 Highlights in the evolution of metallocene-based i-PP catalysts (Me = methyl, Ph = phenyl, and 2-BiPh = a biphenyl group in the 2-position of the indene ring).

time. It was not until the discovery of the MAO cocatalyst and Sinn and Kaminsky's report in 1981 that metallocene catalysts began to be taken seriously in industrial laboratories. One year after the publication of the chiral i-PP Ti-based metallocene catalyst, Kaminsky et al. reported the Zr-analogue [26]. The zirconium analogue had vastly superior activity and lifetime, but was deficient in terms of both molecular weight capability and melting point. Two years later Exxon Chemical [27] and Fina [28] filed patent applications that included the discovery that Hf-based systems showed significant improvements in the molecular weight, but the melting point remained deficient. Having exhausted Group IVB metal options, researchers in several laboratories began to modify the structure of the indene-based ligand system responsible for chirality at the metal center. The organic and organometallic chemistry associated with this effort was challenging. In 1995 Küber et al. of Hoechst reported the 2-methyl-4-phenyl di-substituted metallocene systems [29] shown in Fig. 1.7. These systems allowed for the production of high-melting, high molecular weight i-PP at high activity in solution at 70 °C. It is clear from the patent literature and presentations by the Hoechst scientists that this effort required the synthesis of numerous derivatives (rumored to be hundreds) to identify the optimal substitution pattern on the ligand system. Unfortunately, when these systems were supported on silica, the melting point of the product dropped. In 2002 Hart et al. reported that the 2-isopropyl,4-biphenyl analog to the Hoechst systems retained a high melting point after being supported on silica and screening in a slurry process [30]. The first commercial production trials of i-PP using these advanced systems began in 1995 [31].

Symyx entered the highly competitive field of "single-site" olefin polymerization in 1997 and developed an efficient primary and secondary screening workflow for the synthesis and screening of ligand–metal combinations for the discovery of new, non-Cp catalyst systems. This use of HT-R&D methods in this field requires an even more dramatic paradigm shift for the chemist than in the field of heterogeneous catalysis since in the latter case the chemist is often dealing with complex and poorly defined chemical systems. The organometallic chemist who executes a single-site catalyst discovery program has been trained, for good reasons, to prepare the ligand–metal complex, purify it, and fully characterize it prior to screening for performance. The synthetic procedures for preparing one metallocene often do not work well for other derivatives, and without characterizing and purifying each new ligand–metal complex one runs the risk of screening mixtures and generating data that is difficult to interpret and/or reproduce. The HT-R&D methods developed at Symyx involve the in situ generation of the ligand–metal complex and screening without characterization or purification. We rely on a number of methods that circumvent these issues, including classical synthetic studies to determine multiple viable chemical routes for the synthesis of new classes of ligand–metal complexes, the use of multiple derivatives of each new ligand class, and the use of multiple ligand–metal complexation conditions and activation routes for each ligand in our library designs. The result of this approach is that each ligand and each ligand class is screened under a sufficient number of conditions that the potential for false negatives is extremely low. This

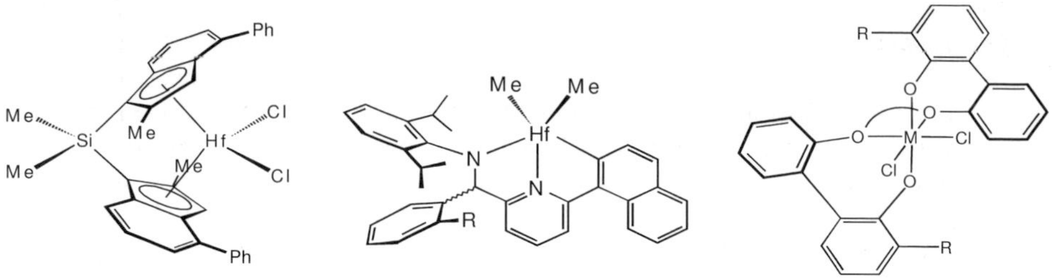

Küber et al. [29] Boussie et al. [34] Boussie et al. [35]

Fig. 1.8 Structural comparisons of i-PP metallocene catalyst precursors and two recently reported new classes of competitive single-site systems discovered using HT-R&D.

approach coupled with efficient and precise secondary screening technologies provides an extremely powerful workflow that allows the team to sort through hundreds of ligand classes and thousands of ligands looking for promising starting points for catalyst systems with targeted properties. The details of an important subset of these workflows and methods are discussed in Chapter 12 and in other publications [32, 33].

Perhaps the most dramatic demonstration of the power of this platform is in the area of non-Cp catalysts for the production of i-PP. As shown in Fig. 1.7, it took 40 years to develop a commercially viable i-PP catalyst from the time of the first metallocene "hit" of Natta and Breslow, and almost 15 years of intense research and competition from discovery of MAO-based metallocene catalysts in 1981. In less than three years the Symyx team screened over 1500 ligands and executed nearly 100 000 experiments targeting non-Cp catalysts capable of producing high molecular weight, high-melting i-PP at high temperatures. The result was the discovery of two completely different non-Cp ligand Group IVB metal catalyst systems that are competitive with some of the most advanced metallocene systems known [34, 35]. The generic structures of these systems and the chiral metallocenes are compared in Fig. 1.8 to help highlight the differences between the three ligand systems responsible for the chirality needed to produce crystalline polypropylene.

It is interesting to imagine how much more quickly and differently the field of olefin polymerization catalysis and the industry that it supports would have evolved had the HT-R&D technologies been available in the mid-1950s when Ziegler and Natta made their historic discovery. The specifics are not possible to determine, but the generalities are that the number of "hit" discoveries would have increased much more rapidly, the rate of maturation of catalyst systems and process technologies would have been greatly enhanced, product quality improvement would have been more rapid, metallocenes would likely not dominate single-site catalysis in commercial processes today, and new bottlenecks in the R&D process would have been identified and new technologies to remove them would have been advanced.

1.4 Conclusions

The advent of HT-R&D catalyst discovery and optimization has begun a revolution in the way R&D is practiced in this industry. To implement such a HT-R&D discovery program requires advanced tools as described above along with advanced data collection, data management, data mining, and machine control software. Thus, the skill set needed to create and use such an infrastructure include synthetic chemistry, physical chemistry, physics, analytical chemistry, chemical engineering, mechanical engineering, electrical engineering, software engineering and database programming expertise. These are not the demographics of a typical R&D organization in the chemical industry. Therefore, the adoption of HT-R&D methods within the industry will largely occur through the purchase of advanced tools and software and through collaborations with HT-R&D technology providers.

The field of HT-R&D technologies as applied to catalysis is young and it will continue to evolve and become adopted within industry due to the demonstrated effectiveness and the competitive advantage it provides. The impact that this young field has already had can be measured by many metrics including the amount of money being spent by large chemical and petrochemical companies on R&D collaborations with HT-R&D technology providers [36], the fact that essentially every major member of the chemical industry has initiated HT-R&D programs internally, externally, or both, the number of conferences, papers, and patents that are dedicated to this subject, and especially the technical success leading to significant commercial development [37, 38]. Specific examples of the implementation of these methodologies are described in the following chapters.

1.5 References

1 X. Xiang, X. Sun, G. Briceno, Y. Lou, K. Wang, H. Chang, W. Wallace-Freedman, S. Chen, and P. Schultz, *Science,* 268 (1995) 1738.

2 T.R. Boussie, G.M. Diamond, C. Goh, K.A. Hall, A.M. LaPointe, M. Leclerc, C. Lund, V. Murphy, J.A.W. Shoemaker, U. Tracht, H. Turner, J. Zhang, T. Uno, R.K. Rosen, and J.C. Stevens, *J. Am. Chem. Soc.,* 125 (2003) 4306.

3 E. Thorsteinson, T. Wilson, F. Young, and P. Kasai, *J. Catal.,* 52 (1978) 116.

4 M. Hatano, and A. Kayo, A10318295 filed in 1988 and assigned to Mitsubishi Kasei.

5 (a) T. Ushikubo, H. Nakamura, Y. Koyasu, and S. Wajiki, 0608838 A2 filed in 1994 and assigned to Mitsubishi Kasei. (b) R.K. Grasselli, J.D. Burrington, D.J. Buttrey, P. De Santo, C.G. Lugmair, A.F. Volpe, Jr., and T. Weingrand, *Topics in Catal.,* 23 (2003) 5. (c) R.K. Grasselli, *J. Chem. Ed.,* 63 (1968) 216.

6 (a) J.C.J. Bart and N. Giordano, *J. Catal.,* 64 (1980) 356. (b) E. Fedevich, E. Krylova, V. Zhiznevskii, M. Grobova, and K. Semenova, *Khim. Promyshlennost,* 6 (1980) 335.

7 H. Borchert, and U. Dingerdissen, WO 98/05619 (1998).

8 The library design software, Library Studio®, is part of a suite of software applications, Renaissance®, developed at Symyx to design experiments, control robotics, integrate and control analytical devices, and capture and search experimental designs, methods, and data.

9 D. Hajduk, J. Bennett, R. Jain, US 6 371 640 B1 assigned to Symyx Technologies filed October 3, 2000 and a continuation of the now abandoned application No. 09/215 417 filed on December 18, 1998.

10 See Chapter 3 and references therein.

11 For example, US 6149882 (2000), US 6410331 (2002), and EP 1001846 (2002), each assigned to Symyx Technologies with additional patents pending.

12 Y. Schuurman, V. Ducarme, T. Chen, W. Li, C. Mirodatos, and G. Martin, *Appl. Catal., A*, 163 (1997) 227.

13 See, for example, Bergh et al., *Top. Catal.*, 23 (2003) 65, and references therein.

14 K. Ziegler, E. Holzkamp, H. Breil, and H. Martin, *Angew. Chem.*, 67 (1955) 541.

15 G. Natta, *Angew. Chem.*, 68 (1956) 393.

16 P. Watson, *J. Am. Chem. Soc.*, 104 (1982) 337.

17 H.W. Turner, and R.R. Schrock, *J. Am. Chem. Soc.*, 104 (1982) 2331–2333.

18 H. Sinn, W. Kaminsky, H. Vollmer, and W. Rudiger, US 4 404 344 assigned to BASF and filed February 27, 1981.

19 H.W. Turner, EP 0 277 004 A1 assigned to Exxon Chemical and filed in January 30, 1987.

20 H.W. Turner, and G. Hlatky, EP 0 277 003 A1 assigned to Exxon Chemical and filed in January 30, 1987.

21 H.C. Welborn, and J.A. Ewen, US 5 324 800 assigned to Exxon Chemical and a continuation of the originally filed specification Ser. No. 501 688 filed June 6, 1983.

22 For selected reviews on metallocenes, see R.F. Jordan, *Adv. Organomet. Chem.*, 32 (1991) 325–387; N. Kashiwa, and J.-I. Imuta, *Catal. Surveys Jpn.*, 1 (1997) 125–142; A.L. McKnight, and R.M. Waymouth, *Chem. Rev.*, 98 (1998) 2587–2598; H.H. Brintzinger, D. Fischer, R. Mülhaupt, B. Rieger, and R.M. Waymouth, *Angew. Chem. Int. Ed.*, 34 (1995) 1143–1170; M. Bochmann, *J. Chem. Soc., Dalton Trans.*, (1996) 255–270; H.G. Alt, A. Köppl, *Chem. Rev.*, 100 (2000) 1205–1221.

23 F. Wild, L. Zsolnai, G. Huttner, and H. Brintzinger, *J. Organometal. Chem.*, 232 (1982) 233.

24 J. Ewen, *J. Am. Chem. Soc.*, 106 (1984) 6355.

25 D. Breslow and N. Newberg, *J. Am. Chem. Soc.*, 79 (1957) 5072.

26 W. Kaminsky, K. Kulper, H. Brintzinger, and W. Wild, *Angew. Chem.*, 24 (1985) 507.

27 H.W. Turner and G.G. Hlatky, EP 0 558 158 A1 assigned to Exxon Chemical and filed January 1, 1988.

28 J. A. Ewen, EP 0 284 707 B1 assigned to Fina and filed September 9, 1987.

29 F. Küber, B. Bachmann, W. Spaleck, A. Winter, and J. Rohrmann, US 5 786 432 assigned to Targor GmbH and filed June 7, 1995.

30 J. Hart, W. Haygood, T. Burkhardt, and R. Li, WO 03/002583 A2 assigned to ExxonMobil Chemical and filed June 14, 2002.

31 F. Küber, M. Aulbach, B. Bachmann, R. Klein, W. Spaleck, and A. Winter, Presentation at MetCon '95, May 17–19, 1995.

32 T.R. Boussie, G.M. Diamond, C. Goh, K.A. Hall, A.M. LaPointe, M. Leclerc, C. Lund, V. Murphy, J.A.W. Shoemaker, U. Tracht, H. Turner, J. Zhang, T. Uno, R.K. Rosen, and J.C. Stevens, *J. Am. Chem. Soc.*, 125 (2003) 4306–4317.

33 H.W. Turner, L. VanErden, G.C. Dales, A. Safir, and R.B. Nielsen, US 6 508 984 B1, filed July 19, 2000 and assigned to Symyx Technologies, Inc.; H.W. Turner, L. VanErden, G.C. Dales, A. Safir, and R.B. Nielsen, US 6 653 138 B1, filed November 28, 2000 and assigned to Symyx Technologies, Inc.

34 T. Boussie, G. Diamond, C. Goh, K. Hall, A. LaPointe, M. Leclerc, and V. Murphy, PCT WO 02/38628 A2 assigned to Symyx Technologies and filed November 6, 2001.

35 T. Boussie, O. Brummer, G. Diamond, C. Goh, A. LaPointe, M. Leclerc, and J. Shoemaker, PCT WO 03/091262 A1 assigned to Symyx Technologies and filed April 24, 2002.
36 For example, the 5-year, $200M Exxon-Mobil-Symyx Alliance announced in July 2003. For details see www.symyx.com.
37 Symyx has announced the commercial development of five catalytic systems for polyolefin and commodity chemical applications. One system is targeted for commercialization in 2004. For details see www.symyx.com.
38 UOP has disclosed at several conferences the successful use of HT-R&D methods for the discovery of an isomerization catalyst that is in commercial development.

2
Mastering the Challenges of Catalyst Screening in High-Throughput Experimentation for Heterogeneously Catalyzed Gas-phase Reactions

Stephan A. Schunk, Dirk Demuth, Alexander Cross, Olga Gerlach, Alfred Haas, Jens Klein, John M. Newsam, Andreas Sundermann, Wolfram Stichert, Wolfgang Strehlau, Uwe Vietze, and Torsten Zech

2.1
Challenges Connected to Catalyst Screening in Gas-phase Catalysis

The production of fuels and chemicals has been subject to major technological changes and improvements over the last century. Especially the development of efficient processes embracing catalysis as their core feature has advanced and brought forward the chemical industry. Today catalysis is the key to more efficient use of feed streams and waste reduction coupled with cost-optimized processes which are key to the wellbeing of society in the third millennium. Still, catalyst and process development are complex challenges that need to be mastered by the respective scientists and technologists. With increasing demand for shorter time to market, including shorter development times in the laboratory and pilot plant stage, tools need be available for the respective development units that hold the promise of accelerated process development. High-throughput experimentation holds this promise and the goal of this chapter is to illustrate with a select number of case studies how this technology can bring forward the development and improvement of catalytic processes in the gas phase.

The history of parallel testing in the gas phase goes back to first attempts of parallel reactor technologies in the 1970s and 1980s [1], or technologies which were further developed in the early 1990s [2]. Today technology is far more advanced and offers more technological features which render high-throughput experimentation an indispensable tool in catalyst development for Stage I [3–5] and Stage II applications [6–8].

The basic challenges for parallel test reactor development for high-throughput experimentation are, apart from technological challenges, related to technical demands that arise with the special issues for parallel test reactors, which are identical with the demands for conventional test reactors for gas-phase reactions. The criteria that must be fulfilled to obtain intrinsic catalyst properties from experimental data relate mainly to mass and heat transfer. A sufficient contact between the reactants and the catalyst must be insured to avoid mass transfer limitations inside and outside of the catalyst particles. Isothermal operation under laboratory conditions and avoidance of heat transfer limitations are also crucial. As a general quality check prior to operation intra- and extra-particle limitations should be

High-Throughput Screening in Chemical Catalysis
Edited by A. Hagemeyer, P. Strasser, A. F. Volpe, Jr.

ISBN: 3-527-30814-8

examined by varying the particle diameter and the flow over the catalyst bed. If extreme hotspots are experienced it may be useful to dilute the catalyst bed with an inert material. It should also be ensured that, due to small particle size of the catalyst, the pressure drop over the reactor does not exceed the expected values.

Last but not least a good description of the reactor characteristics together with a well-defined residence time distribution is essential basics to obtaining relevant test data. Generally the reactor configurations that operate at steady state are preferred over transient reactors or batch wise operated reactor types, mainly due to the ease of operation and comparably lower costs of the steady state operated reactor types. The main use for transient reactor systems, like TAP, lies in obtaining data that can directly be applied for mechanistic model development. A concise review of the advantages and disadvantages of common laboratory reactor systems has been published [9].

Generally, plug flow reactor systems need, as their name implies, plug flow conditions. It is important that the reactor dimensions, the size of the catalyst bed and the catalyst particle size, are adapted towards this technical demand. In many cases plug flow reactors, if carefully engineered, will be used to mimic pilot and technical plant behavior, usually integral operation will include operation at high conversions and respective product yields. There one has to keep in mind that this operation mode is inferior for obtaining kinetic data that are, generally speaking, obtained in better quality from differential reactor systems. Nevertheless, an attractive option is the variation of the space hourly velocity by introduction of varying catalyst amounts of one type of catalyst in the parallel reactor system.

A general distinction for the technological components in Stage I and Stage II technologies is not only useful in terms of technological definition but also helps distinguish and identify the different demands for each stage of catalyst testing and preparation. Tab. 2.1 lists the distinctive features of the two technological stages and their use in catalyst screening.

The general workflow in a high-throughput experimentation program will consist of different phases, starting with a design phase, during which the experiment is planned based on prior knowledge and chemical and engineering intui-

Tab. 2.1 Features and use of Stage I and Stage II technologies for high-throughput experimentation.

Stage I	Maximum sample throughput Reduced information (–/0/+) Analysis of target compounds Used for new discoveries
Stage II	Approaching real conditions Existing system knowledge Detailed analysis of compounds Used for continuous improvements

tion (computational aid is indispensable here), followed by a preparative state where catalyst samples are prepared for the test. This phase then culminates in the test of the materials following the plan established in the design phase [10]. The final phase is dedicated to data evaluation, usually by computational aid. The pinnacle of the evaluation stage is the computational modeling with the aid of the obtained data. For most heterogeneous catalysis reactions most of these modeling efforts will generally be of a rather qualitative stage and rarely include modeling on a molecular basis. Generally the data and qualitative assumptions obtained in this last stage will again be an important indicator in the next design phase.

The core issue for the seamless performance of the whole workflow is the early identification of the most time-consuming steps associated to the full high-throughput experimentation cycle to be able to adjust those bottlenecks to meet the requirements with regard to the overall desired screening speed. Below, we overview the topics addressed in this introduction and discuss the different features together with selected case studies.

2.2 Preparative Aspects

The preparation of catalysts for high-throughput testing is crucial for successful screening. The demands for catalyst preparation are high. Speed of preparation is a key component to the throughput that the screening process is targeted at. The need for such speed is, essentially, the driving force for the use of automated synthetic robots that can contribute to the parallel or fast sequential synthesis of catalytic materials. In addition to the beneficial effect on synthesis speed the use of robotic units is generally also a blessing with regard to reproducibility [11–14]. Synthetic protocols performed by robots are also easily transferable and through detailed electronic control and documentation of the single steps even errors during preparation sequences can be tracked and documented. Still, the major challenges in catalyst preparation for high-throughput experimentation lie within a reasonable scale down and translation of unit operations that are usually performed in the laboratory and on a technical scale to automated synthetic platforms. The term unit operation is in this context used for a preparation procedure or a certain step of a preparation procedure for which technical solutions exist on a laboratory or technical scale [15]. The essence and importance of these unit operations is that they enable the scientist to be sure that, firstly, relevant procedures with regard to standard laboratory procedures and scale up to a technical scale were employed and, secondly, the results obtained by well chosen unit operations for catalyst preparation can in many cases be translated to the technical scale and be a precious aid during scale up.

For many applications, commercial solutions for automated material's preparation are at hand. Basic unit operations like solid handling or liquid dispensing can often be performed with currently available technologies. This leads to a straightforward translation of preparative sequences that are necessary for making impregnated or coated catalysts [16–18]. More complex, but still within the compass of commercial apparatuses, are robotic systems capable of exercising techniques like pH-controlled or co-precipitation [6]. Nonetheless, the experimentator has to keep in mind that such technical solutions essentially present island solutions that hardly offer the perspective of a seamless workflow with regard to full preparation sequences and an integrated data handling. In many cases these island solutions can, in combination with manual work, be a good first step towards the acceleration of certain preparative sequences. Still, a highly desirable goal is the elimination of manual labor in high-throughput preparative sequences – the main reason for this being that manual work will, without disregarding the quality of the obtained results, be strongly operator dependent and not offer the perspective of fully integrated tracking of preparative data. The complexity of certain preparative sequences will always require a manual workforce to elaborate basic synthetic routes that can then be translated into an automated seamless synthetic workflow.

The highest demand with regard to synthetic sequences lies within the fully integrated synthesis stations with a wide range of applications. From a robotic and engineering viewpoint the human hand has an immense capability for sensitive object handling and the human eye is, in accord with the nervous system, a smart solution for system control. Robotic solutions will in many cases have to follow pathways for the synthesis of materials that for more than three centuries of labware development were designed to be performed manually. In few cases will a single robotic system be able to perform a full set of the operations necessary for a complete sequence of synthetic steps that are required, starting from soluble precursors to the final solid product to be employed in the desired catalytic process. These more complex sequences usually require multifunctional robotic systems in accord with the interaction of different robotic units.

The custom design usually applied to these units should be tailored to suit the needs of the user wishing to prepare materials in a high throughput fashion. Specific synthetic needs for tailoring materials towards a certain application combined with the desired throughput will, generally, be the governing principles for the design of these robotic systems. Speed in preparation of samples and flexibility with regard to different unit operations of a given unit are still contradictory principles. For a fully operational high-throughput materials synthesis laboratory it may be useful to have automatic units at both ends of the extremes: to be as flexible as possible with regard to preparation procedures and, conversely, to be able to run validated preparation procedures as fast as possible.

Fig. 2.1 shows an integrated custom made synthesis platform operating at hte Aktiengesellschaft. The gantry system can handle and dose liquids and solids, transport synthesis vials and control several thermostats and agitation units. The system design is tailored for maximum flexibility of the unit with regard to the unit operations envisaged. Thus, the system offers a large platform that can easily accommodate other automated preparative units, which can easily be embedded into the software for fully automated preparation sequences. Proven workflows within project-frameworks that require larger throughput in terms of a given

Fig. 2.1 Integrated synthesis station for liquid, powder and labware handling.

materials class can then be transferred to a robotic platform based on Scara technology. Here, much greater speeds for the single operations can be achieved, but the system "stiffness" with regard to certain procedures means that proven synthetic sequences are still necessary.

There is still much debate as to whether synthetic operations should be divided into Stage I and Stage II operations. This is because synthetic operations at whatever scale they are performed should be scaleable and translate into the later technical application. It becomes clear from the above that the guideline for Stage II synthetic operations will in general be such that the unit operations employed for the material's synthesis will directly lead to scaleable synthetic operations that closely link to laboratory procedures and later the technical production. In extreme cases one may even employ catalyst libraries that stem completely from commercially produced materials to ensure straightforward production of a material in case it proves to be a hit.

For these reasons it is doubtful whether synthetic operations that only fulfill the requirements of Stage I screening with regard to the synthesis of large numbers of samples will make sense in an integrated approach. There is much to recommend the use of these synthetic methods as they can be regarded as an integral piece of the Stage I screening process, but subsequent problems may arise in terms of scalability and the question of relevance with regard to the later technical materials production. As long as an intermediate translational function from the Stage I synthetic process is known, that can be directly linked to larger scale synthesis, the integrity of the screening process is assured and Stage I screening is valuable in the screening process. We shall address this issue again in Section 5.1.

2.3 Analytical Aspects

Analytical precision and the time required for a certain analysis are again trade-offs that often have to be considered as crucial points. Even today, hardened industrial process specialists make a distinction between the yield calculated by GC-analysis, be it as accurate as possible, and the yield calculated from the weight of distilled or crystallized product, with a clear preference for the latter. As Stage I and Stage II screening aim to accelerate conventional testing procedures, seldom will a gravimetric analysis of the separated product be the method of choice. Fast analytical methods are required to maintain the desired screening speed. Thus, the choice is either to push conventional analytical tools to the limits with regard to analysis times, while retaining acceptable margins of error, or to come up with new analytical solutions, either new techniques or the "intelligent" use of known ones. A useful division of the analytical toolsets that can be employed for Stage I and Stage II screening can be made because the two stages require quite substantially different speeds and accuracies.

Tab. 2.2 gives the analytical techniques employed in the two screening stages. All of these techniques have particular advantages and disadvantages. We will show that the technological demands connected to Stage I and Stage II screening differ quite substantially. Clearly, there are a lot of parallel or quasi-parallel analysis techniques but few that are employed in a fast sequential mode (Tab. 2.2). For fast sequential screening, the total analysis time includes not only that of the respective analysis but also the time required for sampling. This is true for sampling via valve arrangements and via robotic systems. As discussed above, these issues have to be considered when analyzing for bottlenecks in the total workflow.

The following demands were discussed by industrial specialists at the NICE meeting in Helsinki in 2001 for the development of a highly desirable analysis

Tab. 2.2 Analytical techniques employed in Stage I and Stage II.

Stage I: Parallel and quasi-parallel analytical techniques	IR-thermography Photoacoustic analysis Photothermal deflection REMPI Adsorption techniques
Stage I: Sequential analytical techniques	Mass spectrometry
Stage II: Parallel analytical techniques	–
Stage II: Sequential analytical techniques	Gas chromatography Gas chromatography coupled with mass spectrometry Multidimensional gas chromatography

tool. The overall demand was for speed of analysis, coupled with a high accuracy. To shorten analysis times the analysis should preferably be parallel or quasi-parallel and non-invasive, preferably an optical technique. The reader may find that many of the demands are still not met and require further developments, yet a major part of that wish list is already fulfilled and serves in select academic and industrial laboratories.

Analysis methodologies for the two different screening modes are discussed, separately, below.

2.3.1 Stage I Screening

Infrared thermography (IRT) and mass spectrometry, rather obviously, find their main application within Stage I screening, whereas the use of gas chromatography is generally associated with Stage II screening. IRT received harsh criticism as a screening tool as it is usually only indicative of the activity of a given sample, regardless of the product composition [19–21]. Naturally, this is not desirable, especially when screening for partial oxidation catalysts as the most active samples will usually be the ones that catalyze complete oxidation; less activity can generally not be taken as indicative for selective oxidation to partial oxidation products. In combination with spectrally resolving detectors the thermographic technique has certainly more potential in terms of accurate product identification and quantification than initially perceived by many groups [22–24]. It should be emphasized that IRT is a true parallel analysis technique with simultaneous analysis of the samples. This is not the case for all the other analysis techniques discussed here. The following techniques can only be subsumed under the term quasi-parallel (photoacoustic, thermodeflection, REMPI) or fast sequential (mass spectrometric, gas chromatographic).

Mass spectrometric analysis systems are advantageous when it comes to product analysis with regard to selectivity to desired or undesired reaction products and can thus overcome the disadvantages connected with IRT [25–28]. Different modes of analysis of the ionized sample can also accelerate the screening process. In general, a scanning mode for the whole mass range will be applied, so as to be able to follow the whole product spectrum and also to identify unexpected hits for products that are not the primary focus of the screening campaign. This mode of operation can be altered to save time during the screening process by tuning the mass spectrometer to filter certain masses (mass/charge ratios) that are indicative of the abundance of a desired species. Again, typically for a Stage I screening, the amount of information is reduced and the main focus is on identifying new compositions for a desired conversion within a short timeframe.

Another apparently undesirable aspect is the general layout of the reactor systems used in combination with the analysis technique [3–5]. The arrangement of samples on flat plates can be an acceptable compromise when it comes to simple reactions with essentially one reaction product, as with CO oxidation to CO_2 or the combustion of H_2 and O_2 to water. In the latter case, diffusion limitations may lead to false activity readouts, product selectivity not being an issue with the type of reaction observed. For more complex chemical reactions, where reaction products that are potentially intermediates to a fully oxidized or partially hydrogenated product are the prime targets of the screening, diffusion limitations can not only lead to false reads on activity but will also give false readouts on product selectivity. In Section 5.2 this design issue is addressed and examples are given of how alternative Stage I reactor systems can be designed to overcome the inherent diffusion limitation due to reactor maldesign.

Senkan et al. [29] introduced the REMPI analysis technique as a Stage I tool and showed its applicability with the example of a dehydrogenation reaction. This method is based on sample ionization via laser light and subsequent detection of the ionized reactor effluent at dedicated electrodes at the reactor exit. Challenges concerning the technique are unambiguous analysis of the ionized fragments, which can become exceedingly complex with complex product mixtures, and the coking of the electrode materials, which may lead to false readouts or total signal loss. REMPI is certainly a highly specialized technique that seems to have found only a minor range of applications. However, it illustrates what can be achieved with optical analysis techniques.

As a Stage I analysis technique photothermal deflection spectroscopy [30] is used by Symyx for the rapid optical identification of product effluents by infrared laser light. Also taking advantage of the infrared adsorption of product molecules is the photoacoustic analysis of reactor effluents with the aid of laser light [31]. The principle of this technique is based on the acoustic analysis of light absorption by vibrational or stretching modes of a molecule in the infrared. The absorption of light will lead to excitation of the molecule and the molecular relaxation to a thermal signal that is detectable as sound via a microphone setup. The technique has limitations with regard to the IR spectral absorptions of the entities to be detected and the availability of laser wavelengths of sufficient power. In some cases, certain absorptions of a product and side-product molecules may overlap (e.g. ethylene and propylene, or acrylonitrile, acetonitrile and HCN) and it may be hard to find alternative absorption bands especially in the fingerprint region. This limits the capabilities of the analysis systems to reaction product mixtures with limited ambiguity. However, the accuracy of photoacoustic analysis certainly exceeds that of commonly applied Stage I analysis techniques. The sensitivity range with regard to product molecules lies within the ppm range and shows a high linearity for a concentration range over several orders of magnitude.

Notably, the analysis method does not necessarily have to be applied in the setup as given in ref. [31]. The essence of this setup is the allocation of a single excitational device (the laser) for the assembly of measurement cuvettes. The detection devices (the microphones) are allocated at an unambiguous position and their signal gives direct feedback of the respective catalyst locus. Two alternative setups can easily be arranged, also taking advantage of the full capabilities of the analysis technique. The simplest approach is the adaptation of a single measurement cell to an automated sample capillary or a valve assembly, which allows stream selection from a given set of samples. This approach is comparable with reactor system setups that have been published for gas-phase analysis of screening reactors and simply takes advantage of another alternative analysis technique. As with the described setups, the catalyst locus is encoded in the position of the valve display or the position of the sampling device allocated over the catalyst assembly.

More challenging is an indirect analysis over an open exhaust reactor system. Here, an array of microphones is allocated close to the reactor exhaust system and a collective excitation of the off-streams of the exhausts of the single catalysts is performed. The array of microphones is used to calculate the locus of the origin of the sound. This setup is challenging in terms of the effort to calculate the sample locus but may offer the opportunity of quasi-parallel screening for arrays of several hundreds of catalysts. The advantage with regard to the analysis in single analysis chambers is that only one sound chamber has to be optimized in terms of noise, resonance frequency and echo.

Product detection via product adsorption and subsequent analysis has also become a proven technique for Stage I screening [24, 32–34]. In general, techniques are used that can readily indicate a physical change, such as a change in color of the adsorbate. Two groups have taken different approaches to this method. While Bergh et al. decided to develop a completely new reactor design that can accommodate the adsorbent in the reactor system, Schüth et al. [24, 32] constructed an adsorber cell that can easily be linked to different reactor systems of different types. Both approaches have their pros and cons. Keeping in mind that Bergh et al. aim at maximum sample throughput (in a reactor system that may contain several hundreds of samples), the detachment of the adsorbent holder makes sense for each change in active compound or change in condition. The separation of the analysis unit by Schüth et al. [24, 32] is advantageous with regard to frequent changes in reaction conditions without changing the catalyst library. Here only the adsorber unit has to be changed, all active components may remain in the reactor, even under reaction conditions.

Although the list of analysis techniques available for Stage I screening is quite notable, there is still great demand for new methods to enhance high-throughput experimentation. Especially high is the need for truly parallel techniques, preferably optical, to enable very short analysis times.

2.3.2
Stage II Screening

Analysis in Stage II has to fulfil totally different tasks to analysis methods for Stage I screening. Of course time is as essential as for the Stage I, but the accuracy of the analysis has to meet the highest demands. As Stage II screening is dedicated to obtaining process relevant data it is essential that no compromises with the errors caused by the analysis method are made. This follows the same ratio as of the engineering of the Stage II reactor unit and makes Stage II screening an essential part of modern industrial catalyst development.

In most cases gas chromatographic analysis methods will be the method of choice for Stage II screening. Gas chromatography is a well established methodology for a broad range of analytical applications and is the workhorse in many analytical laboratories. The history of the use of on-line process GCs goes back decades and major developments have been achieved. High-throughput experimentation, in this case, can take advantage of such development in on-line GC and process GC analyses [35, 36].

Gas chromatography is especially useful as for gas-phase analysis of partial oxidation, hydrogenation or hydroconversion products – in many cases a full carbon balance (educts, products and all side-products) will be required to evaluate sample performance. As the detection and quantification of permanent gases like O_2, CO and CO_2 and also of higher boiling compounds are standard separation problems for gas chromatography, it is wise to employ the method regarding this problem.

First of all, similar to analysis in Stage I, a distinction has to be made between the time required for a GC-analysis (typically one should calculate injection to injection) and the total time required for analysis. This is because all lines, whether sampling the capillary of an automated moving robotic device or a multi-valve assembly, need to be flushed by the sample; it is highly recommended to operate at an excess volume with regard to the tubing diameter of at least a factor of ten to prevent cross-talking. If the time required to flush the lines is larger than the analysis time, it is of little use to accelerate the GC-analysis – rather one should take into account changes in configuration of the sampling capillary or the valve assembly.

There are several ways to arrive at fast separation and thus detection times with GC-units, all of which have their pros and cons. An easy but cost intensive way of accelerating an analysis is the use of several analysis units that all fulfil the same analytical task. As discussed above, this approach only makes sense if the total-run time of the GC-run is much shorter than the time required to flush all of the lines. Instead of increasing the number of analytical units, a smart choice may be to use more columns on which the separations can be performed. With this configuration, as well as the valve display dedicated to the reactor unit, a second valve display for the different columns is needed, in some cases separate detectors may even be necessary. One has to keep in mind that, in general, this analytical setup will require isothermal separation conditions.

Alternative solutions to high-throughput compatible analytical solutions will require the development of fast separations. There are again several ways to shorten the analysis time required for a gas chromatographic separation. The choice of the GC-column is essential but will not be discussed here as this would go far beyond the scope of this chapter. Users will find a large range of different columns for GC-analysis at the respective suppliers and, in general, the suppliers also offer consultation and validation, with some even offering large databases for on-line access.

Having found a prospective GC-column and completed initial runs with conventional separation times, the usual finding is that short separation times come with high temperatures and/or steep temperature gradients. If such separations at high temperatures cannot be run isothermally the cooling time to reach the start temperature of the separation may become critical. In many cases it may therefore be advisable to seek separation conditions that avoid extreme temperature profiles. In addition, high temperatures and drastic temperature programs may lead to rapid deterioration of the column.

The column flow, diameter and length are attractive parameters for shortening chromatographic runs. In comparison with steeper temperature profiles and isothermal runs at higher temperatures one has to remember that such procedures may lead to a more pronounced broadening of the peaks, which is acceptable as long as the total quality of the analysis does not suffer and the respective gain in speed of the separation is sufficient.

An attractive alternative is the use of multidimensional GC-analysis. Figures 2.2 and 2.3 show a typical simplified column setup for multidimensional analytical purposes and an example of the results obtained. The two different columns are connected via a valve that is actuated at the corresponding residence time to give access to the sample fraction of interest, which is then separated on the second column. This setup is especially attractive with regard to high-throughput screening for the analysis of high- and low-boiling compounds on different column types. In many cases back-flushing of the column or the use of park volumes is necessary. The reader is again directed to the literature for further details [35, 36].

Another interesting technological development is micro-GC-units, which also allow short separation times. Their compact design (although the layout of the full GC units is scarcely micro) and attractive pricing are both arguments for their use in high-throughput screening. However, for the time being, owing to the columns typically offered by the manufacturers in these units, their main application lies in the analysis of low-boiling compounds or permanent gases.

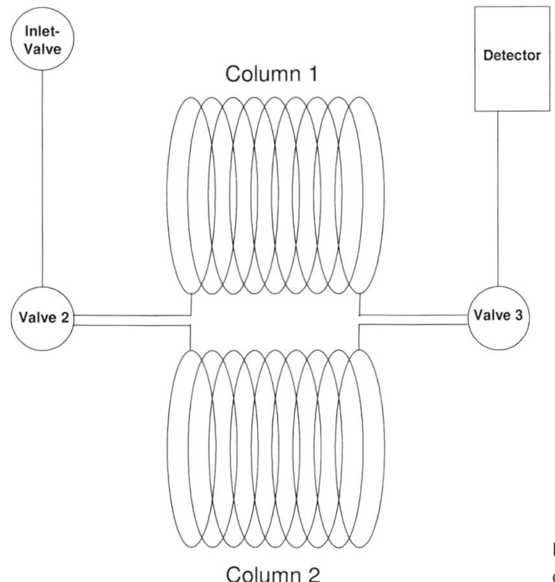

Fig. 2.2 Column setup for multidimensional GC-analysis.

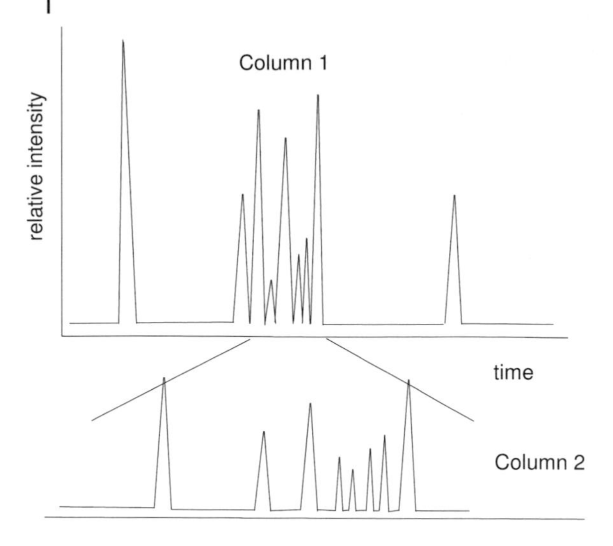

Fig. 2.3 Example separation of a complex product mixture using multidimensional GC-analysis.

2.3.3
King-System: Saving Analysis Time via Intelligent Use of Analysis Techniques

We have discussed several analytical techniques that can be applied in high-throughput screening, with a special focus on accelerated analytical techniques that lead to accelerated screening. Clearly, the fastest screening can be achieved by the use of truly parallel analysis techniques, such as IRT (infrared thermography) [3–5], or quasi-parallel screening techniques like photoacoustic analysis [31]. Techniques that include mass spectrometric or gas chromatographic analysis will deliver more information, but in general also require longer analysis times due to their invasive nature. An approach taken by hte Aktiengesellschaft was the combination of a fast non-invasive analysis technique for material preselection in combination with a more time consuming chromatographic or spectrometric analysis of select samples that are within a certain threshold of the response signal of the preselection method. In our case we chose IRT as a preselection tool and mass spectrometry or gas chromatography as the more time consuming technique applied to a select set of samples [37].

As described in Section 3.1 the crucial aspect in implementing a screening concept that allows the use of a non-invasive optical technique in combination with a spectrometric or chromatographic technique is the central reactor system. The need for the simultaneous or subsequent application of an optical method to determine materials activity in a parallel manner through thermographic visualization of heat release and a sampling capillary connected with a second analytical technique, e.g. MS or GC, for selectivity measurement has two key impacts on the reactor design: Firstly, each candidate material has to be accessible separately and independently through an IR transparent window for the optical method in a way that avoids the measurement of, for example, artefacts by variation of reactor

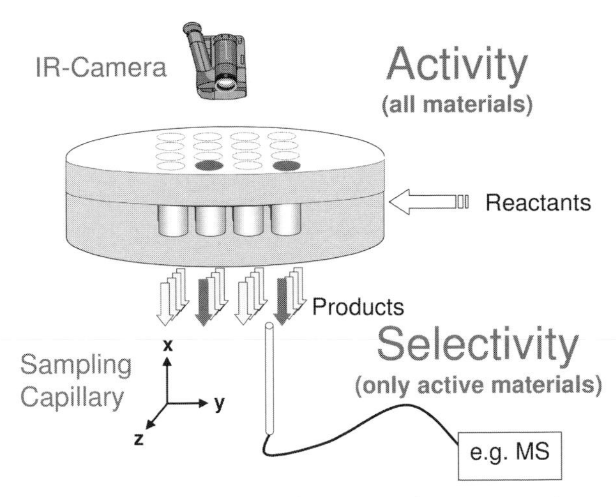

Fig. 2.4 King-System for simultaneous analysis via IRT and mass spectrometry or gas chromatography.

materials emissivities or organic gas-phase reflections and potential cross-talk of neighboring reaction channels. Secondly, each catalyst product flow has to be fed without any cross-talking to a sampling capillary connected to the second analytical technique (e.g. GC, GC/MS, MS, or other techniques). Furthermore, the reaction conditions, e.g. the fluid distribution (concentration, composition and space velocity) and temperature conditions, have to be held constant and equal for each material in the separated reaction channels.

Implementation of the above screening principle requires three main parts: a thermosensitive IR camera capable of recording heat emissions of the catalysts contained in the reactor system, the reactor itself as central part and an *xyz*-positionable sampling capillary, connected to the second analytical tool (MS or GC, etc.) (Fig. 2.4). To simplify matters, a reactor with a 4×4 matrix of reaction channels is illustrated here, the actual reactor formats used at hte Aktiengesellschaft are 96- and 192-fold reactor systems based on the 8×12-MTP (micro-titer plate) matrix.

Through the IR-transparent window on top of the reactor, the IR-camera can detect temperature changes (shown in Fig. 2.4 by two black shadowed spots) of all catalysts in a catalytic gas-phase reaction in parallel. The IR-thermographic image is evaluated automatically and the *xyz*-robotic system directs a sampling capillary to the active candidates, indicated by the dark effluent streams in Fig. 2.4. The positional information of candidates in the appropriate activity range is automatically determined by pattern recognition from the IR-thermogram during the experiment. All parts of the system are fully integrated and automated.

The key hardware for the combined application of, for example, a parallel optical and a subsequent fast sequential analytical method is the reactor. A photograph of the reactor system is shown in Fig. 2.5.

Fig. 2.5 Reactor system of the King-System for simultaneous analysis via IRT and mass spectrometry or gas chromatography.

Fig. 2.5 shows the reactor assembly. The reactor design is based on a modular, sandwich-like construction, which eases the exchange or the maintenance of single modules. The reactor top plate consists of an IR-transparent window, in this case a sapphire or silicon window with a visible sector of sufficient size and wavelength transmittance. The next modular layer consists of a black ceramic plate with an 8×12 matrix of drilled holes, fitted to the positions of the matrix containing the active compounds. The final module is the main reactor part with gas supply, heating system and reaction channels. The resulting "IR camera-"view into the reactor through the sapphire window and the black ceramic surface ensures that only the catalyst materials are in the visual field of the IR camera. The homogeneous emissivity of the black ceramic mask is similar to a black body and allows the above-mentioned precise, high-temperature resolution. The use of a material resembling black body behavior is a technical core feature of any IR-thermographic system intended for the measurement of the true thermal signal of catalytic materials. Without this feature IR-thermographic systems are practically useless as no normalisation to a black-body-like surface is possible and the only signal that can be recorded is the emissivity of the catalytic materials.

Fig. 2.6 is the top view of an open reactor module with a symmetrically distributed materials library in the separated, independent reaction chambers for the validation of the 96-fold screening system. Fig. 2.7 shows a schematic visualization of the symmetrical distribution of equal materials over the 96 reactor positions.

To illustrate the functionality of the system a validation library was prepared and introduced into the reactor system. With the goal of achieving an optimal fluid distribution with a minimal pressure drop over the 96 reactor channels we used multichannel ceramic bodies ("miniliths") as supports, which are impregnated with the corresponding catalyst precursor solutions in an automatic manner (for suitable technical solutions see Section 2). At each of the 96 reactor positions, a candidate material modified by impregnation is available for testing. The shadowed scheme

Fig. 2.6 Materials library in the reactor system.

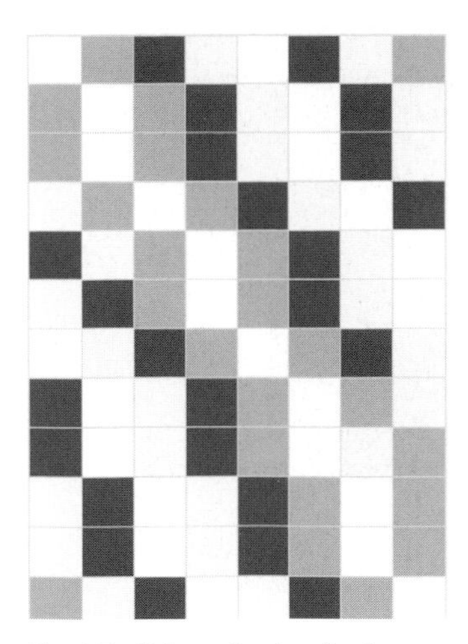

Fig. 2.7 Schematic visualisation of the symmetrical distribution of the materials within the reactor system.

of the material's library positions (Fig. 2.7) helps in understanding Figs. 2.8 and 2.9. Unimpregnated ceramic bodies, which serve as blank materials, are visualized in white. The validation library (Figs. 2.6 and 2.7) consisted of only three different catalytic materials and an inert support material, which are distributed in a pattern of diagonals (Fig. 2.7) over the 96 reactor positions. Materials that are symbolized by the same shading in Fig. 2.7 are synthesized by the same unit operations, but were synthesised separately and not in one batch. The homogeneity of the impregnation procedure is proven and ensured by μ-XRF measurements on different samples of the same elemental composition.

The main aspects of the reactor design are temperature homogeneity and homogeneous gas flow for all 96 reactor positions. In Figs. 2.8 and 2.9, homogeneities of temperature and gas distribution are indirectly verified by an IR-thermogram and a product distribution of a partial oxidation reaction, measured over all 96 catalyst samples.

Fig. 2.8 (a) and (b) shows the temperature homogeneity of the whole reactor system with a symmetrical distribution of equal materials compositions in an 8×12 MTP array.

Fig. 2.8 (a) and (b) are thermographic pictures, recorded with the IR camera above the reactor system (Fig. 2.4) under typical reaction conditions: 1% hydrocarbon in synthetic air, 375 °C and GHSV 3000 h^{-1}. The thermogram is emissivity corrected for these conditions. The homogeneous temperature distribution of the reactor temperature (375 °C, black surface background in Fig. 2.8) is evident. Each deviation from a homogeneous temperature distribution would result in colour gradients in Fig. 2.8. The result of several measurements with thermocouples around the catalyst positions of the reactor system support the finding recorded via IR-thermography on the reactor surface. The maximal temperature deviation found is below ±1 °C.

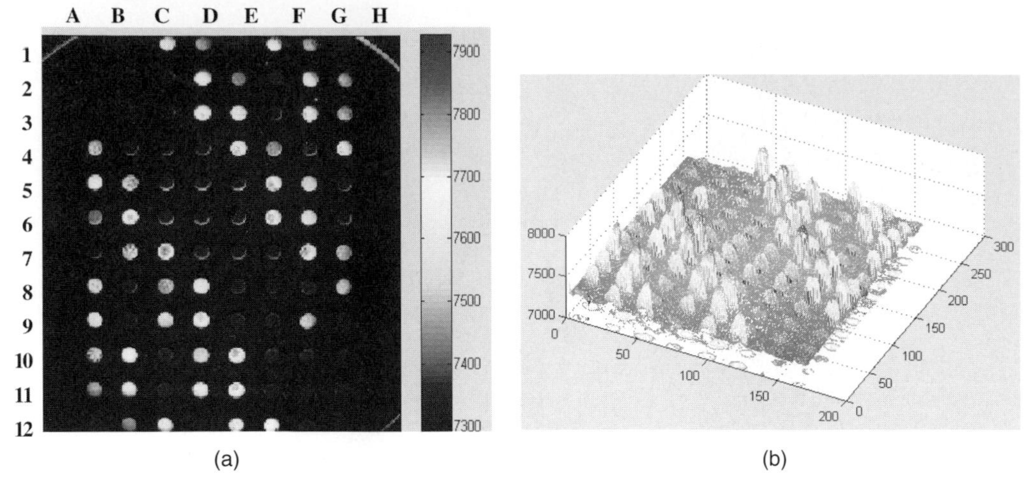

Fig. 2.8 IR-thermograms of the reactor system under reaction conditions.

The activity pattern in Fig. 2.8 matches perfectly with the library pattern in Fig. 2.7; the black-colored catalyst is the most active, followed by the light grey-colored one. The dark grey-codes material shows very low activity and the white materials (unimpregnated ceramic supports) are inactive and serve as a blank reference. This result also supports the finding of the homogeneous temperature distribution as well as the good fluid distribution: equal heat release on equal catalysts indicates obviously identical reaction conditions. The diagonal pattern can be identified in the IR thermogram (Fig. 2.8). Library pattern (Figs. 2.6 and 2.7) and IR results (Fig. 2.8) show very good conformity.

Fig. 2.9 shows the screening results of MS-sampling of all 96 reactor positions with a commercially available mass spectrometer.

The MS intensities for two different oxidation products correlate with the evolved heats of reaction, observed by an IR camera in Fig. 2.8, and reproduce very well the library pattern from Fig. 2.7. Product 1 (Fig. 2.9 a) is produced mainly with the black-colored material. Only minor quantities of product 1 one can be found over the light grey-colored catalyst. The dark grey ones show nearly no activity, as for the unimpregnated support materials (white). For another partial oxidation product (Fig. 2.9 b) the light grey catalysts show high selectivities, whereas the black material is totally inactive. The unimpregnated support ceramics (white) and the dark grey catalyst show a minor selectivity to product 2.

Finding this regular pattern of product distribution in accordance with the validation library (Figs. 2.6 and 2.7) manifests the two main features of the screening reactor set-up: (1) In accordance with the XRF analysis, the sample synthesis fulfils all the requirements of a high-throughput synthesis method. Materials synthesized under the same conditions show very similar, not to say identical, results in the validation reaction. (2) Reaction conditions such as fluid and temperature distribution are the same for all 96 reactor positions. The same materials produce

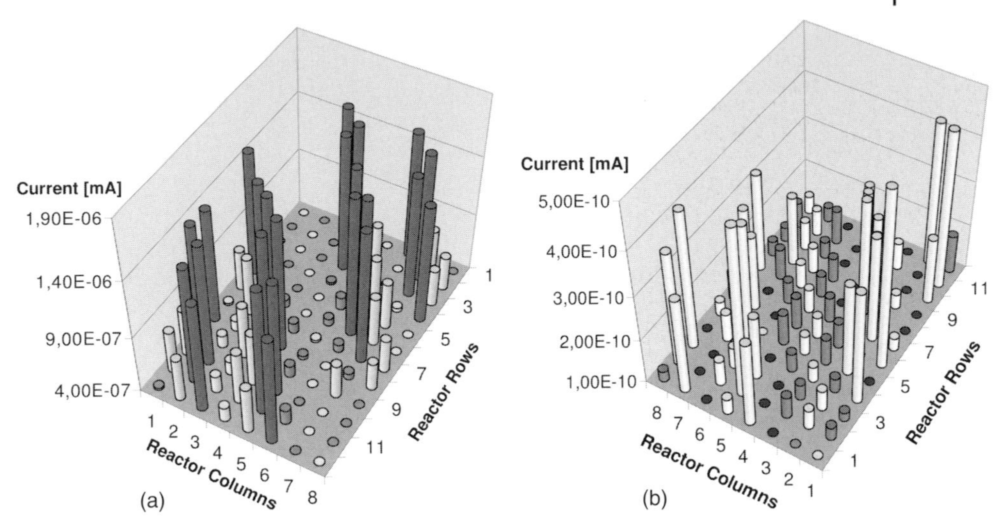

Fig. 2.9 Results of MS-measurements of the different samples. Different masses associated with different product molecules are plotted as a function of the reactor locus.

exactly the same results. Furthermore, the screening principle has a proven value with regard to IRT pre-screening of materials for activity in an appropriate range, with subsequent analysis via a more time-consuming analysis method. As seen from the validation study, materials that do not produce an IR signal can be omitted in the more time-intensive screen for detailed product selectivity. Depending on the threshold chosen for the IR signal this pre-screen can rule out up to 95% of the library under certain conditions and, therefore, help to save essential screening time for lead development.

The so-called King-System is one of the technologies of choice for ambitious screening programs in the gas phase useful for fast lead identification.

2.4 Case Studies of Selected Examples in Gas-phase Catalysis in Stage II Screening

The following case studies are intended to illustrate the technical feasibility and possibilities of the current status of parallel screening technology that has been established during the past decade. The case studies are guidelines for the reader in terms of technical feasibility.

Not all of the case studies have a correspondence to a Stage I screening process and tool. Especially for high pressure applications and fast dynamic studies, there are still opportunities for further developments in Stage I screening. In many of these cases though, Stage I screening can be adapted to allow at least the screening of certain aspects of catalyst behavior that can be used as a discriminator for further lead development.

Apart from detailed studies of the effect of reaction parameters or aging and regeneration behavior, Stage II screening will in many cases be concerned with the exact mapping of parameter spaces of the active components. In some cases the distinguishing parameters may be compositionally related to the samples. However, one has to keep in mind that, usually, Stage II screening is connected with leads obtained either by a Stage I screening process or otherwise. The information connected to the lead may in some cases be a compositional one; in many cases the information may go far beyond this point. Structural details like oxidation states, structure types of the solid compounds, typical pore size distributions or total surface areas, characteristic sequences needed for the preparative success and other factors can be important clues that may not be omitted during the screening process. As in conventional catalysts screening it is important to ensure that the lead structure as such is correctly followed during the screening. This qualitative surveillance of the Stage II screen is essential to arrive at the final goal with high efficiency, namely the handover of a scaleable catalyst formulation into a mini-plant or a technical reactor. Any technique that serves as a qualitative or quantitative check for quality and consistency of the library needs to be compatible with the overall throughput of the process so that the screening speed is not affected, and should not go above a certain level of increase in overall resources needed for a given project.

2.4.1
Bulk Chemicals and Intermediates: Partial Oxidation

Partial oxidation plays a central, if not crucial, role in the chemical industry for the functionalization of, especially, alkanes and olefins. In particular, gas-phase oxidation is attractive with regard to feedstock conversion as high space–time yields can usually be achieved and product separation is often far easier than for product mixtures derived from liquid phase oxidation. Nonetheless, gas-phase oxidation has major challenges that need to be considered when tackling the complex technological challenges. Starting from alkanes, olefins or aromatics, thermodynamically, the formation of CO and CO_2 are favored, which are undesirable in terms of the partial oxidation products of the relevant feedstock. Starting from relatively low-boiling feedstock compounds, increasing boiling points (or decreasing partial pressures) can be observed for hydrocarbon oxidation products with increasing oxidation state of the functionalized hydrocarbon molecule. This will lead to changed adsorption behavior and larger "sticking coefficients" of the functionalized molecule to the catalytic surface and increase the chance of further oxo-functionalization and finally of total combustion. Partial oxidation and total oxidation are sources of heat, and rarely can an industrial process be run isothermally over the whole catalytic bed. In most cases explicit hot spots are formed, which in extreme scenarios can exceed the average temperature of the catalyst bed by 5 to 15 °C. Usually if a hot spot is displayed in a catalyst bed the degree of conversion for single pass-engineered processes is well over 60% with regard to the hydrocarbon feedstock – for multi-pass-engineered processes the degree of conversion can be much lower. Still, this leads to a bizarre scenario: the catalyst is expected to deliver high product selectivity over a temperature range of approximately ±10 °C and a large range of varying partial pressures of educt, oxygen, desired products and side products of the reaction. Keeping these facts in mind it is even more surprising that there are many examples of highly selective gas-phase oxidation catalysts and processes. It is not astonishing that screening efforts with regard to changes in materials' properties, new prospective catalyst candidates and variations in engineering conditions for partial oxidation reactions are highly attractive to the chemical industry.

Optimization of the engineering of a reaction for high-throughput experimentation screening is obviously best tested in Stage II and not in Stage I. Nevertheless, the point at which new materials should undergo Stage II screening is still hotly debated. As discussed before, the issue of lead development is crucial. Yet leads will, in many cases, not be defined by their compositional parameters, but will need structural specification. Therefore the synthetic capabilities of the Stage I screen, also regarding later detailed analysis, have to be considered. The second relevant issue is the degree of differentiation that can be achieved in a Stage I screen, as in many cases even improvements in selectivity of around 10% for the desired product can be indicative of prospective materials candidates. This is a relatively low value in terms of absolute concentration of the respective chemical in the gas phase, considering that in many cases a pre-screen will include the differentiation at conversion levels between 10 and 50% at feed concentrations of 0.5 to 5% with regard to the feedstock employed.

It becomes obvious that, especially once a given lead structure has been identified, careful decisions have to be made as to whether to continue a screening campaign in Stage I or Stage II. In general one may find that Stage II will be the more promising technology for projects in which the above-mentioned issues need a more specific analysis and accuracy of distinction. Especially where a full carbon-balance is demanded, it is usually advisable to go for a test campaign in Stage II. However, Stage I screening can also make sense at that stage of differentiation, although greater care has to be taken with regard to the data obtained. Particularly in Stage I, a good practice can be to include a reference material that produces the target product in known amount and so use these data points as internal references.

We have noted that most gas-phase oxidation catalysts are highly sensitive to the reaction temperature. For any Stage II screening tool aimed at operating under isothermal conditions, for all active materials it is a given that the thermal equilibrium of the reactor and a homogeneous temperature distribution are essential. In Section 3.3 we have seen that IRT can be a valuable tool for verifying the thermal distribution in a reactor system. Still, in this study mass spectrometric data also prove that the catalyst performances for a given set of identical catalysts are in an acceptable range. Although in many cases it is of dubious value to calculate conversion, selectivity and yield from mass spectrometric data, the generally good performance of the reactor system can be deduced from this study. To evaluate the thermal behavior of a standard 48-fold reactor system employed at hte Aktiengesellschaft, a similar approach was taken. The oxidation of propylene to acrylic acid is currently performed in two steps on an industrial scale. In the first step propylene is oxidized to acrolein over a bismuth-molybdate-based catalyst and in the second step acrolein is oxidized to acrylic acid over a molybdenum-vanadium-based mixed oxide. Both steps are very sensitive to the reaction temperature in terms of both catalyst activity and selectivity to the desired product. As a test reaction, the oxidation of propylene to acrolein was chosen, using a standard bismuth-molybdate catalyst [38, 39]. Some 42 identical catalysts were introduced into the reactor system and exposed to 2% hydrocarbon in air at a GHSV of 2000 h^{-1}. The goal of this experiment was to prove that (a) practically identical values for activity and selectivity are obtained for all catalysts introduced into the reactor system and (b) the typical behavior known from conventional testing in laboratory reactors and pilot plant could be reproduced in the Stage II reactor system. Finding (b) also illustrates that good thermal distribution and nearly identical temperatures are achieved in all positions in the reactor as well as a good distribution of the reaction gases over the reactor system.

Fig. 10 shows the behavior of the set of catalysts plotted as a function of the reactor position. Nearly identical degrees of conversion and selectivity were achieved. If the temperature dependence of the conversion/selectivity behavior is plotted, including the relative errors obtained for both values, a curve is obtained that is analogous to the well-known temperature-dependent behavior of this well-known catalyst class (Fig. 2.11).

As discussed above, partial oxidation reactions are in general exothermic and connected with large heat releases. This has several consequences for the reactor

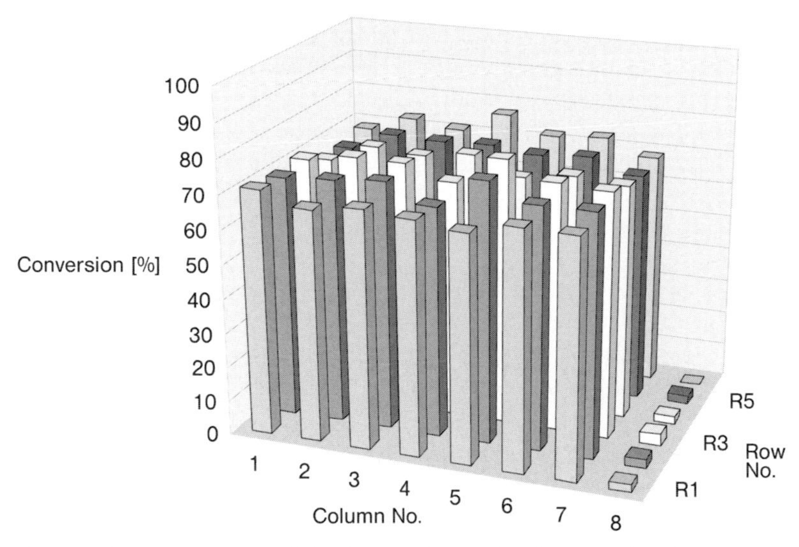

Fig. 2.10 Position sensitivity of the degree of conversion for a set of 48 bismuth-molybdate catalysts in the conversion of propylene into acrolein in a Stage II screening reactor (reaction conditions: 2% hydrocarbon in air at GHSV of 2000 h^{-1}).

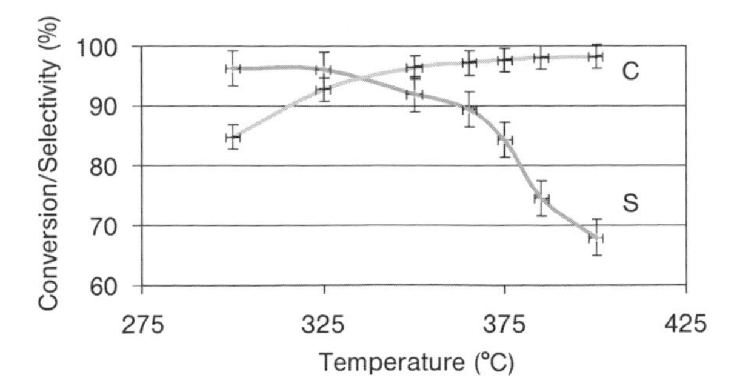

Fig. 2.11 Temperature dependence for conversion and selectivity for the set of 48 bismuth-molybdate catalysts of Fig. 2.10 in the conversion of propylene into acrolein in a Stage II screening reactor (reaction conditions: 2% hydrocarbon in air at GHSV of 2000 h^{-1}).

design that need to be discussed in detail. Usually, a screening will differ from the case study performed here for the isothermal properties of the reactor system in terms of the reactor position as, generally, the different materials introduced into the reactor system will in most cases display different catalytic behavior. This has an enormous impact on reactor design. In the extreme case that several different catalysts may already be fully converting the feedstock into CO_x the maximum heat release will be experienced at these reactor locations. Heat transfer to neighboring reactor positions may become relevant as, depending on the amount of heat released, neighboring reactor positions may be heated up over the targeted average reactor temperature and so give false readings on the respective catalyst activity. Unless the temperature in each catalytic bed is recorded, this is highly undesirable. Simulation may aid in avoiding such cases by looking at borderline cases and reactor design features that prevent such scenarios. Fig. 2.12 shows the possibility of simulating the behavior of these borderline cases for a single catalytic component at a given reactor position. With full conversion of the feedstock introduced to the catalyst bed the evolution of heat does not cause heating of the neighboring reactor positions.

There are elegant ways of designing reactor systems so that thermal contamination is minimised. Fig. 2.13 shows the so-called slice and other reactor systems.

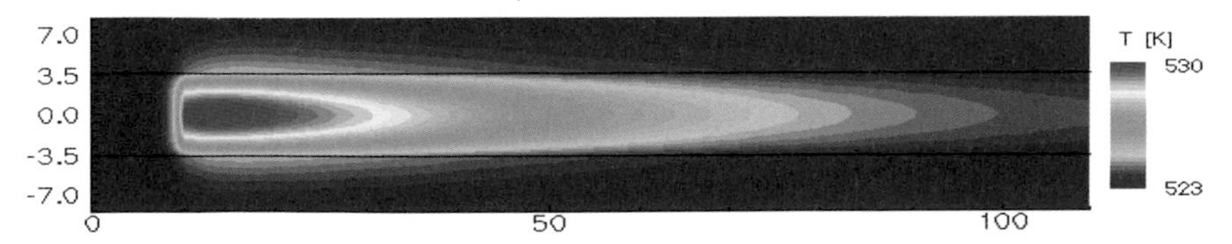

Fig. 2.12 Simulation of the thermal behavior of a catalyst in a multitube test reactor. Though the reaction is highly exothermic, no "thermal contamination" of neighboring reactor sites is experienced. DH >100 kJ mol^{-1}; inlet: 523 K, reactor: 523 K. Porosity: 80%; bed: 3.0 W mk^{-1}; GHSV: 10000 h^{-1}, 0.11×0.007 m.

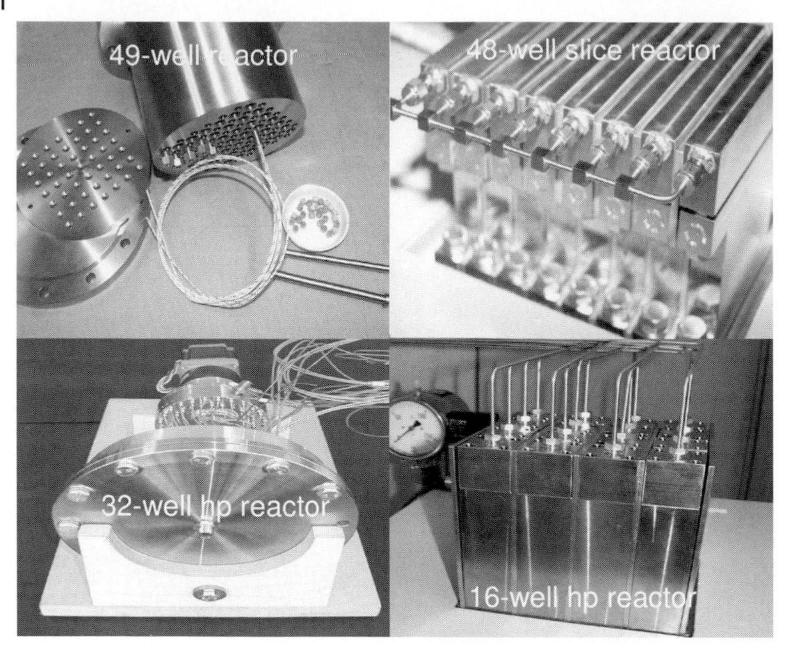

Fig. 2.13 Types of slice reactor systems. Reactor and heat exchanger elements are separated in these designs. This aids in avoiding undesirable thermal cross-talk between catalyst candidates.

For the slice reactor systems the reactor block is divided into reactor and heating exchanger elements. The heat exchanger elements serve as thermal sinks for excess heat produced by catalytic components and insure isothermal operation of the reactor system. The most extreme design of such a slice reactor system is of course the accommodation of a single material of interest in a given reactor slice, this design excludes "thermal contamination" of neighboring catalyst candidates.

Another vital issue to be addressed is the consequences of catalyst formation and aging. Here we emphasise catalyst formation, whereas the subsequent section will address catalyst aging. Typically many oxidation catalysts show a formation behavior, although the time required for formation differs quite substantially for the different material's classes. In some cases, as illustrated for a bismuth-molybdate catalyst for propylene oxidation in Fig. 2.14, formation behavior until steady state operation may take several hours. In other cases, formation may take up to several hundred hours, e.g. VPO-catalysts employed for butane oxidation to maleic anhydride. What can typically be detected during these catalyst formation procedures are essential changes in activity and selectivity. As, usually, during a Stage II screening materials properties are hard to predict it may prove useful to follow two rules of thumb: (1) equilibration under a given set of conditions until no major changes in activity and selectivity are detected prior to the next change in reaction conditions assure a close to steady state operation of the catalyst. (2) A given set of reaction parameters should always be measured at least twice during a cata-

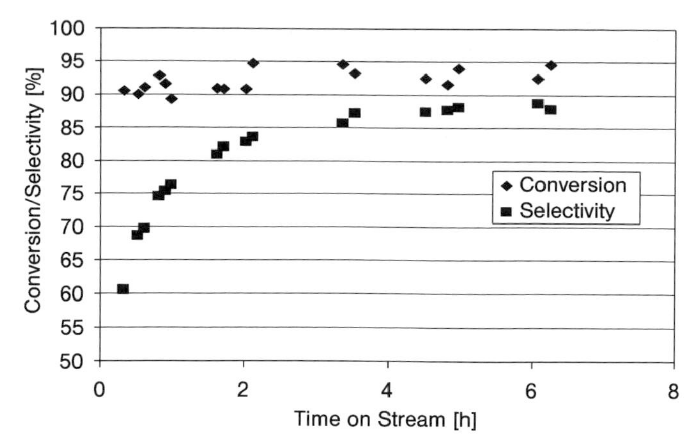

Fig. 2.14 Formation behavior of a bismuth-molybdate catalyst employed in the oxidation of propylene into acrolein (2% hydrocarbon in air, 370 °C, GHSV: 1500 h^{-1}).

lyst screen to detect and evaluate changes in materials that occur during operation. Evidently, it is never useful to screen libraries in a Stage II screen that are too diverse, due to differing behaviors not only with regard to optimum operation conditions but also with formation and aging behavior. In some cases when a material's properties have already been evaluated further during a screen it may be a sensible to employ a separate formation reactor for certain material classes. Such reactors can also usefully be employed as aging or poisoning devices prior to a screen in a Stage II test unit.

We conclude this section on gas-phase oxidation catalysts with an illustrative catalytic study where it can be demonstrated that minor changes in preparation have a major effect on catalyst performance, and a very high degree of sophistication is demanded of the accuracy of analysis to detect those changes.

Silver catalysts employed for the gas-phase epoxidation reactions of light olefins are very temperature sensitive in terms of activity/selectivity, and also show a pronounced sensitivity to details of the preparation procedure. Low surface area materials like α-alumina or steatite are the carrier materials of choice for deposition of the active component. The temperature window for the maximum yield can be fine tuned via the type of dopant and the dopant level, a fact that can be exploited usefully in an industrial process. Most interestingly the preparation procedure has a pronounced effect on optimum catalyst performance. Apart from adding amines to the impregnation solution, an impregnation temperature below 5 °C and the preparation of the solution under the exclusion of light deliver materials with the best performance data in the desired oxidation reaction. Tab. 2.3 compares the maximum yields obtained via the different preparation procedures. From the differences it becomes evident that a high degree of sophistication is demanded of (1) the automated preparation sequences and (2) the accuracy of the analysis of the Stage II screening reactor.

Tab. 2.3 Comparison of the influence of the synthesis procedure on the catalytic performance of silver-containing catalysts in the epoxidation of light olefins.

Ag-precursor	*Ethylene-diamine complex*	*Ethylene-diamine complex*	*Ethylene-diamine complex*	*Ethylene-diamine complex*
Dopant	–	–	Cs or Rb	Cs or Rb
Synthesis conditions	Room temperature/ day light	Ice bath cooling/ darkness	Room temperature/ day light	Ice bath cooling/ darkness
Optimal temperature (°C)	260–280	240–280	230–240	220–240
Maximum yield (%)	24	38	84	87

2.4.2
Refinery Catalysis: High-pressure Reactions

Very different from the application focus of partial oxidation reactions are the demands for screening for new prospective catalyst candidates in the respective test units. Many refinery processes will operate in the mid- to high-pressure regime and in a number of cases reactions will be either run as three-phase reactions or even produce liquid products from gaseous educts. Where reactants and products are present in gaseous and liquid phases a phase separation is needed, and a separate analysis of the gaseous and liquid components is usually required to calculate the conversion, selectivity and establish the carbon balance. Further care has to be taken with the choice of the reactor material. In a number of refining and refinery-related processes carbon monoxide is present in larger amounts at high temperatures and high pressures. Carbon monoxide may lead to so-called metal dusting and corrode the reactor system, and in some cases produce catalytically active reactor walls. A careful choice of the reactor material can avoid such undesirable corrosion phenomena.

As indicated in the last section, catalyst aging and regeneration are phenomena typically encountered in refinery catalysis applications. Improvements in catalyst lifetime in some cases are even of higher value for process improvements than increases in activity and selectivity of a given active component. The times on stream in refinery applications until regeneration is necessary span a wide range in refinery catalysis, from fractions of a second for FCC catalysts up to years for several hydro-conversion reactions. Depending on the way the catalyst deactivates it may prove useful to employ different modes of screening for such conversions. The activity and selectivity curves follow a steep decay during a short period of time (Fig. 2.15). Before this decay the catalyst performance is relatively stable over a longer period of time. With regard to the screening this has an important impact on the level of detail at which activity and selectivity have to be followed and the number of data points required during the time interval of rapid deactivation.

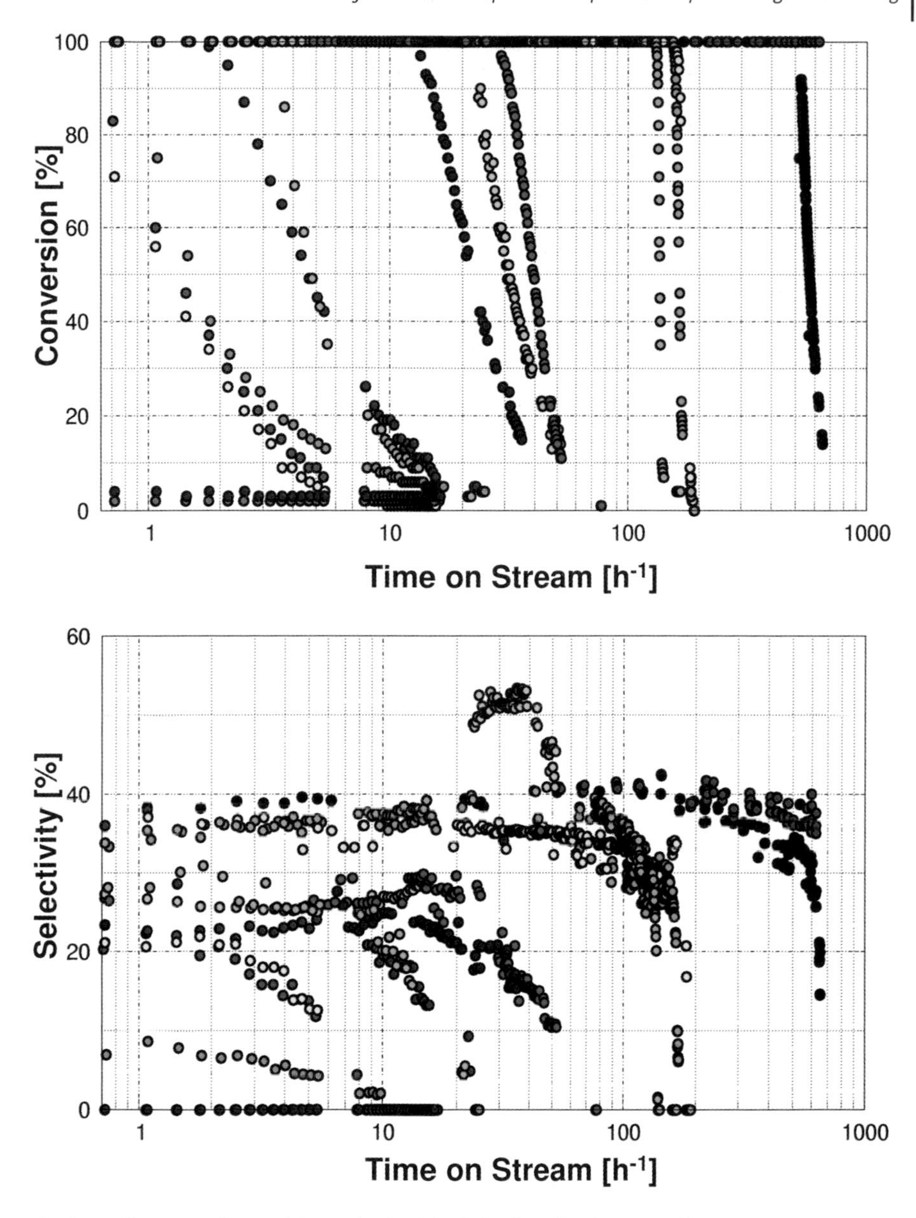

Fig. 2.15 Corresponding activity and selectivity behavior of a given set of catalysts in a petrochemical reaction (top and bottom). Note that the catalysts show pronounced aging with a drastic decrease in activity and selectivity over a very short timeframe.

A feasible solution for this complex challenge is to implement at least two analytical methods with which the course of the reaction can be followed: a fast first method that allows qualitative control of the status of the catalyst performance and a second accurate, and in most cases more time consuming, analysis method that will allow a detailed evaluation of catalyst performance. The two analysis methods can be run on one analytical unit, e.g. a gas chromatograph with two different analysis protocols, or separate analytical units such as a gas chromatograph for accurate performance evaluation in combination with a non-dispersive infrared unit for fast qualitative analysis.

A further level of sophistication is appended if deactivated catalyst samples are intended to be regenerated independently from all other catalyst samples staying under constant reaction conditions. As the technical layout of such a system is far beyond the scope of this chapter.

2.4.3
Environmental Catalysis: DeNOx Catalysis

Stage II testing for environmental catalysis for automotive applications holds several challenges that differ substantially from those of catalysis for partial oxidation and petrochemical applications. Many challenges in automotive catalysis for exhaust applications have been met in recent years: CO-oxidation for diesel-engine applications, three-way technology has become state of the art, and the first successful soot combustion systems have been implemented in new automobile generations. One of the largest challenges that catalyst development still has to face is the development of DeNOx catalyst systems for diesel applications. Generally speaking the difference with regard to three-way technologies implemented in Otto-engines are the much lower exhaust temperatures at which the catalyst has to operate for the conversion of NOx into nitrogen and oxygen (new legislative regulations will not consider the conversion into N_2O as satisfactory). No catalyst-formulation is yet known that can decompose NOx under lean conditions at low temperatures without rapid catalyst aging. State-of-the-art and also currently commercialized technologies for DeNOx for diesel-engines are based on NOx-storage catalysts that typically consist of a storage compound such as barium-oxide, which can store NOx as nitrates, in combination with a noble metal component such as platinum that serves as the center of NOx reduction. Dynamic cyclic operation under rich/lean conditions makes the storage catalyst a technologically advisable solution for diesel-engine applications. Critical, of course, are sulfur-containing fuels as they generally poison the storage component by forming stable sulfates, thereby deactivating NOx storage.

From a Stage II screening perspective it is of the highest importance to mimic the operations, namely the dynamic operation of the lean/rich cycling, which take place in the real engine. The challenge in constructing the screening reactor lies in the ability to construct gas supplies that allow switching between rich and lean cycles over a period of seconds. The general conditions applied during this dynamic cycling are NOx levels between 250 and 400 vppm, up to 50 vppm of propylene,

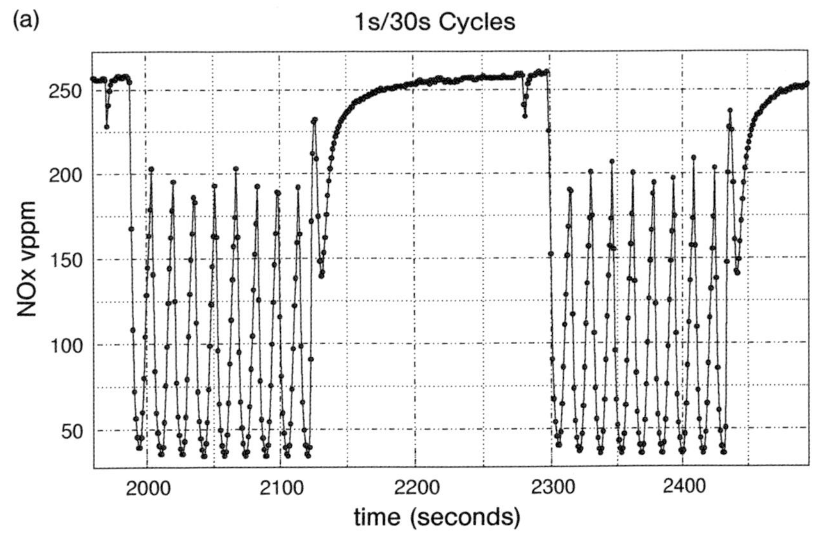

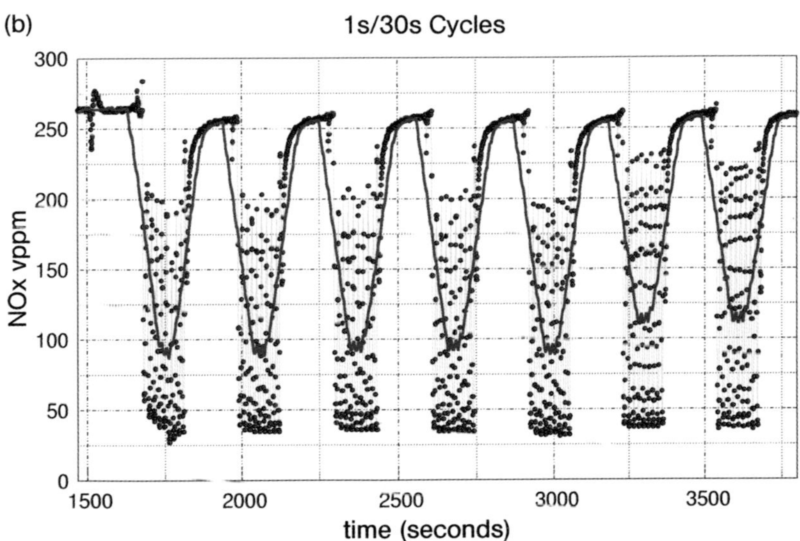

Fig. 2.16 NOx signal as a result of dynamic rich cycling operation. NOx signal as a function of time (top). Averaged NOx signal as a function of time (bottom).

8 vol.% of water, the level of CO may vary between 500 and 10 000 vppm, the level of hydrogen between 2500 and 10 000 vppm, all depending on the oxygen content, which varies from 0 and 7 vol% between the lean and rich phases. Lean phases are generally run between 50 and 300 s, the intermittent rich spikes generally last from 2 to 15 s with the oxygen level going down to 0 vol%. The whole operation is run at gas hourly space velocities between 50 000 and 150 000 h^{-1}.

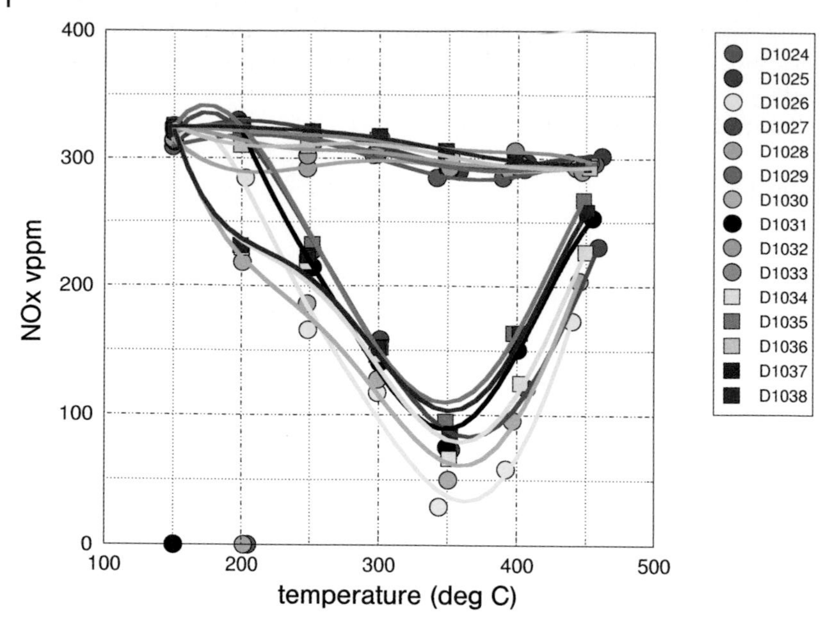

Fig. 2.17 Averaged NOx signal as a function of temperature for Ba-Pt-based catalyst samples.

Fig. 2.16 illustrates the NOx response to the rich/lean-cycling operation, following typically the periodical enrichment of the test gas with a periodical decrease in NOx content. Fig. 2.16 (bottom) shows the averaged NOx signal for the curve in the top part of the figure; clearly, the overall NOx level is reduced.

To compare the performance of the catalysts in the screen, the average NOx-signal as a consequence of the cycling operation is plotted (Fig. 2.17) as a value of the respective catalyst temperature for several different storage catalysts based on the barium/platinum-metal technology. The catalyst performance clearly shows a large scatter over a broad range of two different catalyst classes, one not being very active for NOx conversion the other being very attractive as candidates for further pursuance. The best candidates show a very attractive NOx conversion window with close to 95% NOx conversion below 400 °C.

2.5
The Challenge of Ultrahigh-Throughput Screening

Considering the facts discussed in the previous sections it becomes clear that the central point in catalyst development with high-throughput methodologies is the definition and identification of lead structures. If such lead structures have already been explored by conventional screening or are known from the literature, it is doubtfull whether Stage I screening can still add value to a screening process, unless the search for different lead structures or a fast but qualitative mapping are relevant around certain lead components, shall be addressed during the screening.

When no relevant lead structures are known for a given target conversion, it is of greatest importance, prior to stepping into a Stage II screen, to have a given set of lead compounds at hand that are prospective candidates for further pursuance. Before the definition of such prospective lead compounds a Stage II tool will be degraded, using it for a Stage I screen takes only minor advantage of the high accuracy of its analytical method and has the drawback of comparatively low capabilities regarding catalyst containment and throughput. Evidently, a major effort at the start of a project needs to be dedicated to the fast finding of new leads to provide the Stage II screening with sensible library layouts.

The desire to complete this period of primal lead definition as fast as possible justifies the implementation of architectures in the screening process that can be subsumed under the name of ultrahigh-throughput screening. The domain of ultrahigh-throughput screening is clearly dedicated to Stage I screening and the corresponding system layout is designed for highest efficiency not only with regard to the already discussed features such as accuracy of the analysis method and resulting depth in information density, but with a high focus on streamlined and seamless logistics of synthetic effort, and handling of both active components and reactor formats. We will also draw attention to the inherent and unambiguous tagging of entities of active compounds and the storage of materials libraries in the context of multifunctional reactor formats. In the last two sections we introduce a conceptual approach for efficient materials synthesis and integration of materials libraries in multifunctional reactor formats that allows materials analysis, materials testing and storage in a single reactor format.

2.5.1 Catalyst Synthesis: the Split & Pool Principle

The basis of the single-bead concept is the use of single shaped bodies as the catalytic material of interest. These particles may in principle be of any shape, but spherical particles are usually employed. In accordance with approaches known from combinatorial chemistry, such spherical particles are called "beads", although they fulfil very different functions in comparison to their application in combinatorial setups in organic and bio-chemistry.

Each bead represents one catalyst as a member of a library of solid catalysts. It may consist of an unporous material like α-Al_2O_3 or Steatit or of typical porous support materials – such as Al_2O_3, SiO_2, TiO_2, or the like. These beads can be subjected to different synthesis procedures and sequences like impregnation, coating etc. In addition, full mixed metal oxide catalysts can also be formed as spherical particles.

Using single beads for high-throughput experimentation has several advantages. First, such beads are comparable to well-known fixed-bed catalysts and the synthesis pathways can be the same as for conventional materials, which may facilitate scale-up procedures. A number of common preparation procedures for these beads are available and can be carried out in standard laboratory environments. The second main advantage is that each bead represents a single entity that can be handled independently from other beads or the final reactor configuration. Starting from masterbatches, a large diversity of materials can easily be prepared. Furthermore, different beads may be treated individually, by for instance being subjected to different preparation steps or pretreatments such as calcination or steaming, rather than handling the complete library, as is necessary for substrate-bound thin or thick film catalysts. This allows the use of synthesis procedures different from the parallel approach. The potential use of ex situ synthesis procedures may furthermore present a significant advantage, especially in microchemical systems, as an in situ preparation of the catalysts in the reactor could cause problems with contamination or thermal and chemical stability of the reactor material. From an inorganic synthetic viewpoint it is interesting to see how far the "bead principle" can be exploited for the synthesis of complete combinatorial libraries.

Comparing the chemistries involved in the creation of organic libraries with those of inorganic materials, one can conclude that for the synthesis of organic libraries a series of well-defined synthetic steps lead to libraries of defined chemical entities, whereas for inorganic materials library creation a series of well-defined synthetic steps lead to libraries of complex materials that at first are characterized by the synthetic pathway employed to synthesize the respective library member, but the single synthetic operation and overall pathway may not have the same effect over the whole library of complex materials [40–42]. Is a true combinatorial approach, revising the complexities of inorganic materials synthesis, possible? From a direct transition step of the organic to the inorganic synthesis principles, except for very few examples, the direct answer has to be no. However, if one considers including the intelligent combination of synthetic steps in a combinatorial manner for the generation of complex inorganic materials libraries into the group of combinatorial synthetic chemistries, then synthetic approaches to inorganic materials can also be considered combinatorial. From this point of view the synthetic possibilities offer a wide range of perspectives for variation. Purity of compounds, potential aging of precursor solutions, sequence and decent history of steps during addition of compounds offer a range of synthetic parameters that can be varied and used to create diverse libraries.

As discussed above, among the different methodologies for the efficient synthesis of libraries of a high degree of diversity the Split & Pool principle offsets other synthetic methodologies by far. With relatively few operations, compared with standard parallel operations a large number of variations of library members can be achieved. The essential steps of the synthetic effort are shown in Fig. 2.18; from a principle point of view the synthetic steps of splitting and pooling are identical, the steps wherein a synthetic operation is performed, of course, differ due to the different demands arising from the different chemistries required for inorganic solid state chemistry. Naturally, different considerations have to be taken into account than for organic synthetic strategies.

The synthetic approach of inorganic materials via the Split & Pool methodology is closely connected to the question of synthesizing materials on or within physical entities that will contain the library member throughout the synthetic process. If physical entities are employed as "carriers" that show sufficient chemical inertness towards the screening process, a separation of the final library compound may

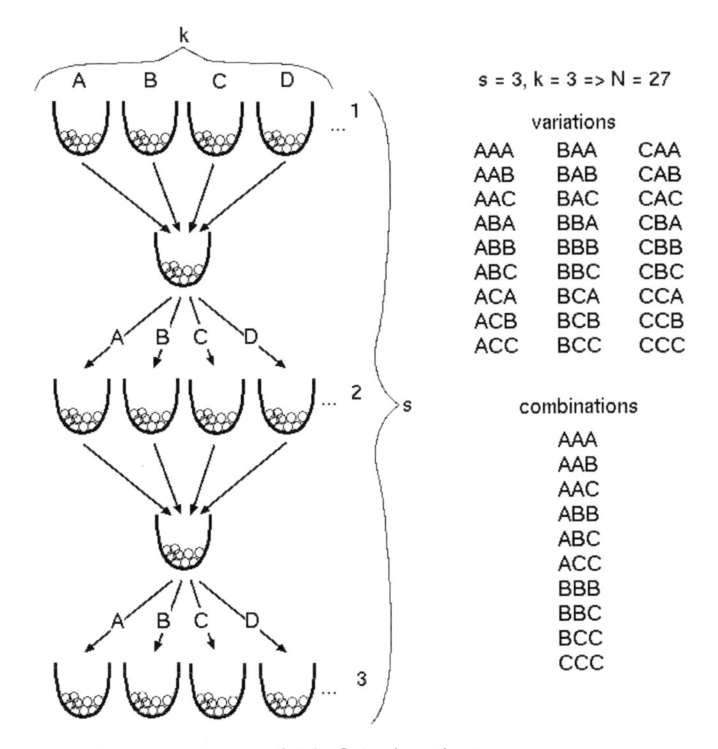

Fig. 2.18 Essential steps of Split & Pool synthesis.

not be necessary at the end of the synthetic steps. We will later see that in some cases it may also be desirable to employ "carriers" that may even become a part of the desired material (and subsequently the "carrier" entity a library member as a whole), so that the separation of the created compound is no longer an issue.

Generally, solution-based approaches for the generation of inorganic split and pool libraries have substantial advantages over approaches where solid phases are introduced as chemical sources during the different synthetic steps. Solution chemistry offers, potentially, a wide range of synthetic opportunities that can be exploited not only for the purpose of parallel synthesis but also for synthetic steps for Split & Pool library creation.

For "carrier"-free synthetic approaches, techniques employing different vessels like cans or alternatively "containers" that can absorb and contain considerable amounts of solutions with regard to their own volume are a central component of the synthetic effort [43]. Fig. 2.19 illustrates a possible approach to a substrate-free synthesis using vessels or cans as the alternative "carrier"-system. Starting from solutions (V_1) these can be "split" into vessels or cans, becoming a library of different identity by either adding components (E1 or E2) or by altering the solution with any form of physical treatment (heat, evaporation of solvent, or the like). The pooling step in Fig. 2.18 becomes, of course, in many cases absurd if all the solu-

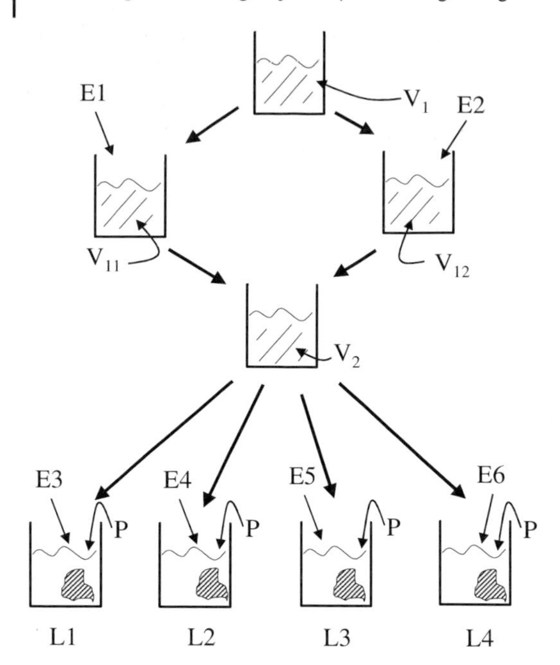

Fig. 2.19 Pathway for obtaining "carrier"-free inorganic solids via a solution/precipitation based Split & Pool methodology.

tions of the first Split step are united again. However, if only some "carriers" or parts of the solutions from the "carriers" are united during the pooling step this step may very well make sense from a synthetic point of view and can be usefully employed, although in many cases the number of pooling steps may be smaller than the number of splitting steps. Through the addition of further components (E4 to E6) the complexity of the mixture is increased and the final performance of a precipitation step P leads to a solid library member L1 to L4. The elegance of the procedure lies in the fact that encoding is eased if "carriers" as vessels or cans are employed which can be equipped with tags. For many applications a shaping step of the resulting library members will be mandatory. Obtaining a powder as a product leaves a range of methods for shaping open, which can be adapted according to the demands of screening procedure.

Another possible option for the application of the Split & Pool principle for obtaining "carrier"-free libraries in the form of powders is the use of combustible "carriers" that can contain solutions. Various organic polymers are available for this purpose; the ones that proved to be of highest value were acrylates or hydrophylized polyolefins, suitable for the needs of an approach based on aqueous chemistries or other highly polar solvents. In this case the similarity with regard to organic synthetic approaches seems very high, from the point of view of the polymer properties discrete functional groups that bind to the entities added during the synthetic steps are obsolete; the only compatibility that is demanded is that the carrier is capable of absorbing the solvent.

Alternative approaches connected to synthetic work based on the use of "carrier" systems have a similar background as the one described above. The synthetic "backbone" can be purely inorganic, organic or of composite nature. As many steps in the generation of functional inorganic materials may involve a thermal treatment of the given material, a clear preference is for materials that tolerate thermal treatment steps without deterioration of the shaped entity. This indicates why fully inorganic materials are often preferred as "carriers".

For a large range of applications impregnation or equilibrium adsorption methods, including subsequent drying or calcination steps for solvent removal, can be employed as a synthetic approach to obtain large libraries of materials [44, 45]. Carriers that can be employed are available or can be customized. At hte Aktiengesellschaft standard "carrier" materials of pure and mixed oxides in the form of beads are kept in stock for synthetic operations. A wide range of chemistries can be performed from a synthetic point of view based on these "carrier" materials. Standard materials include Al_2O_3, SiO_2, TiO_2 and ZrO_2 and various other mixed oxides. Usually, a spherical shape is preferred as it eases transfer operations and minimizes potential damage during transfer.

From a chemical point of view it has of course to be mentioned that these inorganic "carrier" materials are far from inert towards catalytic reactions (apart from select exceptions) and will in many cases become an active component as such via the synthetic pathway followed. This is also an important differentiation with regard to the traditional organic Split&Pool synthesis.

Alternatives to impregnation procedures coating procedures are another possibility for the application of different chemical compositions to a "carrier"-body. A sequential procedure can though lead to an onion like structure of the deposited components, which may not always be desirable, but for some applications one may also exploit this concept. In combination with impregnation procedures, coating procedures may prove to be most powerful and efficient.

To illustrate the Split & Pool synthesis principle a synthetic example of a library based on alumina "carriers" impregnated with different metal solutions was chosen as a case study. The potential application focus on such a library is its use in a catalytic application. The inorganic library of approximately 3000 Mo-Bi-Co-Fe-Ni/γ-Al_2O_3-library candidates was synthesized by applying the following metal salts in aqueous solution as metal precursors in the Split & Pool-synthesis: ammonium-heptamolybdate [$(NH_4)_6Mo_7O_{24} \cdot 4H_2O$] (0.025 M), $Bi(NO_3)_3 \cdot 5H_2O$ (0.075 M), $Co(NO_3)_2 \cdot 6H_2O$ (0.25 M), $Fe(NO_3)_2 \cdot 9H_2O$ (0.25 M), and $Ni(NO_3)_2 \cdot 6H_2O$ (0.1 M). As ceramic "carrier" the sieve fraction with 1 mm diameter of γ-Al_2O_3 beads (CONDEA) was employed. The synthesis started with alumina "carrier" beads (2 g), corresponding to approximately 3000 single, uniform beads. The dividing steps of the starting library (Split step) into four equal portions into porcelain dishes were achieved by weighing on a laboratory bench balance (LA 1200 from Sartorius). The impregnation steps (adsorption impregnation with 80% solution of the water uptake, water uptake of the Al_2O_3 support: 588 μl) were performed on a Multi-PROBE IIEX (Packard) robotic system, which was additionally equipped with an Titramax 100 shaker (Heidolph), on which the porcelain dishes with the appropriate part of the alumina beads were shaken continuously while adding the precursor solutions. After evaporating the solution and a waiting time of 20 min, after each impregnation step each library part was dried for 16 h at 80 °C followed by a final calcination step at 400 °C subsequent to all the impregnation steps. A sketch of the Split & Pool-steps and a synthesis scheme are shown in Figs. 2.18 and 2.19, respectively.

For optical analysis via imaging of the final Split & Pool-library an AX 70 microscope was used (Olympus). The apparatus was equipped with an automated *xyz*-table and autofocus-control. The software package analysis 3.0 (SIS) was used for hardware control and analysis means. Pictures were taken in the multi-imaging alignment mode at a magnification of five.

To verify the success of the different deposition steps, in combination with the Split & Pool methodology X-ray fluorescence was chosen as an analysis tool. Elemental analysis was performed by X-ray fluorescence analysis on an Eagle II μProbe (Roentgenanalytik) with Rh-Kα radiation. An essential feature is the small diameter of the measurement spot: The X-ray beam is focused by a multi-capillary system to a 50 μm spot on the sample surface. XRF analysis of the 8×12 catalyst library selection (Fig. 2.20) was routinely accomplished automatically by an elemental mapping at a pattern of 512×400 points, equally distributed over the rectangular library field, each point (50 μm diameter) was measured for 300 ms.

In the above Split & Pool-synthesis example five different chemical precursor solutions, varying the metal atom (Mo^{VI}, Bi^{III}, Fe^{II}, Co^{II}, Ni^{II}), at four different concentration levels (0, 0.002, 0.1, 1 wt%) were used. From the underlying mathematics the resulting Split & Pool library obviously includes $4^5=1024$ different members, covering all possible element combinations of the above-mentioned metal oxides and concentrations. As the pool of 3000 library members exceeds the number of chemically different members (only 1024 compositionally different materials), the library has a certain redundancy. Each material is approximately represented by three library members of the S&P-library. For ease of discussion we decided to specify this factor, which represents the redundancy of the library as a function of chemically identical library members, as the factor of over-determination for a certain library. For the total number of ca. 3000 Al_2O_3-beads, the factor of over-determination of the materials library in this case was 2.9. Realistically, in many cases of routine Split & Pool synthetic operation where redundancy of the library can be assumed, a factor of over-determination between 1.2 and 1.5 from a statistical point of view versus screening efficiency is sufficient, assuming a reproducible synthesis procedure. Still an over-determination of a factor of 3 is in good accord with the theoretical considerations described above.

For materials characterization and a consistent proof of the synthetic principle we chose μ-XRF with a spot size of 50 μm as the analytical tool, allowing acceptable speed, resolution and data quality for the synthesis approach, designed for primary screening (stage-1-screening). Elemental mapping over an 8×12 matrix with a 512×400 resolution (0.3 s per point) produced a high-resolution XRF mapping in ca. 17 h. This data quality is not required for routine analysis of Split & Pool libraries. An elemental mapping over a 96-fold or 384-fold single-bead reac-

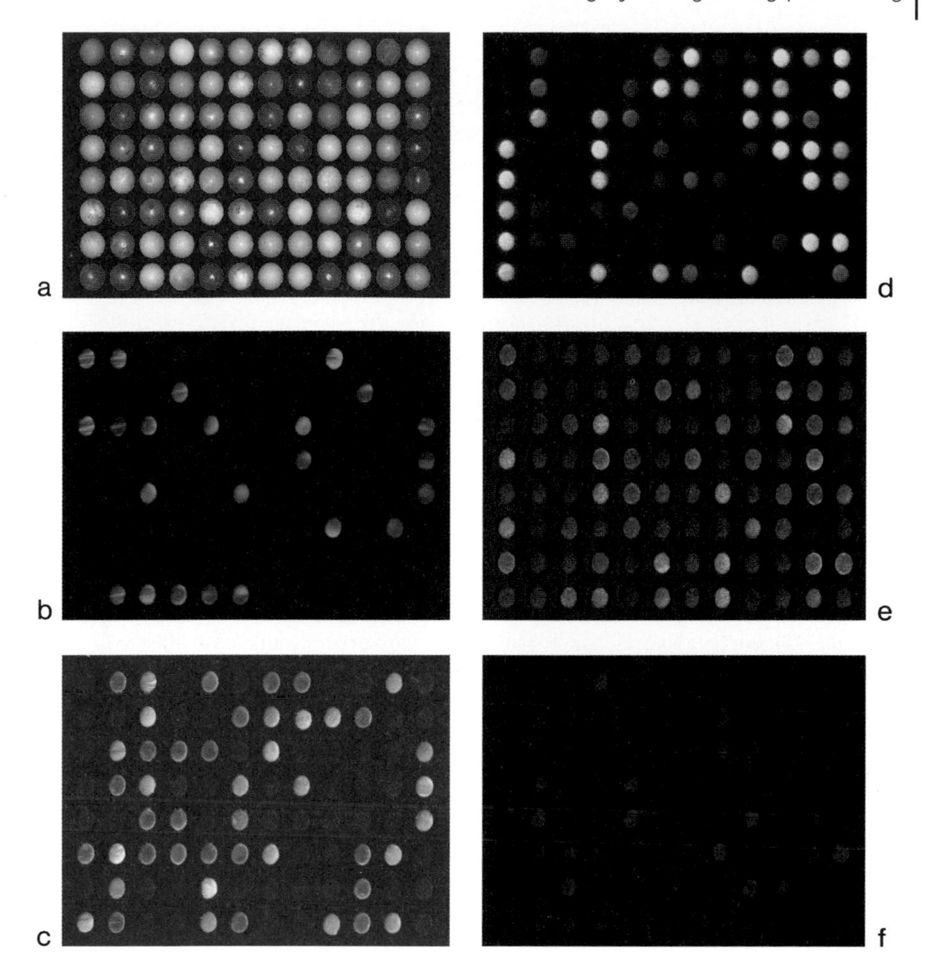

Fig. 2.20 Optical microscope picture (a) and μ-XRF results of an 8×12 array of randomly selected materials (diameter 1 mm), coincidentally selected from the library, described in the text, (b) Ni distribution, (c) Fe distribution, (d) Mo distribution, (e) Bi distribution, and (f) Co distribution.

tor (SBR) used for storage and analysis of the library members with sufficient data quality can be performed overnight. Fig. 2.20 represents the XRF mapping results (Fig. 2.20b–f) for 96 randomly selected materials as well as a picture taken by optical microscopy (Fig. 2.20a), composed of several frames from the above-mentioned selection of beads. The photograph of the 96 beads already illustrates the homogeneity of the multiple impregnation procedure, showing a homogeneous coloring of all the library members. The reddish beads correspond to materials with a high iron content, the bluish beads are materials with a high cobalt content. For other library members, no obvious correlation between color and composition could be observed.

For ease of viewing, the XRF elemental mapping results were arbitrarily color coded (Fig. 2.20 b–f); the more intense the color the higher is the amount of the corresponding metal in the material. Fig. 2.20 shows that each material has a different chemical composition. The four concentration levels (0, 0.002, 0.1, 1 wt.%), as described in the experimental part, could be estimated from the mapping results (Fig. 2.20 b–f); differentiation between 0 and 0.002 wt% could only be realized with our equipment though with very careful calibration. Thus, only three different levels in concentration can be recognized in Fig. 2.20 (b–f). Recognizing the level of precursor concentration is obviously sufficient to determine the synthesis history of interesting library members for the given synthetic example.

The routine workflow to identify outstanding candidate compositions of library members of a Split & Pool-library is fairly straightforward: a detailed post-identification of single leads or hits takes place after the screening in our single-bead reactor system (SBR) [46]. Fig. 2.21 displays the routine workflow, integrating the newly developed Split & Pool-synthesis, screening the materials in the single bead reactor system, identifying materials with the best catalytic performances, optionally storing or archiving the whole library in the reactor as a "materials chip" for further treatment (activation, pre-treatment, or testing in another target reaction), and evaluating the whole data set, i.e. with a combination of evolutionary algorithms and neural networks for modeling and for predicting a new library generation, to be synthesized in the next step. Next to the presented new synthesis tool, the central part of this workflow is our SBR system. This multifunctional architecture fulfils several needs, which are described elsewhere [46]: besides the basic features such as, of course, screening reactor, it acts as "zero-background" carrier for the characterization part (i.e. μ-XRF, μ-XRD, ...etc.), because of its chemical inertness it can be used for synthetic pre-treatment steps (activation, hydrothermal treatment, reduction, etc.), and due to its advantageous format it serves as storage and archiving device with the positional information of each material.

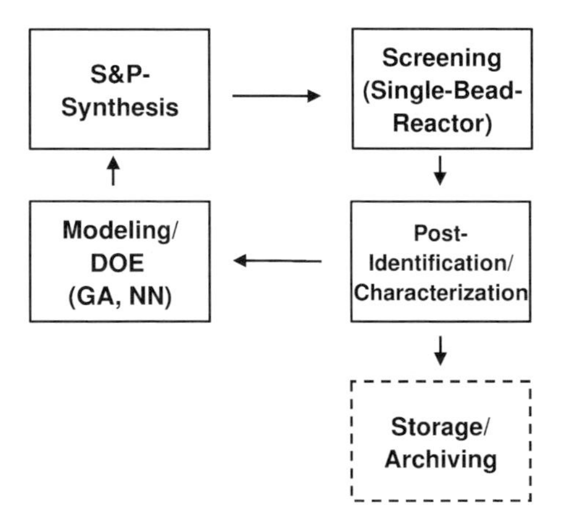

Fig. 2.21 Routine workflow for Split & Pool combinatorial synthesis and testing and archiving.

2.5.2
Catalyst Testing: Integrated Reactor Formats as Critical Key Components

Fig. 2.22 shows the routinely employed 384-fold and the prototype 105-fold single bead reactor. As a size comparison, a conventional match is shown; the 384-fold single bead reactor is approximately half the size of a credit card. The flexible concept of the single bead reactor allows the adaptation of several analysis techniques. Analysis can take place either by sequential techniques that retrieve samples from single library members, typically via a transducing capillary, or via integral techniques monitoring all samples at a time. Suitable fast sequentially analysis techniques are conventional MS, GC/MS, GC, or dispersive or non-dispersive IR. Though mostly sequential techniques are used in current screening setups, it is also feasible to adapt integral analysis techniques that allow true parallel analysis by applying, for example, IR-thermographic or PAS (photoacoustic spectroscopy) techniques. Notably, the IR-transparent architecture of the silicon-based reactor contributes to the versatility in conjunction with IR-thermography.

Several sets of experiments were carried out to evaluate the performance of the different single-bead reactors in continuous flow catalytic experiments. 2% Pd/Al_2O_3-beads (1 mm diameter) were prepared from α-Al_2O_3 by wet impregnation and subsequent calcination at 450 °C. The mass of one catalytic bead is ca. 700 µg.

Such Pd-catalysts are active in hydrogenation reactions. Together with inactive reference beads, a 105-parallel single-bead reactor is charged and tested in the partial hydrogenation of 1,3-butadiene at 60 °C. Using a suitable, heatable flange system, base and top part are pressed together and mounted into a test rig. The reactants are continuously supplied by mass flow controllers and preheated to the reaction temperature before entering the reactor. Product analysis is carried out in a fast sequential mode by spatially resolved mass spectrometry, as described in Section 3.1. Fig. 2.23 shows a section of the reactor base part together with the sampling capillary in front of a specific product outlet.

Exemplary results of a screening are shown in Fig. 2.24. At a residence time of about 10 ms, conversion degrees of ca. 50% are achieved on the active catalysts (Fig. 2.24 a). In contrast, only very low conversions are measured for the inactive

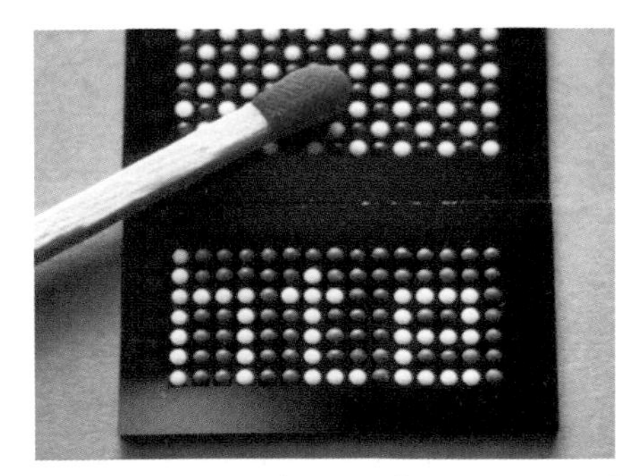

Fig. 2.22 Integrated reactor formats employed at hte Aktiengesellschaft.

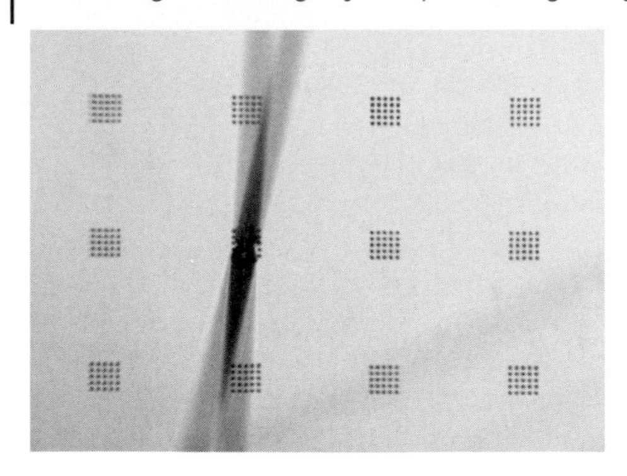

Fig. 2.23 Sampling capillary talking a sample at the reactor base part.

samples. This indicates that cross-talking between adjacent members of the library can almost be neglected. Furthermore, only slight differences in the cumulative selectivities to the n-butenes are measured (Fig. 2.24 b). Fig. 2.24 (c) shows the results for a residence time of 32.5 ms. Compared with a residence time of 10.8 ms, the resulting conversion degrees for the active catalysts are higher, while a still very low conversion is measured for the inactive reference catalysts. Nevertheless, a slight trend for the conversion degree from position B 15 to F 1 can be observed for both residence times. As identical Pd-catalysts from the same preparation batch were used, it could be shown that these trends arise from the reactor configuration. To exclude a potential reactant misdistribution induced by the flange or heating system, the library was rotated by 180° and the experiment reproduced under the same reaction conditions (10.8 ms) (Fig. 2.24). Once again, a trend for the conversion is observed, now in the opposite direction. The reason for this trend could be identified by measuring the pore sizes of the single pores of the pore membranes across the reactor. Rather large deviations in the pore size and the resulting pore area that is directly responsible for different resistance to fluid flow were observed. As indicated from the experiments shown in Fig. 2.24 (a) and (b), longer residence times lead to an increase in conversion. Thus, the observed trend in the conversion degree from one region of the reactor to another can be explained by the differences in the pore sizes of the pore membranes.

A major improvement in terms of flow distribution could be achieved by a more accurate fabrication process that was used for the development of 384-parallel single-bead reactors with a 24×16 format. Prior to applying this reactor to catalytic reactions, the pore sizes were determined by optical microscopy. The pore size distribution (Fig. 2.25) indicates excellent uniformity of the pores. The 384-parallel single-bead reactor can, for instance, be applied in a partial oxidation reaction. In an initial experiment, the reactor was filled with inactive and active catalysts according to the pattern shown in Fig. 2.22 upper right side in a chessboard-like arrangement. A continuous flow screening experiment is carried out at 400 °C and a reactant flow of $V=1$ ml min^{-1} per bead, using again spatially resolved mass spectrometry for sequential product analysis. The normalized results for the conversion degree are shown in Fig. 2.26. According to the mass spectrometric in-

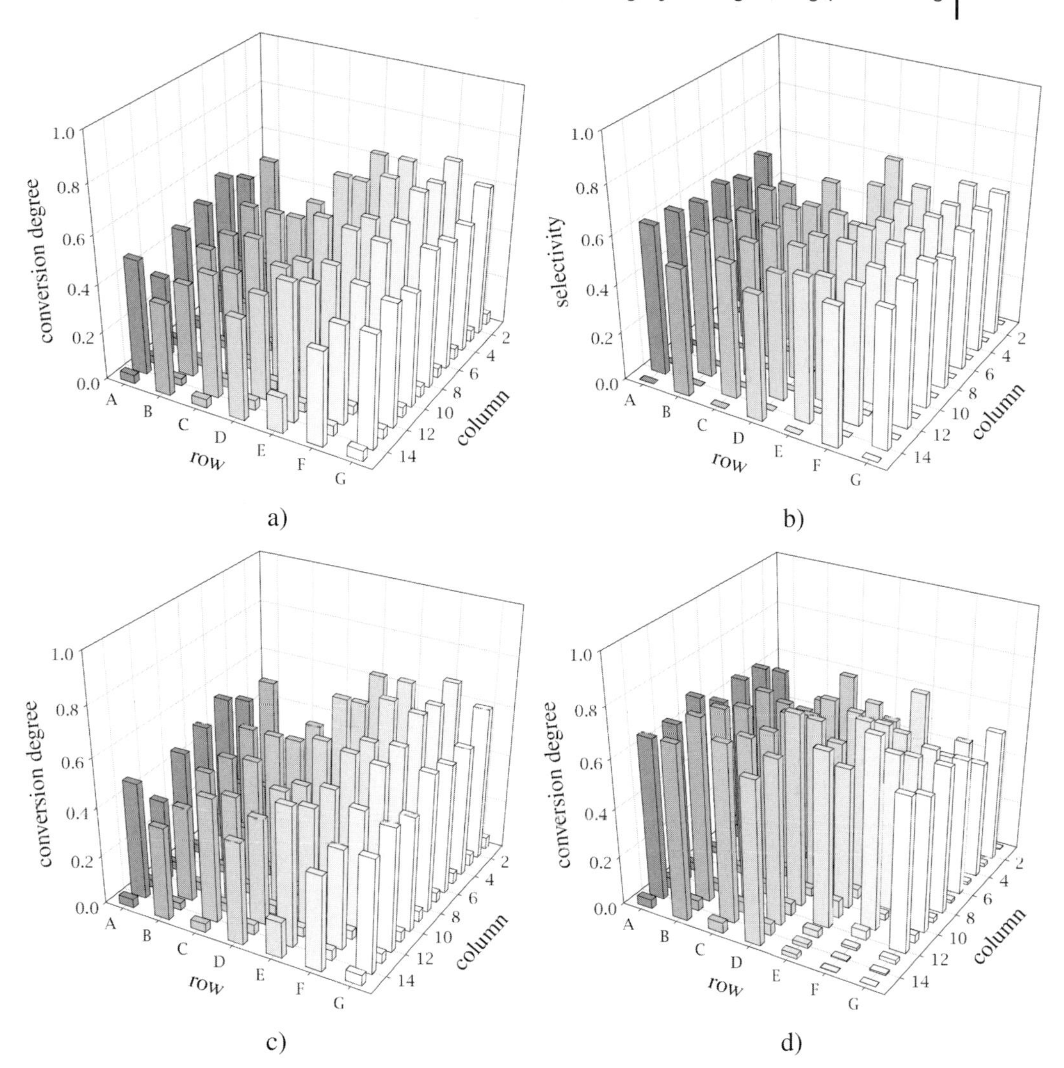

Fig. 2.24 Results from a library of Pd-α-Al_2O_3 catalysts and inert materials in the partial hydrogenation of butadiene.

tensities, white represents low conversions, while dark colours represent high conversions of the hydrocarbon. The maximum conversion in this experiment is ca. 60%. As can be seen, the results for the conversion degree correspond very well with the filling pattern of the reactor. Cross-talking between adjacent reaction chambers can be observed to some minor extent, although not limiting the applicability of the single-bead reactor for primary screening purposes. Furthermore, it is expected that cross-talking can efficiently be reduced by bonding the base and the top part of the reactor, instead of just pressing them together as presented here.

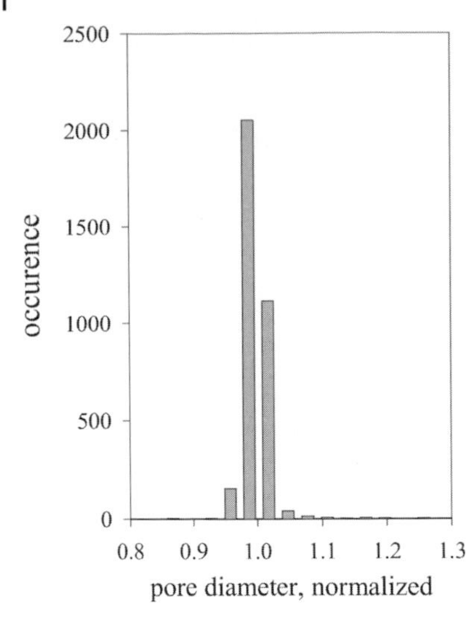

Fig. 2.25 Pore size distribution of the 384-fold reactor system.

In conclusion, the single-bead reactor concept presents a novel system architecture for the high-throughput primary screening of solid catalysts. Using single beads as catalytic materials and a scalable method for the fluid distribution, array densities of up to 60 catalysts per cm^2 have successfully been applied. The array density vastly exceeds those used by other groups. Thus, it is possible to realize parallel reactors for heterogeneous catalysis with a much higher degree of parallelization, as shown successfully with the 384-parallel single-bead reactor. The developed multifunctional reactor architecture enables handling, identification and storage of the catalysts and the automation and integration of the complete high-throughput workflow. This includes the possibility to adapt different sequential or parallel analysis techniques for the detection of interesting products. The versatility of the system facilitates the application in different reaction classes. The operation at conditions relatively close to conventional and/or secondary catalyst test systems can be ensured.

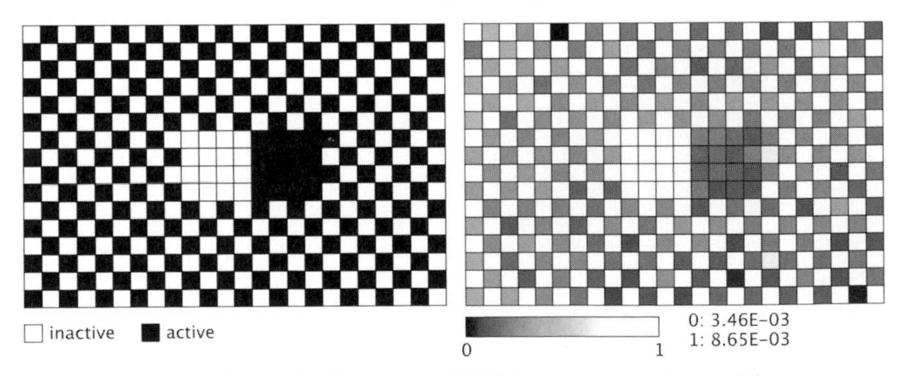

Fig. 2.26 Conversion degree of a library in the 384-fold reactor system in a partial oxidation reaction (left: scheme of catalyst library; right: conversion degree).

2.6 Summary and Outlook

We have illustrated the importance of integrated approaches for Stage I screening and technically relevant Stage II screening methodologies. Especially in a Stage I screen the logistics of synthesis and screening play a very important role with regard to desired throughput and finally actual screening speed. In concordance with this goes the loss in information that is experienced due to the boundary conditions of the Stage I screening. Careful decisions have to be taken when to continue a project on to the Stage II level in order to still be able to keep up the pace of screening and also not lose the level of accuracy needed to bring forward the catalyst development. For Stage II gas-phase screening we have discussed the current state-of-the-art screening for catalysts for partial oxidation, refinery catalysis and environmental catalysts applications. Technically the reason for screening in single reactor systems is no longer a given, as essentially all of the features that can be implemented in classical laboratory reactor systems are also available for Stage II reactor systems.

Future challenges for the development for Stage I systems will for sure be mainly in the area of fast and reliable analytical methods. There are great incentives for the future development of new non-invasive techniques which allow truly parallel screening. Primary exploration of optical analysis methods which hold a great promise in that regard has only just begun and will remain an interesting field in the future. The value of integrated reactor formats is apparent and the future will show how far the concept can be expanded. Especially here fully automated units for synthesis and screening which only need minor manual work steps is a short term goal. Integrated reactor formats with the surrounding architecture are potentially also the first technologies that can make it to the instrument market as true bench top units.

For Stage II screening a main issue will also be the improvement of analytical capabilities. Although full carbon balances for reactions can be established and run in a routine screening mode, especially separations for complex mixtures occurring in petrochemical applications are still challenging and may still slow down speed of screening. Also in the future Stage II system design will remain a domain of custom specifications and adaptation to customer needs and process demands.

Based on the above made conclusions it becomes clear that high-throughput experimentation has kept the promise of becoming an important additional tool for catalysis research in academia and industry.

2.7 References

1 US 4099923, Sohio, E. Milberger (1977).
2 L. Singoredjo, R. Korver, F. Kapteijn, J.A. Moulijn, *Appl. Catal. B1* (1992) 297.
3 S. Taylor, J. Morken, *Science* 280 (1998) 267.
4 A. Holzwarth, H.W. Schmidt, F. Maier, *Angew. Chem. Int. Ed. Engl.*, 37 (1998) 2644.
5 P. Cong, R.D. Doolen, Q. Fan, D.M. Giaquints, S. Guan, E.W. McFarland, D.M. Pooraj, K. Self, H.W. Turner, W.H. Weinberg, *Angew. Chem., Int. Ed. Engl.*, 38 (1999) 484.
6 C. Hoffmann, A. Wolf, F. Schüth, *Angew. Chem.*, 111 (1999) 2971.
7 C. Hoffmann, H.W. Schmidt, F. Schüth, *J. Catal.*, 198 (2001) 253.
8 S. Thomson, C. Hoffmann, S. Ruthe, H.W. Schmidt, F. Schüth, *Appl. Catal. A: General*, 230 (2001) 253.
9 F. Kapteijn, J.A. Moulijn in: G. Ertl, H. Knözinger, J. Weitkamp (Eds.), *Handbook of Heterogeneous Catalysis*, Vol. 3, VCH, Weinheim (1996).
10 J. Newsam, F. Schüth, *Biotechnol. Bioeng.*, 61 (1999) 203.
11 www.gilson.com
12 www.chemspeed.com
13 www.zinser-analytik.com
14 www.accelab.de
15 A.B. Stiles, *Catalyst Manufacture-Laboratory and Commercial Preparations*, Marcel Dekker, New York (1983).
16 U. Rodemerck, P. Ignaszewski, M. Lucas, P. Claus, M. Baerns, *Chem. Ing. Technol.*, 71 (1999) 872.
17 O. Buyevskaya, D. Wolf, M. Baerns, *Catal. Today*, 62 (2000) 91.
18 O. Buyeskaya, A. Brückner, E.V. Kondratenko, D. Wolf, M. Baerns, *Catal. Today*, 67 (2001) 341.
19 S.J. Taylor, J.P. Morken, *Science*, 280 (1998) 267.
20 F.C. Moates, M. Somani, J. Annamalai, J.T. Richardson, D. Luss, R.C. Willson, *Ind. Eng. Chem. Res.*, 35 (1996) 4801.
21 A. Holzwarth, H.-W. Schmidt, W.F. Maier, *Angew. Chem.*, 110 (1998) 2788.
22 C.M. Snively, G. Okarsdottir, J. Lauterbach, *Angew. Chem.*, 113 (2001) 3117.
23 C.M. Snively, G. Okarsdottir, J. Lauterbach, *Catal. Today*, 67 (2001) 357.
24 F. Schüth, O. Busch, C. Hoffmann, T. Johann, C. Kiener, D. Demuth, J. Klein, S. Schunk, W. Strehlau, T. Zech, *Top. Catal.*, 21 (2002) 55.
25 P. Cong, R.D. Doolen, Q. Fan, D.M. Giaquints, S. Guan, E.W. McFarland, D.M. Pooraj, K. Self, H.W. Turner, W.H. Weinberg, *Angew. Chem., Int. Ed. Engl.*, 39 (1999) 484.
26 S. Senkan, S. Ozturk, *Angew. Chem., Int. Ed. Engl.*, 38 (1999) 791.
27 M. Orschel, J. Klein, H.W. Schmidt, W.F. Maier, *Angew. Chem., Int. Ed. Engl.*, 38 (1999) 2791.
28 P. Claus, P.D. Hönicke, T. Zech, *Catal. Today*, 67 (2001) 319.
29 S. Senkan, *Nature*, 394 (1998) 350.
30 P. Cong, A. Dehestani, R. Doolen, D.M. Giaquinta, S. Guan, V. Markov, D. Poojary, K. Self, H. Turner, W.H. Weinberg, *Proc. Natl. Acad. Sci. USA*, 96 (1999) 11077.
31 T. Johann, A. Brenner, M. Schwickardi, O. Busch, F. Marlow, S. Schunk, A. Brenner, *Catal. Today*, 81 (2003) 449.
32 DE 19830607 (1998), EP 9911311, hte Aktiengesellschaft, A. Brenner et al. (1999).
33 WO 00/51720 S. Bergh et al. Symyx Technologies (1999).
34 S. Bergh, J.R. Engstrom, A. Hagemeyer, C. Lugmair, K. Self, P. Cong, S. Guan, Y. Liu, V. Markov, H. Turner, W.H. Weinberg, High-Throughput Screening of Combinatorial Heterogeneous Catalyst Libraries, IMRET 4, 4th International Conference on Microreaction Technology, Topical Conference Proceedings, Atlanta, Georgia: AICHE National Spring Meeting (2000) oral presentation.
35 K. Grob, *Split and Splitless Injection for quantitative Gas Chromatography*, Wiley-VCH, Weinheim (2001).
36 H.-J. Huebschmann, *Handbook of GC/MS*, Wiley-VCH, Weinheim (2001).

37 J. Klein, W. Stichert, W. Strehlau, A. Brenner, D. Demuth, S.A. Schunk, H. Hibst, S. Stork, *Catal. Today*, 81 (2003) 329.

38 DE 19622331 (1996), A. Tenten et al., BASF AG.

39 WO 001997046506 (1997), A. Tenten et al., BASF AG.

40 Á. Furka, F. Sebestyen, M. Asgedom, G. Dibo, *Abstr. 14th. Int. Congr. Biochem.*, Prague, 5 (1988) 47.

41 F. Balkenhohl, C. von dem Busche-Huennefeld, A. Lansky, C. Zechel, *Angew. Chem.*, 108 (1996) 2436.

42 D. Demuth, K.-E. Finger, J.-R. Hill, S.M. Levine, G. Löwenhauser, J.M. Newsam, W. Strehlau, J. Tucker, U. Vietze, ACS Symposium Series 814, *Combinatorial Materials Development*, Ed. R. Malhotra, Oxford University Press, Oxford, p. 147 (2002).

43 DE 10053890 (2000), J.M. Newsam et al.. hte AG.

44 J. Klein, T. Zech, J.M. Newsam, S.A. Schunk, *Appl. Catal.*, A254 (2003) 121.

45 Y. Sun, B. Chan, R. Ramnayaranan, W. Leventry, T. Mallouk, *J. Comb. Chem.*, 4 (2002) 569.

46 T. Zech, J. Klein, S.A. Schunk, T. Johann, F. Schüth, S. Kleditzsch, O. Deutschmann, *High Throughput Analysis: A Tool for Combinatorial Materials Science*, Eds. A. Radislav and E.J. Amis, Kluwer Academic/Plenum Publishers, New York (2003).

3
High-Throughput Workflow Development: Strategies and Examples in Heterogeneous Catalysis

H. Sam Bergh

3.1
Introduction

High-throughput research methods have been applied and developed in the field of heterogeneous catalysis at a rapid rate over the past few years [1–46]. The use of high-throughput methods aims to increase the rate of experimental learning and discovery and to reduce the time-to-market of new or optimized catalyst technologies. The experimental process consists of sets of activities that can be grouped into several categories: experimental design, materials synthesis, materials characterization, materials testing, and data analysis. These are typically executed in a sequential fashion, and then repeated iteratively with the results of previous rounds of experiments feeding into the experimental designs of subsequent experiments. In a high-throughput experimental process, each of these activities requires significant data handling as each activity interacts with all other activities, generating a large amount of data. Data handling is accomplished through a database (Fig. 3.1). The hardware and activities required to undertake a specific high-throughput research effort comprise a workflow, and before initiating a discovery program, the workflow is developed and validated. A workflow is made up of various hardware and software components that together function with a high degree of integration. This chapter will review some of the strategies for high-throughput

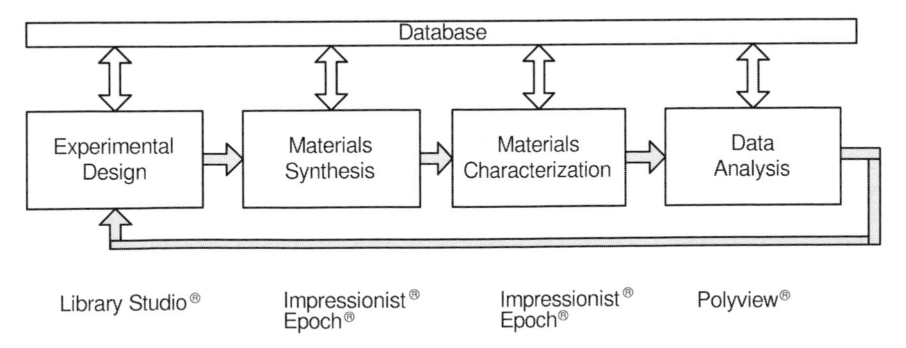

Fig. 3.1 High-throughput workflow block-diagram.

High-Throughput Screening in Chemical Catalysis
Edited by A. Hagemeyer, P. Strasser, A. F. Volpe, Jr.

ISBN: 3-527-30814-8

experimentation and the corresponding hardware that has been developed to enable these strategies. Finally, specific examples of how these components have been used together in specific chemistry research workflows and selected results from these research efforts will be reviewed.

3.2
High-Throughput Methods

High-throughput research efforts are analogous to traditional research efforts in that hypotheses are formed based on current knowledge, experiments are designed to test these hypotheses, and the results of these experiments are analyzed to support or disprove hypotheses. For example, consider a search for an M1-M2-M3 ternary catalyst with improved activity and selectivity compared with catalyst ABC for reaction Y at condition Z. Depending on the set of metals to be screened and the number of compositional levels, the required number of experiments can be quite large. In addition, variables such as metal precursors, loading, promoters, pre-treatments, post-treatments, supports, etc. can be extremely important, influencing such properties as dispersion and particle size, phase and phase segregation, acidity, surface area and pore structure, etc. Including multiple levels of synthesis variables increases the number of experiments further and requires intelligent experimental design to cover the desired experimental phase space efficiently and effectively. Screening key variables such as temperature, pressure, space velocity, feed composition and in situ activation/deactivation must be studied in many cases.

One of the main strategies employed in high-throughput methods to increase throughput is the use of hierarchical screening. This is the use of multiple levels of synthesis and testing as depicted in Fig. 3.2. The primary scale affords the highest number of experiments, typically greater than 50000 experiments per year. Secondary scale research typically allows greater than 5000 experiments per

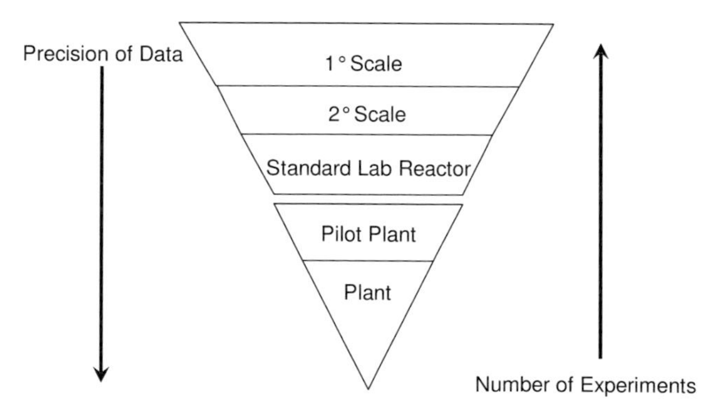

Fig. 3.2 Hierarchical screening stages.

year. This scale is closer to traditional laboratory-scale hardware in form (as compared with primary scale), but with smaller dimensions, higher levels of integration and parallelization, and increased automation. The best materials from the secondary screen then are run in a standard laboratory reactor scale, or tertiary scale. This hierarchical approach has proven very effective at rapidly and efficiently screening catalysts. Each of the experimental stages assists the other. The primary screen identifies catalysts or conditions that are unsatisfactory, eliminating the need to screen them further. In addition, the secondary screening stage allows a certain level of false positives in the primary screen to be acceptable. Clearly, false negatives are much less desirable than false positives and experimental strategies have been developed to bias towards false positives.

The use of hierarchical screening fundamentally affects experimental design. For example, the use of gradient libraries can reduce the probability of false negatives if the response surface is sufficiently smooth. In addition, such gradient libraries can have their element locations randomized in synthesis and screening and then reassembled in the software for analysis. This reduces sensitivity to systematic errors in synthesis or screening hardware. Since the synthesis chemistry and process variables may be non-optimal in the primary screen, it can be prudent to select regions of local performance maxima for secondary testing even when their performance is inferior to other locations in phase space (or compared with a standard). This strategy helps avoid false negatives by passing potentially interesting candidates to the secondary screen. Because combinatorial methods quickly lead one to very large numbers of experiments, it is important to understand rate-limiting steps, or bottlenecks, in workflow components and to develop approaches to removing or reducing them. Throughput limitation can occur in any stage: the experimental design, the materials synthesis, the materials testing, and/or the data analysis. Increasing the throughput can be accomplished by improving experimental design strategies (decreasing the numbers of experiments), increasing the speed of each activity, and, more preferably, through numbers and speed. As new methods and hardware are developed and used, the bottleneck can move between activities. A goal in high-throughput workflow development is to match the throughput of the various workflow activities so that on average no one component is the primary bottleneck. Below is a description of some of the approaches for each experimental stage. These can be applied within each of the primary or secondary stages.

1. Experimental design: software tools for intelligent experimental design including:
   - a priori statistical information
   - chemical, economic, and/or hardware constraints
   - prior knowledge from previous experiments

2. Materials synthesis and characterization: hardware technology and methodology:
   - automation
   - parallelization and pipelining
   - error checking and quality control

3. Data analysis: software tools for processing large quantities of data effectively:
   - visualization
   - modeling and parameter estimation
   - filtering, sorting, hypothesis testing

3.2.1q
DOE – Designing Experiments Based on Statistics

The challenge of developing good experimental designs can be addressed in part by the use of experimental design software tools and statistical methods to determine good ways to sample phase space and organize experiments, e.g. organizing experiments into blocks to reduce uncontrolled factors, using repeats to understand statistical variations in results, and using randomization to reduce systematic errors. Many of these methods are scientific common sense, but having software to help implement such methods across large numbers of experiments in high-throughput research makes it practical. A large body of experimental protocols and design strategies to incorporate simple to complex a priori statistical information into designs has been developed [47]. Some of these methods are based on identifying experimental factors and factor interactions. High-throughput methods can be very helpful in doing this quickly.

3.2.2
Constrained Optimization – Independent Variables

Independent experimental variables, such as elemental components and process conditions, are often constrained by economic, safety, or physical limits. Incorporating these constraints into the library design can significantly reduce the number of samples to be screened, and in the case of safety are essential. Such constraints are ideally suited to be incorporated into software library design tools and data analysis tools so that sufficiently unattractive experiments can be given lower priority than more attractive experiments.

3.2.3
Constrained Optimization – Dependent Variables

Conversely, factors involved in evaluating a new catalyst for commercial viability often include many dependent variables such as catalyst cost, space–time yield, conversion, selectivity, capital costs, separation costs, catalyst lifetime, etc. For example, the objective of a catalyst research program may be to replace an existing catalyst with a new material exhibiting a higher activity, or longer lifetime, keeping all other variables fixed. In this case a single performance metric is needed to rank catalysts. In other cases, modification of plant hardware or process flow is possible and the number of parameters that should be measured or calculated for each potential catalyst will be multiple and the performance metric is actually a vector. Once the metric(s) of fitness has been identified, constraints on these metrics can be incorporated into the data analysis and viewing software.

3.2.4
Methods to Include Synthesis Hardware Constraints

In practice, synthesis and screening hardware can have significant constraints, and optimizing experimental designs to include synthesis constraints can lead to significant throughput gains. Constraints can be throughput constraints, batch constraints or process constraints. One example of how process constraints can be included in the experimental design is through Monte Carlo methods [48]. In one example, Monte Carlo methods were used to optimize the synthesis of arrays of materials based on the desired compositional elements and the physical constraint imposed by the dynamic masking synthesis hardware. The sequence of depositions and shutter motions for each library were optimized over the set of desired compositions. This method can be applied quite generally and effectively to high-dimensional experiments where complex synthesis or processing constraints exist.

3.2.5
Process Simulation for Hardware Bottleneck Identification

Library synthesis bottlenecks can be due to the physical time required to synthesize arrays of materials. These bottlenecks can be addressed by identifying the rate-limiting steps in the synthesis and eliminating them. However, if the number of steps in a synthesis is large and variable, identifying bottlenecks can be a challenging stochastic problem. One approach we have used is to analyze a workflow using a process simulation software package to estimate capacity requirements and queue requirements for different hardware components and processes. Such analyses can be used to help specify hardware requirements in early development and can help identify where automation should (or should not) be applied. Within a single hardware component, the throughput can be increased through several routes, including increasing the speed of the process at hand, improving the process itself, by increasing the parallelization within the stage, and by numbering up the hardware.

3.3
Workflow Components

To implement high-throughput experimental strategies, the hardware is prerequisite. A range of parallel and high-speed hardware has been developed for creating heterogeneous catalysis workflows. Gas–solid-fixed bed reactors are applicable to a wide range of chemistry and will be the focus here. Reaction hardware can take on a number of different configurations and scales. At the primary screening level, wafer-based catalysts can be used to form small scale-fixed beds whereas at the secondary scale, sized bulk catalyst materials are used. Listed below are some of the hardware that has been developed to date at Symyx for the synthesis and screening of catalysts. These workflow components are then assembled as appropriate to address a specific chemistry. Some details of these systems are presented in the following paragraphs.

Primary synthesis:	1. Sol-gel and evaporative synthesis 2. Impregnation synthesis
Primary screening:	1. Scanning mass spectrometer 2. Massively parallel microfluidic reactor
Secondary synthesis:	1. Evaporative synthesis 2. Impregnation synthesis 3. Hydrothermal synthesis
Secondary screening:	1. A 48-channel fixed-bed reactor
Tertiary screening:	1. Lab-scale reactors.

3.3.1 Primary Synthesis

Proprietary Symyx software packages, Library Studio® (Fig. 3.3) and Impressionist®, are used to design experiments and control synthesis hardware based on those designs [37, 38]. 76 mm and 100 mm diameter quartz wafers and 3″×3″ quartz plates were bead blasted through a physical mask to produce arrays of 11×11, 12×12, or 16×16 wells, each well ca. 3 mm in diameter. Details and examples of the synthesis procedure have been described in Bergh et al. [44] and references within. The wafer surface prior to bead blasting is silanized to form a hydrophobic surface. Bead blasting creates a hydrophilic/hydrophobic surface pattern that helps in containing liquid aliquots in their wells during synthesis.

3.3.2 Primary Synthesis: Wafer-based Sol-gel and Evaporative Synthesis

A liquid-handling robot was customized for both sol-gel [46] and evaporative wafer-based syntheses [44]. The appropriate precursors are mixed from stock solutions into micro-titer plates and then volumetrically transferred to the support wafers. Key variables in these syntheses are the drying environment and drying temperature and post treatments. The liquid or gel is dried under controlled conditions and then thermally processed in a tube furnace exposed to desired process gases. After synthesis is complete, the wafers are analyzed by XRD, SEM, and XRF to monitor structural phases, morphologies, and compositions respectively.

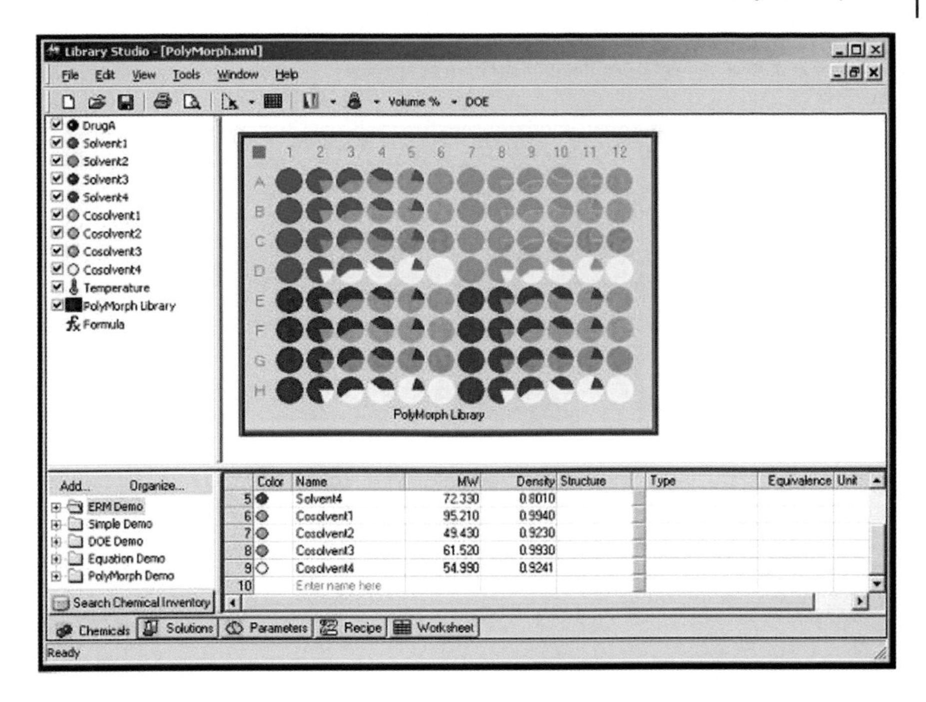

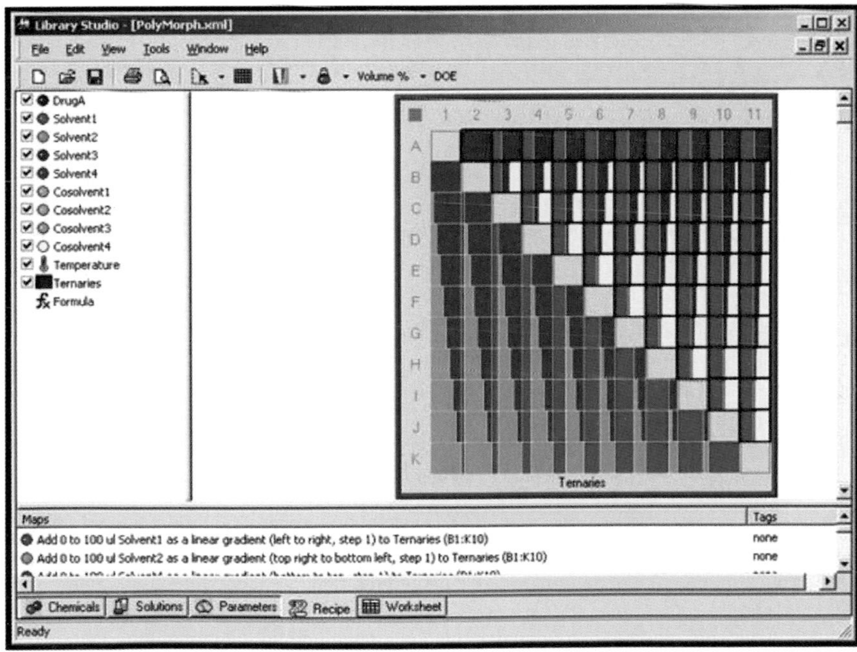

Fig. 3.3 Library Studio® software library design interfaces.

3.3.3
Primary Synthesis: Wafer-based Impregnation Synthesis

A Cavro liquid-handling robot has been customized to automate the synthesis of wafer-based impregnated catalysts. Quartz or glass substrates are prepared as described above and a slurry of support material is deposited into the substrate wells using the liquid-handling robot resulting in an array of thick film support patches. Metal precursor stock solutions are made and the liquid-handling robot is used to create solutions of these precursors based on the desired library design in a micro-titer plate. Aliquots of this micro-titer plate array of solutions are volumetrically impregnated into the support wafer. Subsequently the impregnated supports are dried, followed by calcination, reduction, and/or additional impregnation and thermal cycles. In the thermal processing step, precursors decompose to form metals, alloys, or oxides distributed on the support. The catalytic properties of the resulting catalysts can be highly dependent on the thermal processing variables. Fig. 3.4 shows hardware components for impregnation synthesis, including the liquid-handling robot, parallel liquid transfer tool, and processing furnace.

The film thickness and density of the support on the wafer can be varied in analogy to particle thickness and density. Also, the heat transfer on such format is much better than in a vial. This improves control of drying conditions across each thick-film patch as well as among patches. The support–precursor interaction affects the distribution of materials throughout the support particles in the wetting and drying processes. Drying conditions and times can be tailored, based on these

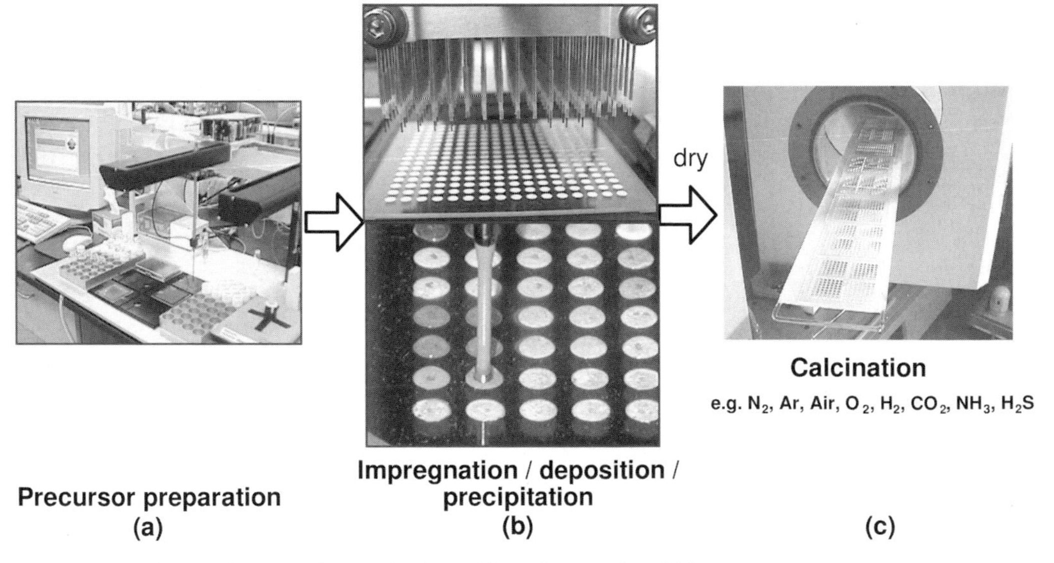

Fig. 3.4 Impregnation synthesis workflow photographs of (a) precursor preparation station, (b) parallel precursor transfer unit; serial precursor transfer tip, and (c) thermal processing oven.

interactions, to influence the distribution. Initial method development to optimize the precursors, the solvent mixtures, impregnation time, and drying times helps to insure quality catalysts.

3.3.4
Primary Screening: Scanning Mass Spectrometer

This system is a rapid-serial primary screen. It uses a probe to scan from sample to sample across a catalyst wafer, analyzing reaction products at each location via mass spectrometry and/or optical absorption [1, 2, 45, 46]. The mass spectrometer is a commercial quadruple system with a custom ionizer interface. The optical absorption cell is a custom device and uses the method of photo-thermal deflection (mirage) to measure very low levels of a specific analyte.

The sample wafer is loaded onto an x–y translation stage that registers the desired sample under the probe. The probe can move in the z-axis, allowing the head to be placed at an appropriate distance from the catalyst wafer. A CO_2 infrared laser locally heats the catalyst under the probe while neighboring samples remain at low temperature to insure neighboring catalysts remain inactive and unaffected. The sample temperature is controlled using an IR thermometer. A feed gas mixture is created using a set of digital mass flow controllers and can be mixed with vaporized liquids fed via Isco liquid pumps. This reactant stream is introduced into the reaction region (Fig. 3.5). The residence time in the reactor is determined by the gap between the probe head and the wafer and by the reactant feed rate. Both touch-down and floating nozzles have been built and used. The gap distance in the floating nozzle is controlled using an optical PID control loop. Reactant gas flows through the reactor and is pumped away via a region outside the reaction volume. This additionally insures that neighboring samples remain in their initial state until they are tested. A small fraction of the gas within the

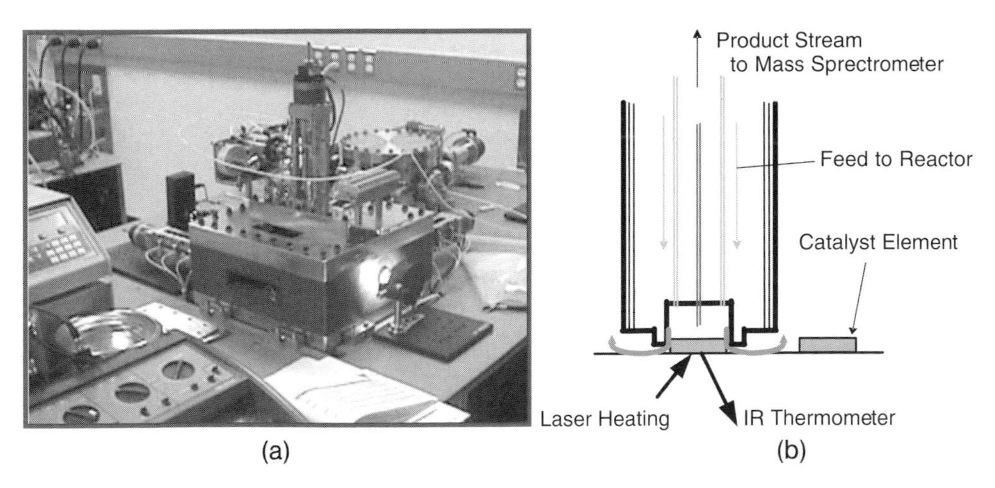

Fig. 3.5 (a) Scanning mass spectrometer; (b) schematic of scanning probe tip and its operation.

reaction volume is sampled using a capillary and fed to the detector. This system can typically run at a rate of one sample per minute corresponding to approximately 4 h for a 256 element library.

Differences in product yields can be amplified when low conversions are screened because the initial kinetics are monitored rather than integral productivities at high conversions. This can strategically be used to facilitate catalyst comparisons.

The operation of the Mirage detector can be understood as follows. A sample of reactor effluent is introduced into the small volume optical analysis chamber where a high power laser beam tuned to a unique adsorption line of the analyte of interest passes through. The result is that the gas is locally heated in the beam and the steady-state temperature distribution of the gas in the cell will have a cylindrical shape, the magnitude of the temperature gradient depending on the concentration of the analyte present and the heat loss out of the adsorbing region. The index of refraction of a gas is temperature dependent and by passing a second probe laser perpendicularly through the main beam it is refracted by the heated gas. A position-sensitive detector is used to accurately measure small changes in gas concentration. To increase the sensitivity a multi-pass geometry is employed and, additionally, the cell is pressurized with an inert gas to increase sensitivity.

3.3.5
Primary Screening: Massively Parallel Microfluidic Reactor

This system is a parallel primary screen. It uses a wafer-based primary screening reactor system for parallel reaction of catalysts and parallel detection of the reaction products (Fig. 3.6). The system consists of three hardware components: a reactor, a spray station, and an imaging station. The reactor operates by pre-mixing a feed composition and then using an upstream flow distributor to split the feed equally to each of 256 reactors. The flow restriction in this distributor is set to be at least 20 times higher than all downstream flow restrictions so that flow splitting will be independent of downstream geometry variations (e.g. catalyst loading). The flow splitting is accomplished with a microfluidic distributor plate (Fig. 3.7). The binary tree structure of the fluidic path provides identical distance and number of turns between the inlet and each outlet. This distributor uses silicon microfabrication techniques that can provide tolerances on all the critical channel dimensions of less than 0.5 μm. This is achieved by using silicon-on-insulator wafers with very high tolerance device layers. Devices fabricated with this technique have been measured and provide a flow splitting accuracy of better than 1% (min to max).

The distributor plate is sandwiched against the catalyst wafer using a gasket of polymer or graphite and the volume above each catalyst patch defines a reaction volume. A cross-section view of one of the 256 reaction/detection channels is shown below (Fig. 3.8). Feed gas enters the reaction volume from the side. For residence times greater than 100 ms the reactor behaves as a diffusionally mixed

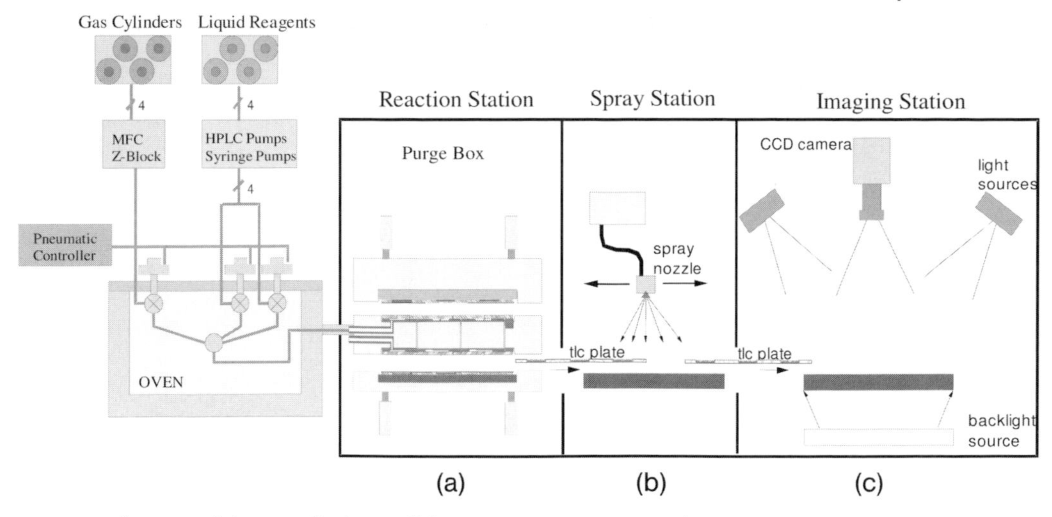

Fig. 3.6 Schematic of the microfluidic parallel screening reactor system showing (a) reaction station, (b) spray station, and (c) imaging station.

Fig. 3.7 A micromachined flow distribution manifold.

continuously stirred tank reactor (CSTR). Additionally, the small cross-sectional area of the reactor inlet channel provides a sufficiently high gas velocity at the entrance of each reactor to prevent back-diffusion. Gas flows out of this reaction volume through a hole in the distribution wafer and travels up through one of 256 independent channels in a thermal insulation block to a trapping plate which is held at an appropriate temperature to selectively trap the analyte or analytes of interest. A trapping wafer, such as a TLC plate with appropriate stationary phase and possibly pre-treated with a chemically active species, is employed. A binary-

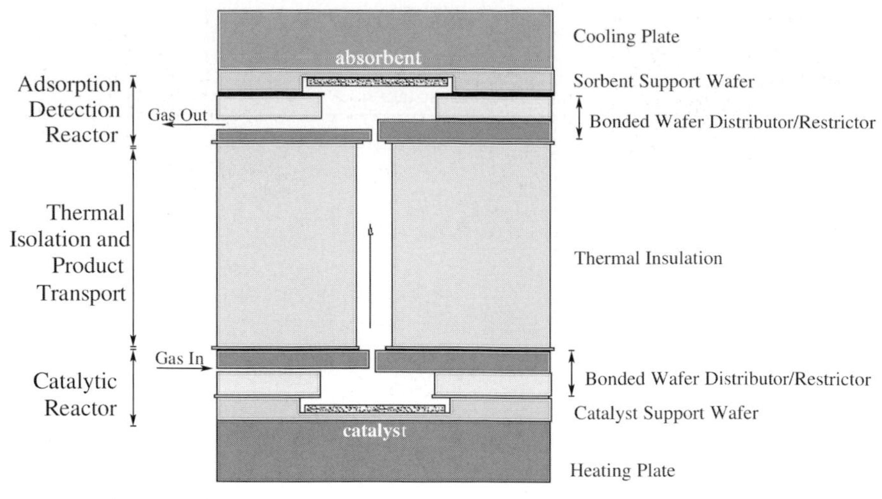

Fig. 3.8 Schematic of the Microfluidic Parallel Screening Reactor System showing a single reaction/detection channel.

tree silicon distributor with low flow restriction is used to form the adsorbent reactor and outlet manifold. As the reaction is run, the adsorption plate is held in contact with the effluent stream for a time appropriate to collect sufficient product. Plates can be run for different amounts of time to provide a wide range of sensitivities. Adsorption plates can be exchanged while the reaction is running and in this way the reaction can be monitored as a function of time or for multiple analytes. After the plate is removed from the adsorption cell, it is transferred to the spray station and uniformly coated with a dye that selectively reacts with the analyte to form a colored or fluorescent moiety. After spray development the adsorbent plate is imaged using a high-resolution CCD camera. The images are quantified using image processing and calibration standards. Transmission images are processed as absorbance units – directly related to analyte concentration. Detection of multiple analytes is possible using one plate per analyte and appropriate development and imaging of each plate. A mass spectrometer is attached to the exit manifold of the reactor and used to confirm that reaction products are as expected.

3.3.6
Secondary Synthesis: Bulk Impregnation

An impregnation synthesis workflow for secondary screening consists of five stations: precursor preparation, support dispensing, impregnation, washing, and thermal processing (Fig. 3.9). The metal precursor solutions are prepared in an array of vials from a set of stock solutions with their composition dictated by the experimental design. The supports are pre-sized, typically by grinding and sieving larger scale commercial supports. These supports are dispensed into an array of

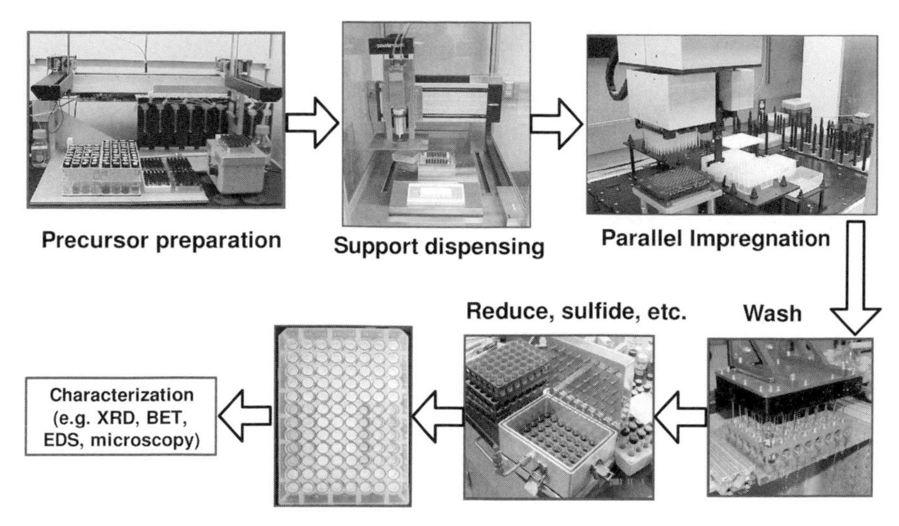

Fig. 3.9 Secondary impregnation workflow showing the precursor preparation station, support preparation, impregnation station, washing station, thermal processing hardware, and impregnated catalyst array.

vials using a powder handling robot. Obtaining good liquid–solid contact and controlling drying conditions are important for obtaining high catalyst quality and uniformity. An effective way to contact the liquid and the solid supports held in the array of vials is to mechanically fluidize the supports while adding the precursor solutions. Spraying into a fluidized bed and mechanical mixing have also been used for liquid–solid contacting. After drying and calcining, the catalysts can be optionally washed with water or other solvent and optionally thermally treated under gas flow for oxidation, reduction, sulfidation, etc. Each of these stations uses the same array-of-vials holder to allow convenient handling of bulk sample libraries. Optical microscopy inspection of catalyst particles as well as methods like SEM/EDS are used to investigate internal catalyst distributions on particles split in half.

3.3.7 Secondary Synthesis: Bulk Evaporation/Precipitation

A 24-channel parallel rapid evaporation/precipitation workflow has been developed and is shown below; it consists of automated precursor preparation, parallel solvent evaporation/precipitation, automated solids washing, parallel particle agglomeration, and parallel crushing and sieving (Fig. 3.10). Evaporation synthesis is a precipitation process initiated and sustained through an evaporation-driven concentration change in a solution of precursors. Rapid evaporation can favor the production of small crystal sizes and reduces segregation during drying. The liquid-handling robot prepares the precursor solutions within the precipitation vials. The vials are then loaded into the four heated blocks and interfaced to one of four

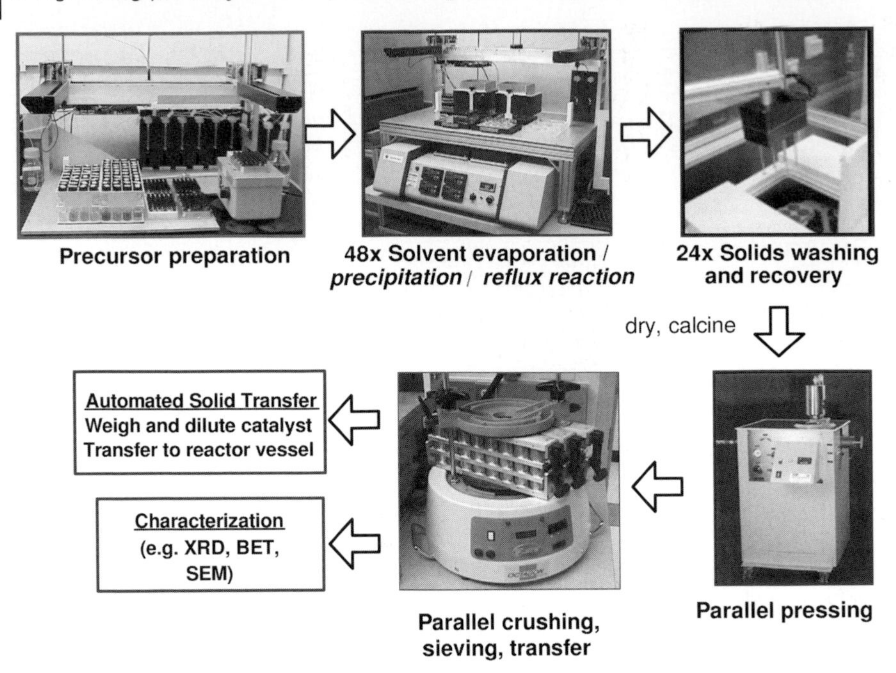

Fig. 3.10 Secondary evaporation/precipitation workflow.

modular cold traps. The cold traps are then evacuated and the solvent is cryopumped from the precursor solutions. The vials are orbitally shaken to provide a large surface area vortex to evaporate the solvent and conductively heated from below in a controlled manner to prevent liquid bumping. The dried materials are removed from their vials, calcined, optionally washed using an automated centrifuge washing station, and dried – after which they are sized for the reactor. Sizing is done by first parallel pressing the powders in elastomeric molds in an isostatic press to form dense pellets. These pellets are then sized in a 24-channel parallel grinder/siever to form controlled particle size catalysts in the range 50–250 μm. The final sized catalyst particles are transferred into vials from which they are characterized with standard analytical methods or transferred into reaction wells for catalytic evaluation.

3.3.8
Secondary Synthesis: Hydrothermal

The hydrothermal synthesis system for secondary scale screening is based on multi-well-tumbled autoclaves modules [49]. Each module contains 16 PTFE vials; each vial has an internal volume of 7 mL. The reactors can operate up to 200 °C. Many chemical and physical factors can affect the final hydrothermal synthesis outcome, making it a good application for high-throughput methods. In addition to the gel ingredients, these factors include the order of reagent addition, gel mixing, gel ageing, reaction thermal history, mixing quality during crystallization, etc. At the end of the hydrothermal synthesis the yields vary, for the specific case of MoVNb catalysts these were typically in the range of 0.5 mL. After synthesis, catalysts are collected and washed, sized and prepared for analysis. The washed materials are transferred to a wafer for analysis by XRD and SEM.

3.3.9
Secondary Screening: 48-Channel Fixed-bed Reactor

Secondary screening is carried out in parallel fixed-bed reactors sized for catalyst loadings between 15 and 500 mg. The system is a 48-channel reactor consisting of six modules of eight reactor wells. Six channels are analyzed in parallel and stream selection valves provide fluidic multiplexing between reactor vials and analytical hardware. A typical system is shown in Fig. 3.11.

Reactant feeds are generated by vaporizing liquid flows from HPLC pumps with manometric pulse dampeners or high-pressure syringe pumps and mixing this vapor with gas components metered through mass flow controllers. This reaction feed is then divided between the 48 channels equally by using flow restrictors such as silica capillaries or micromachined channels (Fig. 3.12). The capillaries feed into the inlet stand-offs of the reactor modules.

The reactor wells are pre-loaded with catalyst (diluted with silicon carbide if desired) and inserted into the module. Two graphite seals are used per well, one to pressure seal each vial and one to prevent bypass flow around the vial. A single central bolt in tension supplies the sealing force onto these seals (Fig. 3.12 b).

Each of the eight reaction product streams leaving the reactor module are fed into a Valco stream selection valve where one of the eight effluents is selected and sent to a GC for on-line analysis. The remaining channels, combined, are sent through a back pressure controller to a waste. The selected channel from each

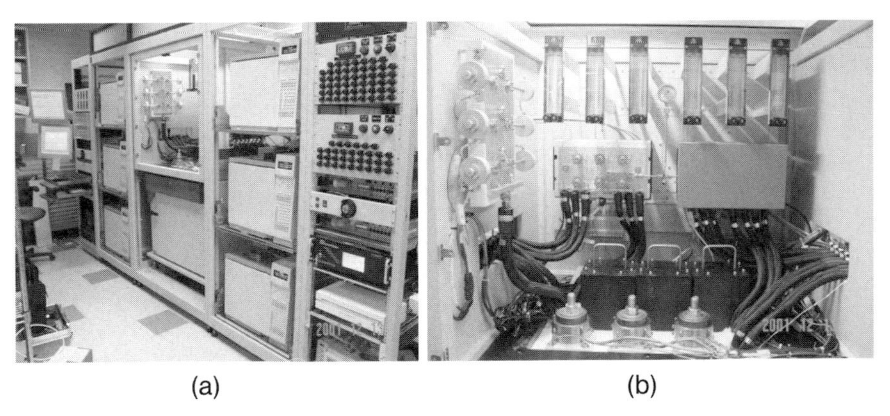

Fig. 3.11 (a) A 48-channel-fixed bed reactor system; (b) close-up of reactor plumbing and valving.

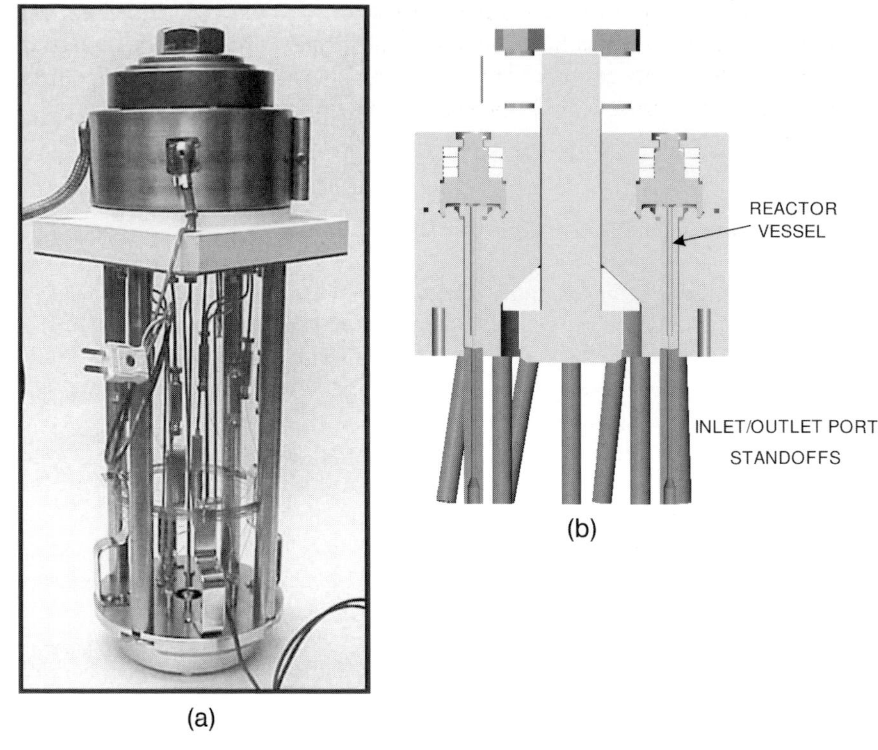

Fig. 3.12 An 8-channel-fixed bed reactor module (a) and a cross-sectional diagram of a reactor module (b).

module enters into a back pressure controller which then feeds into a GC sampling valve within an Agilent 6890 GC.

An important aspect of running a 48-channel MCFBR successfully (and parallel high-throughput hardware in general) is to ensure high service factor and data integrity. In this direction several features and subsystems have been developed and integrated into the system. After the catalysts are loaded into the reactor and the modules are sealed the system performs a leak check to insure that any leaks are below an acceptable level. If a leak is detected, then additional valving is used to isolate the location of the leak to expedite repair. After leak checking has successfully completed, the hardware executes a flow check, measuring and databasing the flow rate through each of the 48 channels to insure that they are within tolerance and allowing any deviations to be tracked. Once the system is flow and leak checked the reaction gases are mixed and stabilized. If flows don't reach set point an error is flagged. Reaction temperatures are also set and must reach a set point or an error is flagged and the run is halted. Each analytical selection valve has a bypass input to allow the feed to be analyzed on each GC. This allows analytical measurements to be normalized to reactant feed compositions and allows tracking of any compositional changes. During the screening process several parameters such as temperatures, pressures, and flow rates are monitored and if they ever deviate outside a predetermined limit the system flags an error and the reactor goes into a safe mode. At the end of a run, the flows rates can be measured again to insure that all flows are still within range. All data are stored in a database for future reference.

3.3.10 High-Throughput Catalyst Characterization

It is often of interest to characterize properties of catalysts synthesized by various high-throughput methods. For example, pore volume, acid site density, and metal surface area can be obtained by gas adsorption. Additionally, particle size distribution via light scattering, morphology via SEM or TEM, and phase and structure by XRD, and elemental composition via EDS and XRF have been implemented for use on arrays of catalysts. A Bruker GAADS XRD system and proprietary Symyx Software package, SpectraStudio®, is used to cluster, sort, and view large numbers of XRD patterns (Fig. 3.13, see page 80).

3.3.11 Tertiary Screening

Tertiary screening or traditional laboratory-scale reactors are used to evaluate the secondary leads. Depending on the richness of the data generated in the secondary screen, the tertiary screen may be focused on more complete analysis, or may be focused on other aspects of catalyst performance such as lifetime, susceptibility to poisons, or hydrodynamics.

3.4 Example: Ethane to Ethylene

The goal of this research was to develop high activity and high selectivity catalysts for the oxidative dehydrogenation of ethane to ethylene. A high-throughput workflow was assembled as follows: a primary stage consisting of a wafer-based impregnation synthesis and a scanning mass spectrometer with mirage detection was employed for screening, and a secondary-scale automated bulk-catalyst impregnation synthesis and a multi-channel fixed-bed reactor. Specific examples of the wafer-based synthesis and bulk syntheses have been reported [46]. The state-of-the-art catalyst prior to this research was a mixture of molybdenum-vanadium-niobium oxides. The workflow was validated by measuring the performance of this system and reproducing literature results.

A catalyst design strategy based on ternary metal oxides with catalytic and chemical constraints limited the desired set of catalyst compositions to about 100 000, a number that could be tested in the scanning mass spectrometer [46]. A feed gas mixture of C_2H_6, O_2, and Ar (4:1:5) was created using a set of digital

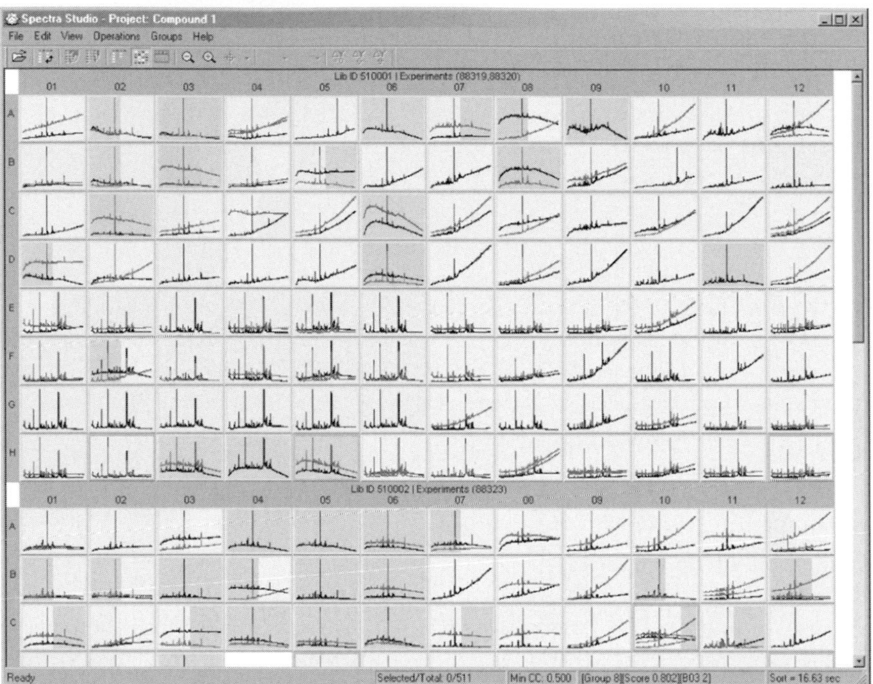

Fig. 3.13 XRD system and output from proprietary Symyx Software package, SpectraStudio® used to cluster, sort, and view large numbers of XRD patterns.

mass flow controllers, and libraries were systematically screened. A number of new nickel-based catalysts were identified. Of specific interest were Ni-Ta-Nb oxides. Scanning mass spectrometer results from an array of samples within this composition are shown in Fig. 3.14. Of specific note, the production of ethylene was 50 ppm for the Mo-V-Nb oxide standard whereas for the best sample in this class ethylene production was around 1800 ppm [46]. Samples (50 mg) were synthesized in bulk and tested in a 48-channel-fixed bed reactor. The conversion

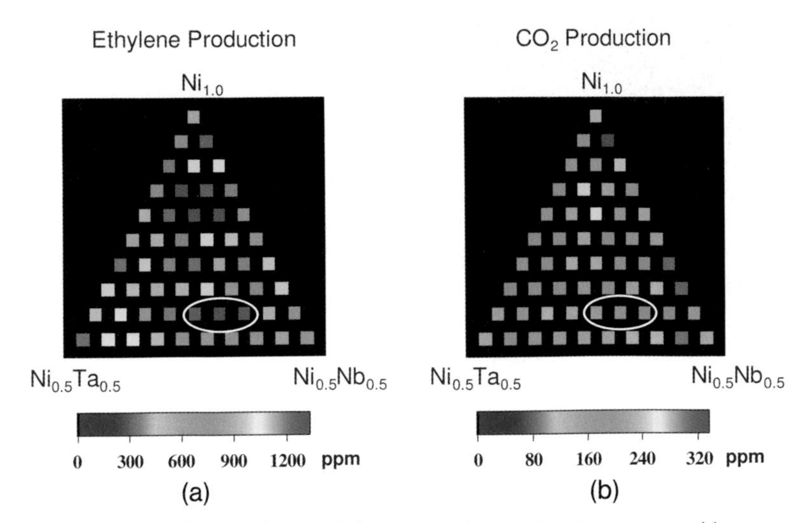

Fig. 3.14 Plots of (a) ethylene and (b) CO_2 production (ppm) as measured by photothermal deflection spectroscopy and mass spectrometry for the NiTaNb oxide library.

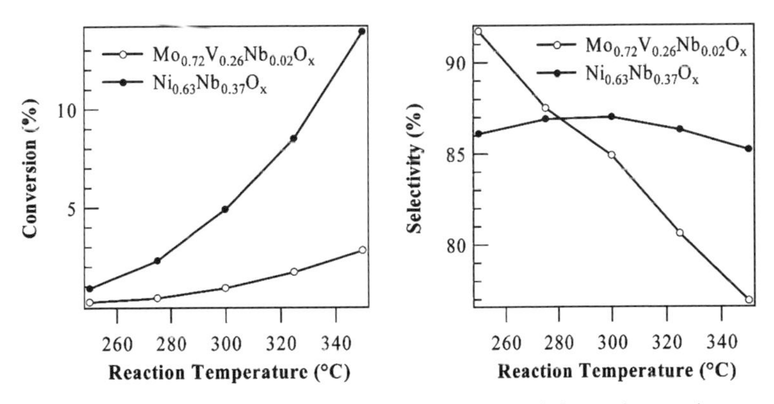

Fig. 3.15 Comparison of (a) ethane conversion and (b) ethylene selectivity between $Ni_{0.63}Nb_{0.37}O_x$ and $Mo_{0.72}V_{0.26}Nb_{0.02}O_x$ as a function of temperature.

and selectivity plots of a representative sample from the new nickel catalyst class and the Mo-V-Nb oxide are shown in Fig. 3.15. This catalyst has a markedly different temperature dependence, suggesting a different reaction mechanism. Also, the Ni-based catalysts produced low levels of CO_2, but no CO, in contrast to the Mo-V-Nb oxide catalysts, which produced both, which could be advantageous in reducing separation costs in production. To confirm results from the secondary screen, catalysts such as $Ni_{0.62}Ta_{0.10}Nb_{0.28}O_x$ were scaled up and run in a 5 g fixed-bed reactor. The 50 mg sample in the 48-channel fixed-bed reactor measured an ethane conversion of 16% and an ethylene selectivity of 80%; the 5 g reactor obtained an ethane conversion of 16% and an ethylene selectivity of 84%.

3.5
Example: Ethane to Acetic Acid

Although the direct oxidation of ethane to acetic acid is of increasing interest as an alternative route to acetic acid synthesis because of low-cost feedstock, this process has not been commercialized because state-of-the-art catalyst systems do not have sufficient activity and/or selectivity to acetic acid. A two-week high-throughput scoping effort (primary screening only) was run on this chemistry. The workflow for this effort consisted of a wafer-based automated evaporative synthesis station and parallel microfluidic reactor primary screen. If this were to be continued further, secondary scale hardware, an evaporative synthesis workflow as described above and a 48-channel fixed-bed reactor for screening, would be used.

For the microfluidic screening reactor, a detection method was developed for acetic acid based on pH detection using methyl red indicator [45, 50]. The absorption spectra of the acid and base forms of Methyl Red dye are shown in Fig. 3.16 (a). The dye is yellow in the presence of base and is red in the presence of acid. In the screening process, reaction products were absorbed into the TLC trapping plate and accumulated over time. After an appropriate exposure to the 256 reactor effluents the TLC plate was removed from the reactor and uniformly sprayed with a water–methanol solution of Methyl Red dye and KOH using the spray station. The KOH ensures that the dye is in its base form and sufficient acid accumulation on the plate will change the color from yellow to red. The sensitivity of the detection system can be modified by varying the KOH concentration in the dye solution. The acid production was quantified by imaging the base form of the dye present using a high-resolution CCD camera with a 550 nm band pass filter. Image intensities were corrected for the darkfield response and illumination field inhomogeneities using standard methods. Image array spot absorbance values were integrated to provide a measure of acid present. The detection system was calibrated using acetic acid solution standards manually spotted onto a TLC plate and processed with the spray station and imaging station, and produced a linear response as shown (Fig. 3.16 b). Interference from

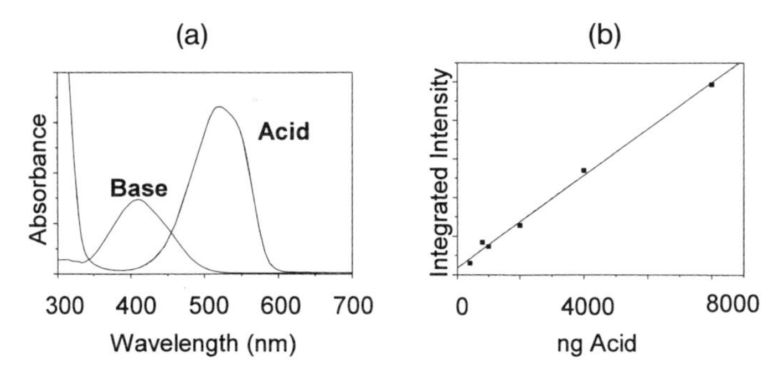

Fig. 3.16 (a) Absorption spectra of the acid and base form of the Methyl Red dye and (b) integrated intensity vs. acid amount on absorbent plate calibration curve.

CO_2 was not significant because the absorption capacity for acetic acid that condenses on the plate is much greater than for CO_2. The mass spectrum of the reactor exhaust did not exhibit peaks other than those associated with ethane, CO_2, acetic acid, and a CO (and/or C_2H_4 cracking fragment of ethane).

Initial work consisted of screening approximately 40 catalyst libraries of binary, ternary, and quaternary compositions. One of the screening strategies used was to first identify the best binary combinations of redox metals out of the set of V, Mo, Cr, Mn, Fe, Co, Ni, Cu, Ag, Re, Sn, Sb, Ti, Bi. The most promising binary combinations were then expanded to ternaries by adding a dopant metal selected from the main group metals, transition metals, and rare earth metals. Promising ternaries were expanded into "focus libraries" through up-sampling and finally the most promising focus libraries were doped uniformly with noble metals using impregnation methods and the effect of noble metal concentrations were studied.

This screening strategy is graphically depicted with selected examples in Fig. 3.17 (see page 84). The wafer of binary gradients indicated the most active binary composition was MoV. Some 18 fifteen-sample ternaries were then created as shown and a number of dopants including Nb, Ni, Sb, and Ce, and other metals showed increased acetic acid productivity. A subset of these was made into a focus library of 8 twenty-nine-sample ternaries. Pd doping on a Mo-V-Nb ternary significantly increased acetic acid productivity (Fig. 3.17 d), as reported in ref. [51].

3.6 Example: Propane to Acrylonitrile

Acrylonitrile is commercially produced from propylene by a molybdate-based catalyst that has been optimized to produce a yield of around 80% acrylonitrile. Utilizing a less-expensive feedstock, the selective ammoxidation of propane to acrylonitrile has significant potential in reducing acrylonitrile production cost. The workflow for this chemistry consisted of a primary scale evaporative synthesis station and 256-channel parallel screening reactor using a proprietary optical-based detection method. For the initial work shown here, secondary screening was done on a six-channel fixed-bed reactor.

Reaction conditions for the Massively Parallel Microfluidic Reactor were a feed of 6% C_3H_8, 7% NH_3, 17% O_2, and 70%, a reaction temperature of 420 °C, and a residence time of 0.4 s. Integrated responses from the detection plate spots were non-linear and a calibration curve was generated from standards to quantify acrylonitrile content in the detection plate spots (Fig. 3.18).

The state-of-the-art catalyst system is a Mo-V-Nb-Te mixed oxide [52]. This catalyst is quite sensitive to its synthesis and process parameters and the automated catalyst synthesis tools described above were capable of synthesizing these and other challenging mixed metal oxides successfully. The workflow was validated by synthesizing a region of the known Mo-V-Nb-Te catalyst system phase space in the primary scale and secondary scale (Fig. 3.19 a and b). Very good agreement between primary, secondary, and literature optima were obtained. One of the pri-

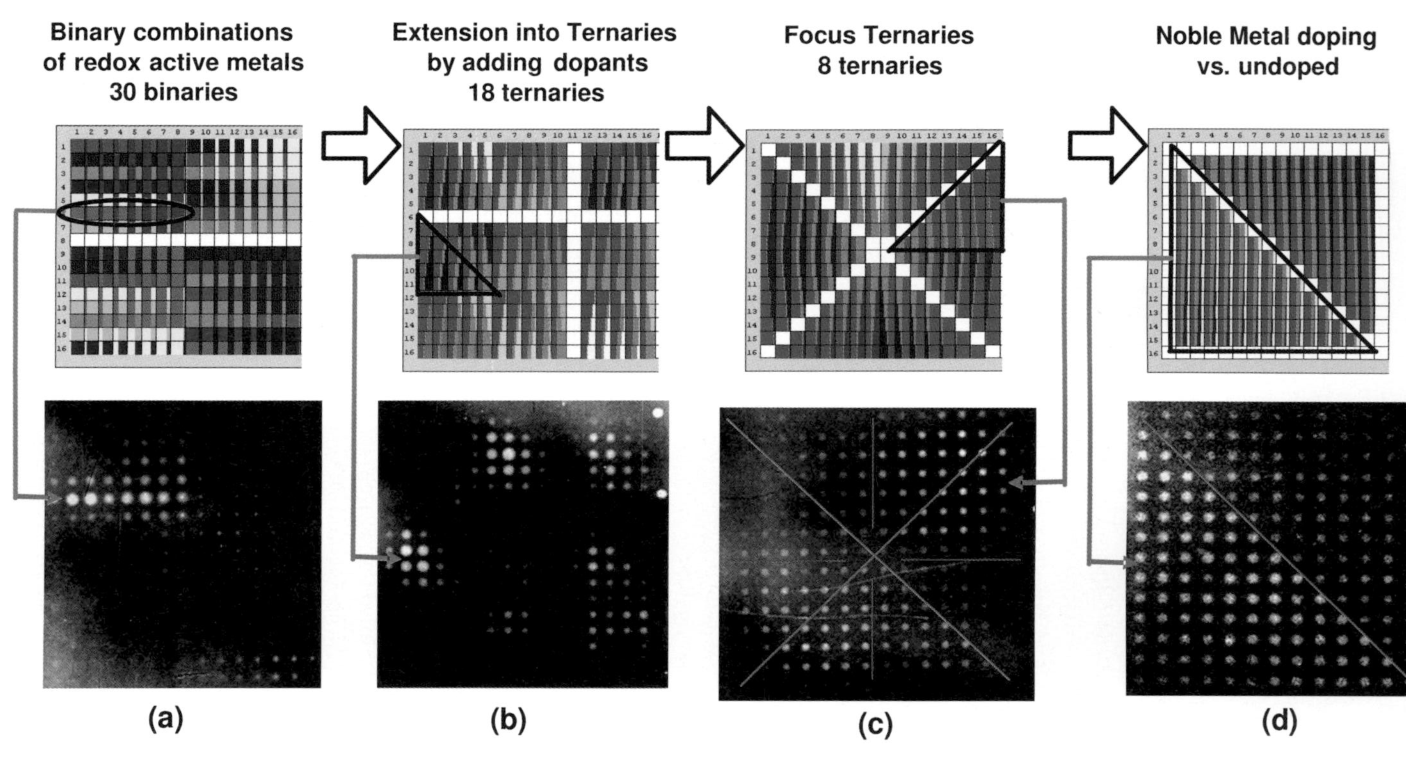

Fig. 3.17 Screening protocol showing library designs (top) and post reaction images of TLC detection wafers (bottom). Note that the white TLC plates appear black and the red spots appear white in the photo. (a) Binaries of redox active metal oxides; (b) extension of binaries into ternaries by adding dopants; (c) focus ternaries of best hits; (d) noble metal doping of MoVNb ternary. Compositional details are given in the text. Reaction temperatures for (a–c) = 375 °C, (d) = 325 °C.

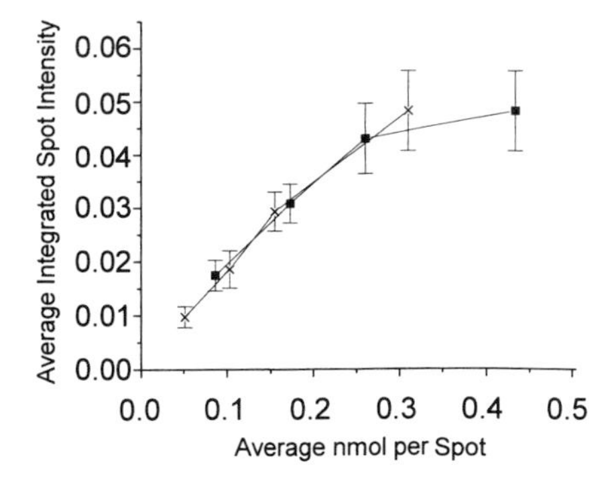

Fig. 3.18 Average integrated spot intensity vs. average nmol per spot for two gas phase acrylonitrile concentrations (× 0.00517 nmol min^{-1} per spot, ■ 0.00867 nmol min^{-1} per spot). Error bars 1σ for the average of 256 spots.

mary screening strategies employed was to take arrays of an interesting ammoxidation catalyst and screen thermal processing conditions and dopants. In one case, four catalyst wafers were prepared with a $VSbO_4$ on Al_2O_3-SiO_2 [52] in each well, and then each of the four wafers were "pre-calcined" at different temperatures (250, 350, 530, and 610 °C). Each of these was then impregnated with 31 dopants in gradients of 8-levels each and then calcined at 620 °C (Fig. 3.20). This showed a clear interaction of some dopants and the pre-calcination temperature. Some dopants were deactivating and others such as Cr, W, Fe, and Sn increased acrylonitrile yield in agreement with literature reports [53–56]. Other active samples in these experiments have not been previously reported and cannot be disclosed here.

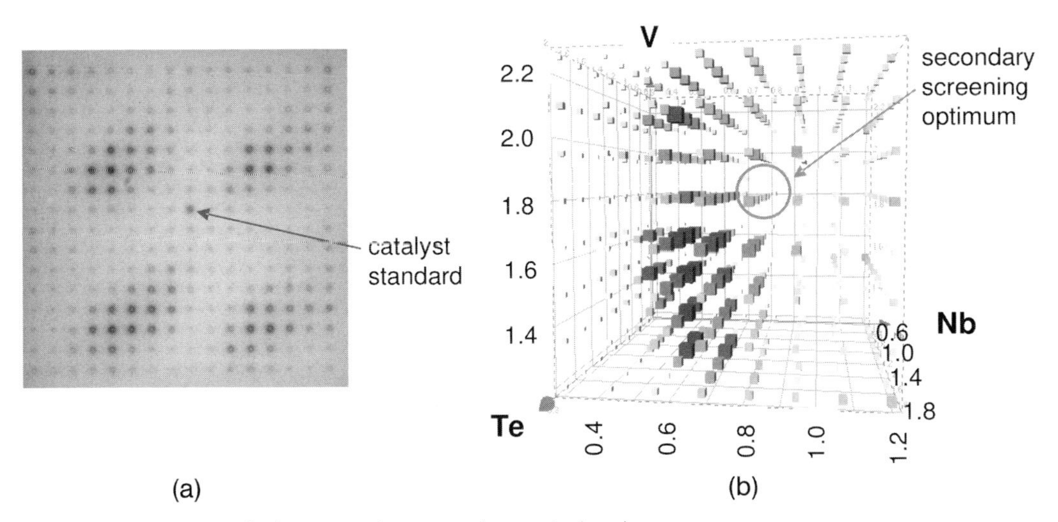

Fig. 3.19 (a) Raw microfluidic reactor data. Note the standard at the center of the plate. (b) Plot of acrylonitrile yield vs. composition of $Mo_6V_xNb_yTe_z$. Yield increases with increasing size and darkness of marker. The circle indicates optimal composition as determined by the secondary screen.

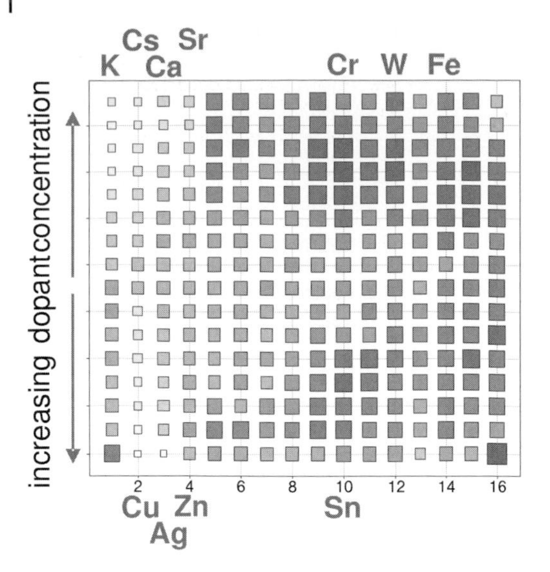

Fig. 3.20 Compositional optimization of $Mo_6V_{1.8-4}Sb_{0.8-2.0}Nb_{0.6}(H_2O_2)_{1-3}$ using hydrothermal synthesis. Yield increases with increasing darkness of marker.

3.7 Summary

High-throughput methods in heterogeneous catalysis have driven the development of new methods for experimental design, hardware for catalyst synthesis, hardware for chemical and physical analysis, and software tools for processing and visualizing data. Various synthesis and screening hardware have been presented and details of how these were used in specific research programs have been summarized.

These developments feed off each other. In an integrated workflow the constituent components should be throughput matched in a statistical sense, and as different steps become bottlenecks they are where improvements in throughput should happen.

3.8 References

1 A. Hagemeyer, B. Jandeleit, Y. Liu, D.M. Poojary, H.W. Turner, A.F. Volpe, Jr., W.H. Weinberg, *Appl. Catal. A: General*, 221 (2001) 23, and references cited therein.

2 B. Jandeleit, D.J. Schaefer, T.S. Powers, H.W. Turner, W.H. Weinberg, *Angew. Chem., Int Ed.* 38 (1999) 2495.

3 A. Hagemeyer, R. Borade, P. Desrosiers, S. Guan, D.M. Lowe, D.M. Poojary, H. Turner, H. Weinberg, X. Zhou, R. Armbrust, G. Fengler, U. Notheis, *Appl. Catal. A*, 227 (2002) 43.

4 US 20020042140 and US 20020014546, each assigned to Symyx Technologies.

5 A. Nayar, R. Liu, R.J. Allen, M.J. McCall, R.W. Willis, E.S. Smotkin, *Anal. Chem.*, 74 (2002) 1933.

6 J. Ren, J. Feng, D.-M. Sun, W.-Y. Li, K.-C. Xie, *Gongye Cuihua*, 10 (2002) 1.

7 E.G. Derouane, V. Parmon, F. Lemos, F. Ramôa Ribeiro, NATO Science Series, II: Mathematics, Physics and Chemistry 69 (*Principles and Methods for Accelerated Catalyst Design and Testing*) (2002) 3, 101, 245, 449, 469.

8 F. Schueth, O. Busch, C. Hoffmann, T. Johann, C. Kiener, D. Demuth, J. Klein, S. Schunk, W. Strehlau, T. Zech, *Topics Catal.*, 21 (2002) 55.

9 H. Su, E.S. Yeung, *Appl. Spectrosc.*, 56 (2002) 1044.

10 D. Demuth, K.-E. Finger, J.-R. Hill, S.M. Levine, G. Lowenhauser, J.M. Newsam, W. Strehlau, J. Tucker, U. Vietze, ACS Symposium Series 814 (*Combinatorial Materials Development*) (2002) 147.

11 Y. Sun, B.C. Chang, R. Ramnarayanan, W.M. Leventry, T.E. Mallouk, S.R. Bare, R.R. Willis, *J. Combinatorial Chem.*, 4 (2002) 569.

12 T. Zech, Fortschritt-Berichte VDI, Reihe 3: Verfahrenstechnik 732 I-ix (2002) 1.

13 S. Ozturk, S. Senkan, *Appl. Catal. B*, 38 (2002) 243.

14 P. Desrosiers, A. Guram, A. Hagemeyer, B. Jandeleit, D.M. Poojary, H. Turner, H. Weinberg, *Catal. Today*, 67 (2001) 397.

15 Y. Yamada, T. Kobayashi, N. Mizuno, *Shokubai*, 43 (2001) 310.

16 T. Hanaoka, *Shokubai*, 43 (2001) 321.

17 V.V. Guliants, ed., *Catal. Today*, 67 (Special Issue on Current Developments in Combinatorial Heterogeneous Catalysis) (2001) 307–409.

18 H. Su, Y. Hou, R.S. Houk, G.L. Schrader, E.S. Yeung, *Anal. Chem.*, 73 (2001) 4434.

19 J.M. Newsam, T. Bein, J. Klein, W.F. Maier, W. Stichert, *Microporous Mesoporous Mater.*, 48 (2001) 355.

20 F. Schuth, C. Hoffmann, A. Wolf, S. Schunk, W. Stichert, A. Brenner, in: *Combinatorial Chemistry*, ed. G. Jung, Wiley-VCH, Weinheim, 1999, 463.

21 C. Hoffmann, H.-W. Schmidt, F. Schuth, *J. Catal.*, 198 (2002) 348.

22 S. Thomson, C. Hoffmann, S. Ruthe, H.-W. Schmidt, F. Schuth, *Appl. Catal. A*, 220 (2001) 253.

23 J.M. Dominguez, E. Terres, A. Montoya, H. Armendariz, *Prepr. Am. Chem. Soc., Division Petrol. Chem.*, 46 (2001) 51.

24 S. Senkan, *Angew. Chem., Int. Ed.*, 40 (2001) 312.

25 U. Rodemerck, D. Wolf, O.V. Buyevskaya, P. Claus, S. Senkan, M. Baerns, *Chem. Eng. J.*, 82 (2001) 3.

26 O.V. Buyevskaya, D. Wolf, M. Baerns, *Catal. Today*, 62 (2000) 91.

27 U. Rodemerck, P. Ignazewski, M. Lucas, P. Claus, *Chem. Eng. Technol.*, 23 (2000) 413.

28 C. Mirodatos, *Actualite Chim.*, (2000) 35.

29 A. Holzwarth, W.F. Maier, *Platinum Metals Rev.*, 44 (2000) 16.

30 P. Desrosiers, S. Guan, A. Hagemeyer, D.M. Lowe, C. Lugmair, D.M. Poojary, H. Turner, H. Weinberg, X.P. Zhou, R. Armbrust, G. Fengler, U. Notheis, *Catal. Today*, 81 (2003) 319.

31 J. Klein, W. Stichert, W. Strehlau, A. Brenner, D. Demuth, S.A. Schunk, H. Hibst, S. Storck, *Catal. Today*, 81 (2003) 329.

32 G. Grubert, E. Kondratenko, S. Kolf, M. Baerns, P. van Geem, R. Parton, *Catal. Today*, 81 (2003) 337.

33 P.P. Pescarmona, J.C. van der Waal, T. Maschmeyer, *Catal. Today*, 81 (2003) 347.

34 A. Muller, K. Drese, H. Gnaser, M. Hampe, V. Hessel, H. Lowe, S. Schmitt, R. Zapf, *Catal. Today*, 81 (2003) 377.

35 S. Geisler, I. Vauthey, D. Farusseng, H. Zanthoff, M. Muhler, *Catal. Today*, 81 (2003) 413.

36 J.M. Serra, A. Corma, D. Farusseng, L. Baumes, C. Mirodatos, C. Flego, C. Perego, *Catal. Today*, 81 (2003) 425.

37 W. Li, F.J. Garcia, E.E. Wolf, *Catal. Today*, 81 (2003) 437.

38 T. Johann, A. Brenner, M. Schwickardi, O. Busch, F. Marlow, S. Schunk, F. Schuth, *Catal. Today*, 81 (2003) 449.

39 J.A. Moulijn, J. Perez-Ramirez, R.J. Berger, G. Hamminga, G. Mul, F. Kapteijn, *Catal. Today*, 81 (2003) 457.

40 L. Vegvari, A. Tompos, S. Gobolos, J. Margitfalvi, *Catal. Today*, 81 (2003) 517.

41 H.M. Reichenbach, P.J. McGinn, *J. Mater. Res.*, 16(4) (2001) 967.
42 H.M. Reichenbach, P.J. McGinn, *Appl. Catal. A*, 244 (2003) 101.
43 H.M. Reichenbach, H. An, P.J. McGinn, *Appl. Catal. B*, 44 (2003) 347.
44 S. Bergh, P. Cong, B. Ehnebuske, S. Guan, A. Hagemeyer, H. Lin, Y. Liu, C.G. Lugmair, H.W. Turner, A.F. Volpe Jr,. W.H. Weinberg, L. Woo, J. Zysk, *Top. Catal.*, 23(1–4) (2003) 65.
45 C. Kiener, M. Kurtz, H. Wilmer, C. Hoffmann, H.-W. Schmidt, J.-D. Grunwaldt, M. Muhler, F. Schuth, *J. Catal.*, 216(1–2) (2003) 110.
46 V. Murphy, A.F. Volpe Jr., W.H. Weinberg, *Curr. Opin. Chem. Biol.*, 7 (2003) 427–433.
47 D. Montgomery, *Design and Analysis of Experiments*, John Wiley & Sons, New York, 2001.
48 WO 02/48841 (2002), assigned to Symyx Technologies.
49 WO 96/11878 (1996), assigned to Symyx Technologies.
50 D.R. Dorsett Jr. 222nd ACS National Meeting, Chicago, IL, August 26–30, 2001. Patents pending to Symyx Technologies.
51 K. Yaccato, A. Hagemeyer, A. Lesik, A. Volpe, H. Turner, H. Weinberg, *Topics in Catalysis*, Vol. 30/31, 2004, 127.
52 EP 1080435 (2002), assigned to Symyx Technologies.
53 US 5985356, US 6004617, US 6326090, EP 1080435, and EP 1175645, each assigned to Symyx Technologies, additional patents pending.
54 US 6605470 (2003) assigned to University of Houston; US 6410332 (2002) assigned to Symyx Technologies, Inc.
55 US 6623967 (2003) assigned to University of Houston.
56 US 6627571 (2003) assigned to Symyx Technologies, Inc.

4
Integrated Microreactor Set-ups for High-Throughput Screening and Methods for the Evaluation of "Low-density" Screening Data

Andreas Müller and Klaus Drese

Index of formulae (all dimensions in SI units unless otherwise noted)

A	surface area of channel cross-section
a	channel width
b	channel depth
c	ratio between maximum and average channel velocity
c	concentration
c_A	concentration of substance A (kg m^{-3})
D	diffusion coefficient
E	activation energy (kJ mol^{-1})
$f(\varepsilon)$	extensive function of the aspect ratio given in ref. [37]
ΔH_r	reaction enthalpy
k	rate coefficient of a first-order reaction (s^{-1})
k_0	pre-exponential factor of a first-order reaction (h^{-1})
k_r	dispersion coefficient of a rectangular channel
k_ν	apparent velocity contribution of the reaction (m s^{-1}) given in ref. [37]
L	channel length
M, M_A	mass of substance A
R	universal gas constant (kJ mol^{-1} K^{-1})
T	temperature
t	temporal coordinate
u_0	maximum fluid velocity
u_1	velocity of the moving frame of reference
u_2	velocity of the moving frame of reference for a heterogeneous reaction
X	fixed distance on the x-axis (pulse length in m)
x	axial channel coordinate
x_1	axial coordinate in the moving frame of reference
x_2	axial coordinate in the moving frame of reference for a heterogeneous reaction
ε	aspect ratio
ϕ	Thiele modulus $\phi = [ab/(a+b)]\sqrt{(k/D)}$
θ_A	dimensionless concentration
τ	mean residence time

High-Throughput Screening in Chemical Catalysis
Edited by A. Hagemeyer, P. Strasser, A. F. Volpe, Jr.

ISBN: 3-527-30814-8

4.1 Introduction

Microreactor technology as a new branch in chemical processing has become an efficient tool with significant advantages over conventional reactors [1–7]. One important area is heterogeneously catalyzed gas-phase reactions, which are very common in reaction engineering on an industrial scale [8–17]. Catalyzed processes often provide important compounds used in a large variety of different processes and products. In addition, the catalysts and their carriers have become the focus of scientific investigation in terms of their preparation, morphology, porosity, composition etc. This new technology offers new routes in chemistry by utilizing laminar flow reactors [18, 19]. These reactors open up new methods of reactor operation, such as the possibility to execute a reaction in the explosive regime, as shown by the synthesis of ethylene oxide [9] or the H_2O_2 reaction [20], and also to facilitate higher product yields [21].

Laminar flow reactors are equipped with microstructured reaction chambers that have the desired low Reynolds numbers due to their small dimensions. Mass transport perpendicular to the laminar channel flow is dominated by diffusion, a phenomenon known as dispersion. Without the influence of diffusion, laminar flow reactors could not be used in heterogeneous catalysis. There would be no mass transport from the bulk flow to the walls as laminar flow, in contrast to turbulent flow, cannot mix the flow macroscopically.

To consolidate the experimental screening data quantitatively it is desirable to obtain information on the fluid mechanics of the reactant flow in the reactor. Experimental data are difficult to evaluate if the experimental conditions and, especially, the fluid dynamic behavior of the reactants flow are not known. This is, for example, the case in a typical tubular reactor filled with a packed bed of porous beads. The porosity of the beads in combination with the unknown flow of the reactants around the beads makes it difficult to describe the flow close to the catalyst surface. A way to achieve a well-described flow in the reactor is to reduce its dimensions. This reduces the Reynolds number to a region of laminar flow conditions, which can be described analytically.

The catalysts are deposited on microstructures and will thus have the appropriate environment for an exothermic reaction, enabling fast quenching of the reaction with near isothermal conditions. By choosing the width of microstructures below the range of the explosion quench diameter the reaction can also be executed in the explosion regime. The residence time is adjustable from some milliseconds to seconds.

The reaction is also influenced by the heat of reaction developing during the conversion of the reactants, which is a problem in tubular screening reactors. In microstructures, the heat transport through the walls of the channels is facilitated by their small dimensions, which allows the development of isothermal reaction conditions. Thus, by decoupling the heat and mass balance, an analytical description of the flow in the screening reactor is achievable.

Combinatorial screening, at first glance, might appear superficial as a number of identical experiments are executed under the same conditions in parallel fashion. This is certainly too simple a contemplation. As soon as fluid dynamic information is added to the pure experimental data, more complex aspects of catalysis are derivable from overall conversion data, such as the intrinsic reaction kinetics. This is a different approach to established screening methods, which normally concentrate purely on the experimental collection of data on the conversion or the selectivity of a certain catalyst.

Several strategies can be applied for catalyst development. A very basic approach utilizes the in situ characterization of a catalyst during the reaction to design more efficient versions of a catalyst. These in situ methods give a basic understanding of surface phenomena between single molecules. A surface reaction always follows adsorption. Adsorption is described by advanced simulation techniques like the Monte-Carlo method or the methods of molecular dynamics, which calculate the so-called configurational diffusion of molecules adsorbed to surfaces [22]. These statistical methods increase the knowledge of surface interactions but are not widely used in catalyst development. This is partly because these surface studies on mono-crystalline surfaces are executed under low vacuum conditions and are not readily transferable to realistic reaction conditions. Another approach is the so-called microkinetics of heterogeneous catalysis [23]. Here the reaction is divided into its elementary steps of adsorption, reaction and desorption and studied for the selectivity of intermediate products using parameters estimated from chemical bonding theorics. Despite these promising strategies, empirical strategies are still the most commonly applied tool in industrial catalyst development. This is certainly due to the rough industrial environment, which requires robust equipment for screening. Such a more empirical approach was enabled by the reactors of Adler et al. [24]. They recommended the use of a unit construction system based upon modularized single channel reactors applicable to integral or differential reactor operations. The dimensions are close to conventional tube reactors.

4.1.1 Pellet-type and Ceramic Reactors

Various reactor principles are recommended for fast catalyst screening. Some authors recommend the use of pellets. Among the pioneers Senkan has to be mentioned [25–27] who has developed a parallel gas-phase reactor with 80 parallel channels that are equipped with impregnated γ-alumina tablets. The gas samples are delivered to a mass spectrometer by a sample robot (Fig. 4.1).

A collection of 66 catalyst combinations was operated for 24 h to check the long-term behavior. Fig. 4.2 demonstrates that all combinations show deactivation, especially the most active catalysts. This clearly demonstrated that screening also depends very much on the selection of the appropriate time scale for a reaction (in this study at least 24 h).

A 15-fold glass tube parallel-packed bed reactor has been described [28–30], which is close to conventional catalyst testing equipment. The same authors also reported a 64-fold ceramic block reactor and a ceramic monolithic reactor for the screening of up to 250 catalysts in parallel. The individual catalysts were coated

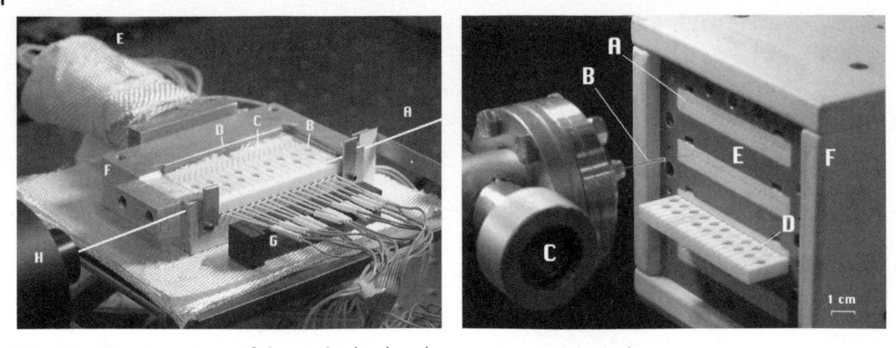

Fig. 4.1 Reactor set-up of the multiple chamber reactor system at the University of California (left: upper three chambers removed; right: detail of sample gas delivery via capillary B to mass spectrometer C, lower drawer moved forward to show the catalyst pellets D).

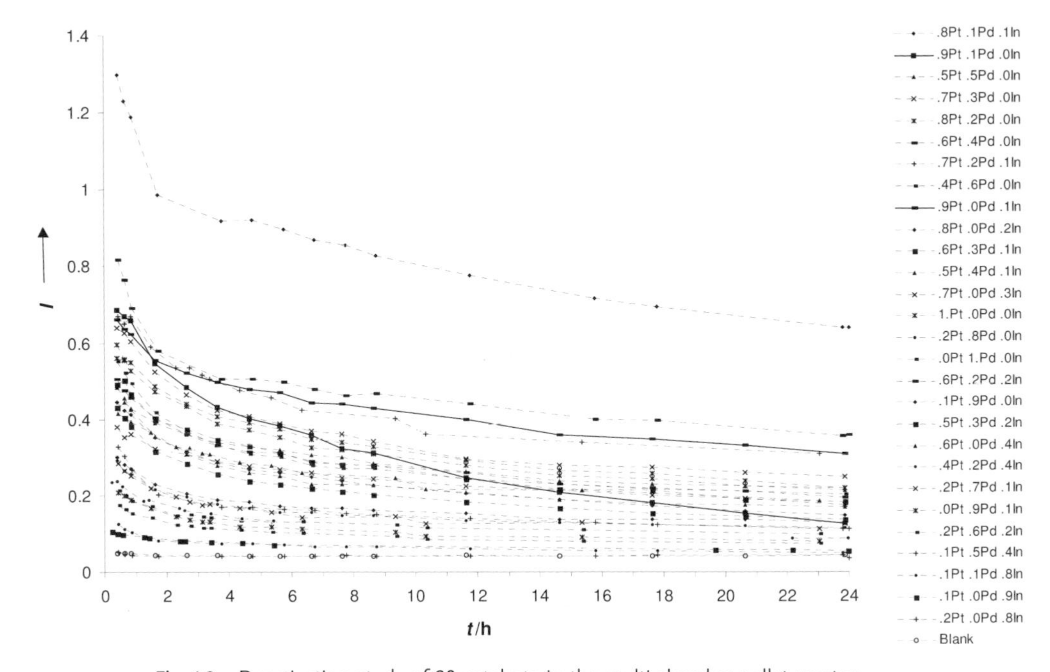

Fig. 4.2 Deactivation study of 29 catalysts in the multi-chamber pellet reactor of Senkan et al. [26].

onto the ceramic walls of the monolith and the gas samples were sent to a mass spectrometer by a sample robot after the reaction (Fig. 4.3).

The group of Schüth has developed a number of reactors close to conventional testing methods with different degrees of sample integration. For multiphase reactions a 25-fold stirred vessel reactor has been developed [31] and for heterogeneous gas-phase reactions a 16-fold fixed-bed reactor has been presented [32],

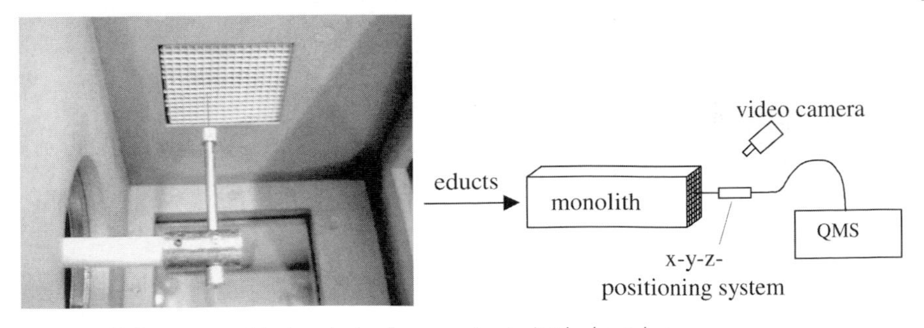

Fig. 4.3 Right: *x-y-z* positioning device for scanning individual catalyst-coated micro-channels of a ceramic monolith; right: plant set-up for high-throughput experiments at the ACA Berlin-Adlershof.

which was later followed by a 49-fold parallel reactor [33]. An improved version of the 49-fold reactor described in ref. [33] has been used for methanol production from Syngas (up to 50 bar).

Some of their methods have been commercialized by hte AG, a German supplier of services in combinatorial catalysis. High-pressure applications have not been reported often due to their expense. A 14-fold stirrer autoclave has recently been presented [34] for liquid–gas reactions.

4.1.2
Multiple Microchannel Array Reactors

A characteristic of this type of reactor is the steel substrate, which is preferably used as the reaction chamber (it can also be titanium or aluminum). This allows the use of microstructures under robust experimental conditions such as high temperatures.

The reactor system of Zech et al. [35] is a good example of an integrated approach as it combines devices from different suppliers into a complex screening system. The reactor was manufactured at IMM (Fig. 4.4) and the sampling device delivered by AMTEC in Chemnitz. The catalyst preparation method was developed at the TU Chemnitz. The latter consists of an *x*/*y*-positioning robot supervised by a CCD camera. The sampling capillary was connected to a quadrupole mass spectrometer. This set-up has been used to screen up to 35 catalysts a day.

Kolb et al. from IMM [36] have developed a 10-fold parallel reactor that includes an exchangeable distribution section for the reactant delivery to the single platelets, and internal cartridge heaters (Fig. 4.5). This reactor can be operated as a 10-fold parallel reactor as well as a single channel reactor. In the latter case the 10 single channels are combined to a single channel by changing the distribution section. This allows the adjustment of the residence time without changing the mass flow and thus enables it to maintain the flow conditions.

A new approach into reactor modularity has been introduced by Müller et al. [37, 38]. They subdivided the screening process into often used process operations.

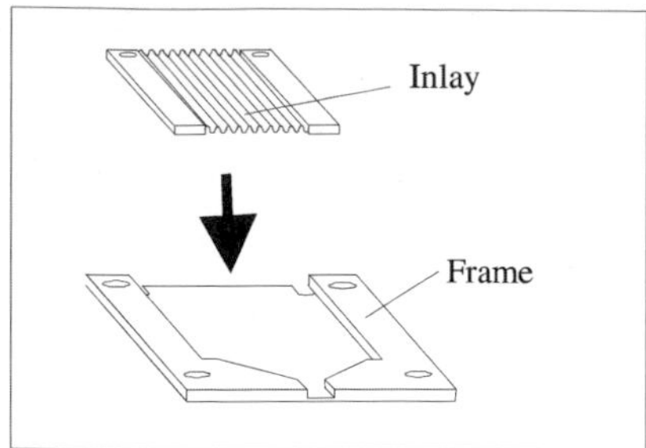

Fig. 4.4 35-Fold multiple microchannel reactor used at the TU Chemnitz.

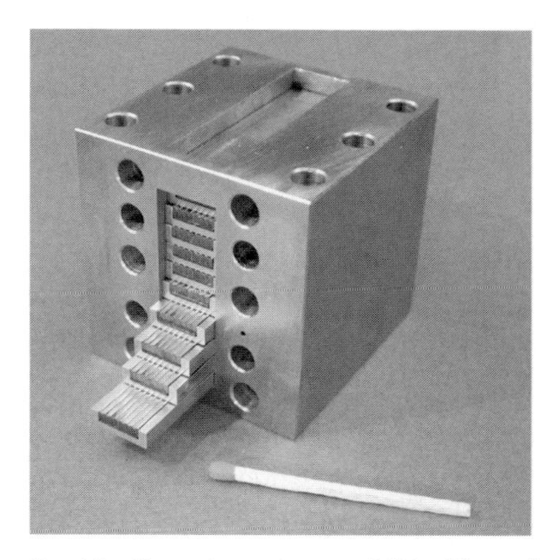

Fig. 4.5 Screening reactor at IMM with catalysts in separate drawers (gas distribution section removed).

In chemical process engineering these so-called unit-operations are components of every complex plant. As catalyst screening contains many different processes, such as heat exchange, flow distribution, sampling, analysis and reaction, such a subdivision in unit operations is justified. An overview of the modules that are part of the screening set-up is given in Fig. 4.6.

The flexibility of such a system has been demonstrated with two example configurations, one for transient studies and another for steady-state experiments (Fig. 4.7).

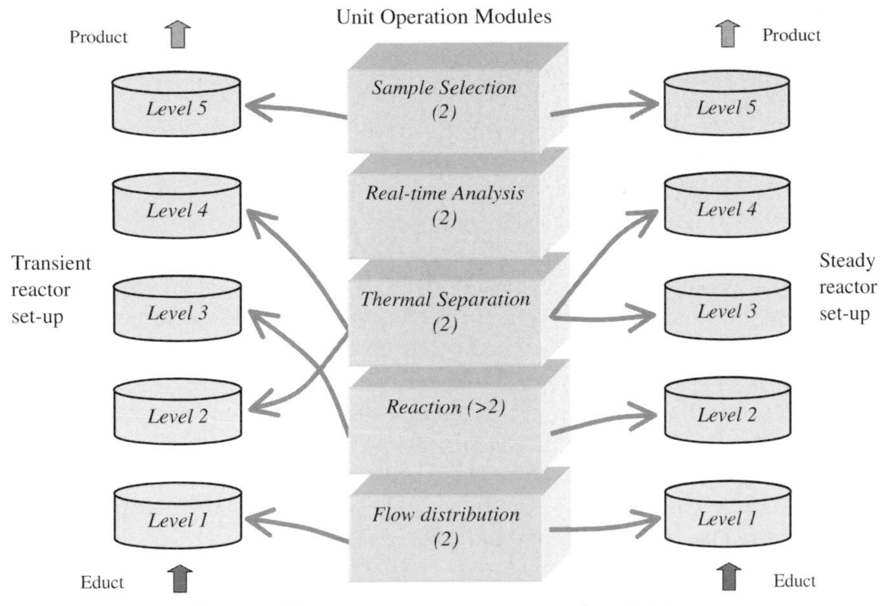

Fig. 4.6 Overview of the modular screening set-up at IMM and its division in unit operation modules (available number of modules in brackets).

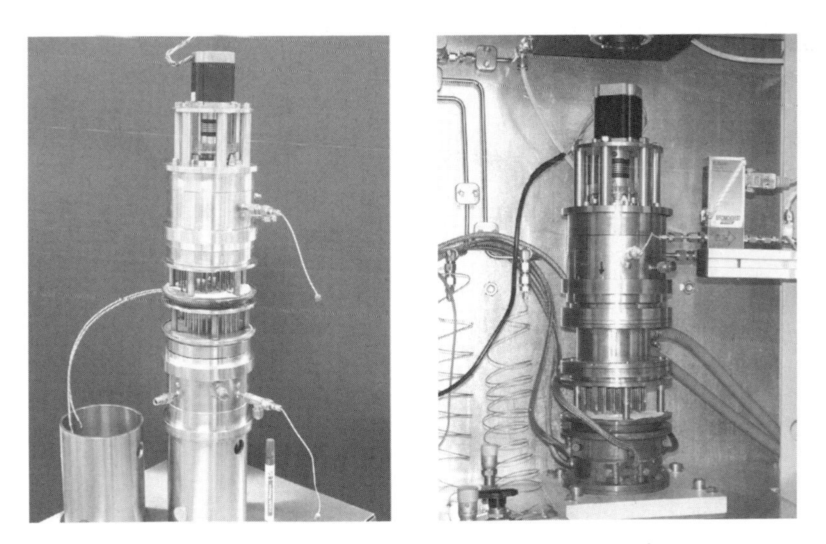

Fig. 4.7 Two versions of the unit construction system built at IMM for transient (left) and for steady screening (right).

4.1.3
Chip-type Reactors

Sample integrations similar to pharmaceutical approaches were already examined in 1997 [39]. Here, a chip-like microsystem was integrated into a laboratory automaton that was equipped with a miniaturized micro-titer plate. Microstructures were introduced later [40] for catalytic gas-phase reactions. The authors also demonstrated [41] the rapid screening of reaction conditions on a chip-like reactor for two immiscible liquids on a silicon wafer (Fig. 4.8). Process conditions, like residence time and temperature profile, were adjustable. A third reactant could be added to enable a two-step reaction as well as a heat transfer fluid which was used as a mean to quench the products.

One step closer to up-scaling to industrial environments is the multiple-bead reactor shown in Fig. 4.9. Here pellet-type catalyst carriers, so-called beads, are positioned in square containers. The beads are made of alumina and are 1 mm in diameter. Gases are passed over these beads through microstructured pore membranes in the cover and the base plate of the containers.

Without being itself a screening device, the reactor of Jensen et al. [6, 42, 43] also has to be mentioned because they opened up a completely new field in catalysis by combining MEMS (micro-electro-mechanical systems) technology with a chip-based catalytic reactor (Fig. 4.10). A mixing-tee was equipped with heaters and temperature and flow sensors, thus giving on-stream information about the reaction conditions.

Symyx, one of the pioneers in combinatorial screening, presented a chemical processing microsystem in 2000 for the screening of 256 catalysts on a silicon or quartz glass wafer [44]. The reactant flow was distributed by a microstructured manifold etched into a silicon wafer (see Bergh, Chapter 3).

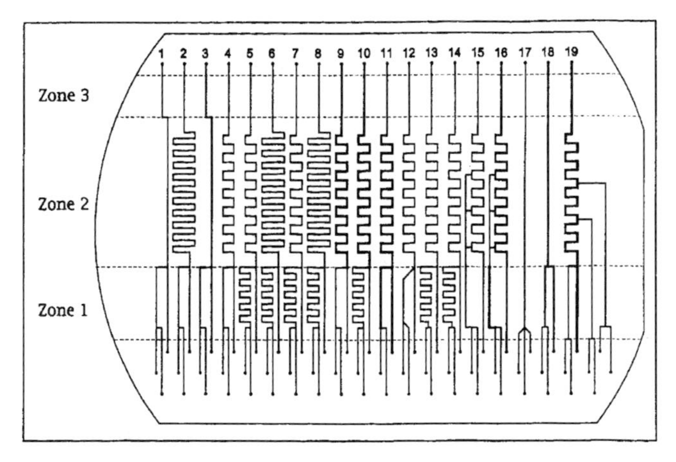

Fig. 4.8 Nineteen different channel configurations on a silicon chip reactor to screen process conditions from the TU Chemnitz [41].

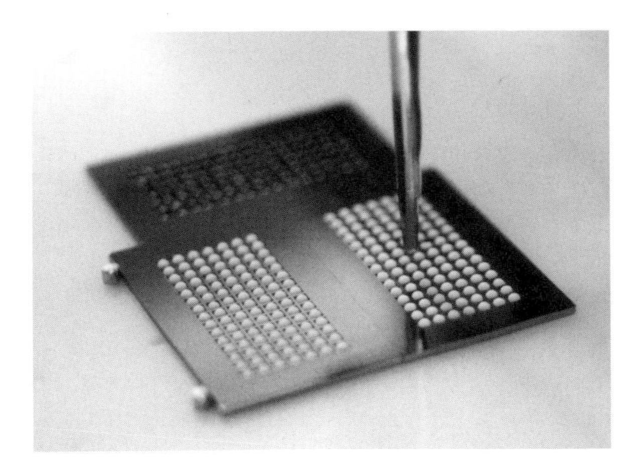

Fig. 4.9 105-Fold bead-type chip reactor from the TU Chemnitz [41].

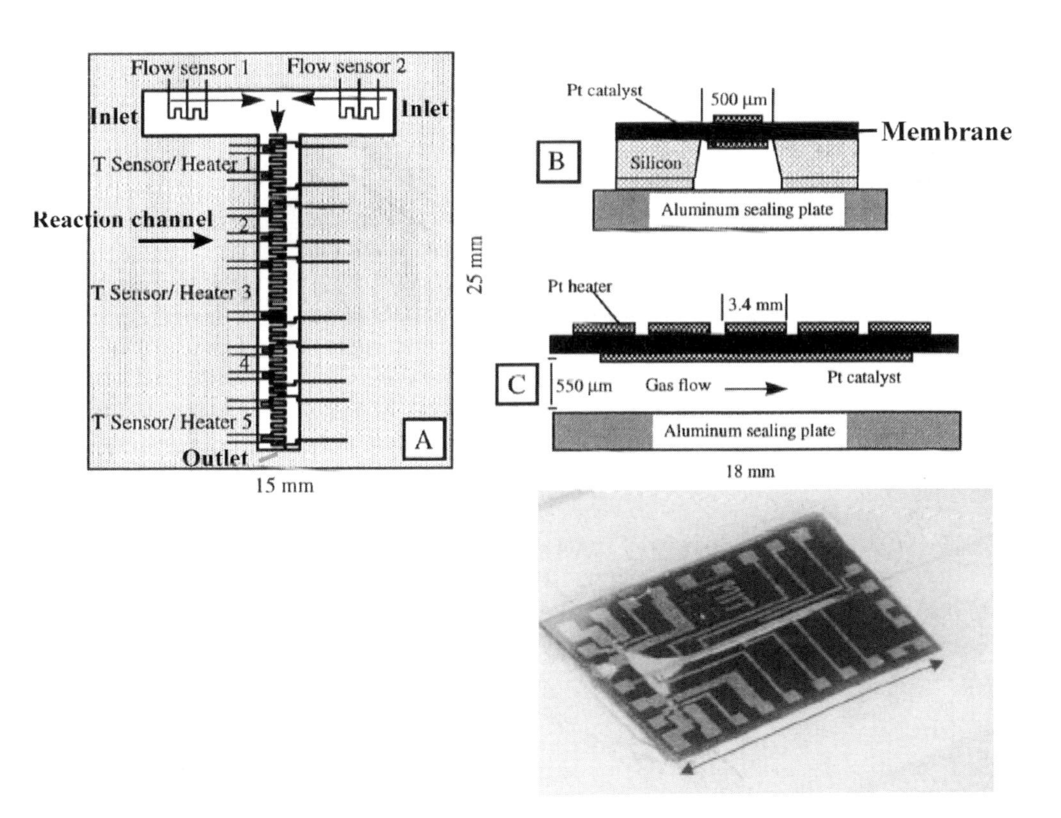

Fig. 4.10 Chip-type microreactor equipped with flow and temperature sensors from the Massachusetts Institute of Technology [6].

4.1.4
Well-type Reactors

Reactions which introduce a temperature difference either by the emission of heat or by consuming heat are accessible to infrared-thermographic methods. Reetz et al. [45] have reported the screening of enantioselective reactions on a modified micro-titer-plate. A time-resolved picture of a sector of this titer-plate showed the increased activity of the catalysts in three individual wells.

At the same Max-Planck-Institut, the group of Klein [46, 47] has distinguished catalysts for the hydrogenation of 1-hexyne and the oxidation of iso-octane by their temperature increase during the reaction. The catalysts were deposited in small borings on a slate by the sol-gel technique. After the evaporation of the solvent the catalysts were calcined and tested through an IR-transparent window. Such an infrared technique requires a correction for the emission of different materials, otherwise the indicated temperatures are misleading. The same group also developed a multi-chamber micro-autoclave for hydrothermal synthesis, with catalyst samples deposited on cavities on a silicon substrate. In 1999 this group examined oxidation reactions by locally resolved mass spectrometry. For this purpose the injection needle of a laboratory sample robot was exchanged for the gas inlet pipe of a mass spectrometer and a reactant gas supply pipe. This arrangement of pipes was inserted in the cavities of a heated slate to examine the activity of different catalysts.

Cong and co-workers [54] have prepared a ternary library of transition metals by sputter deposition on a quartz wafer. The catalyst samples were supplied with reactants through a concentric tube that also delivered the product gas flow to a sensor for spectroscopic analysis (see Chapter 3). The catalysts could be activated by a CO_2 heating laser from the backside of the wafer.

Fluorescence spectroscopy as a means of judging process conditions in a polymerization reaction has been reported [48]. The authors optimized parameters such as the reactant ratio and catalyst amount to reach the smallest variability of material properties. The samples were deposited on a 96-micro reactor array and examined with a spectro-fluorometer during the reaction.

An improvement in the efficiency of photochemical splitting of water was the incentive for Morris [49]. A parallel optical screening method was developed to select photocatalytically active catalysts by their adsorption spectra (UV and visible light).

Despite recent promising strategies, the principle of micro-process engineering is still not widely used in combinatorial catalysis. One drawback certainly is the increasing distance from industrial applications with decreasing dimensions. However, the small structures possess laminar flow conditions that are fully accessible by analytical as well as numerical macroscopic descriptions. This offers the chance to describe thoroughly the fluidic, diffusive and reactive phenomena in catalysis to find intrinsic kinetics on using, for example, non-porous sputtered catalysts.

That fast screening does not necessarily mean parallel screening has been convincingly demonstrated by the group of de Bellefon [50]. They combined a micromachined mixer tube reactor and applied pulsed injections of the respective catalysts for liquid–liquid and gas–liquid phase reactions. The throughput testing frequency exceeded 500 d^{-1}. In 2002 some authors reported parallel transient methods to study kinetics. Rothenberg [51] recommended the use of an improved sampling strategy to reduce the number of data points and so increased the screening throughput. Lauterbach [52] introduced FTIR hyperspectral imaging for fast transient studies.

A convenient method to produce porous surfaces is the anodic oxidation of aluminum plates. Such microstructured aluminum platelets have been coated by wet impregnation with Pt-, V- and Zr-precursors [35], and tested under catalytic methane combustion conditions. The conversion rate of oxygen followed directly the platinum content in the catalysts. These data were well reproducible even after five different runs.

Besser et al. [53] have studied the hydrogenation of cyclohexane over a platinum catalyst. They used microstructured chips with channel widths of 100 µm and 5 µm and showed that the conversion in the smaller channels is larger, as expected, due to a larger surface–volume ratio. The conversion data were consistent with data from macro-scale reactors.

CO oxidation on a quartz wafer has been examined by the group of Weinberg [54]. They prepared catalyst samples by co-sputtering of the individual targets and by sol-gel techniques on a quartz wafer. The time for the preparation of a sputtered library is said to be only 1 h. A ternary library consisting of Pd, Pt and Rh exhibited sensitivity to CO_2 production, with the highest product yield in Pd-rich catalysts.

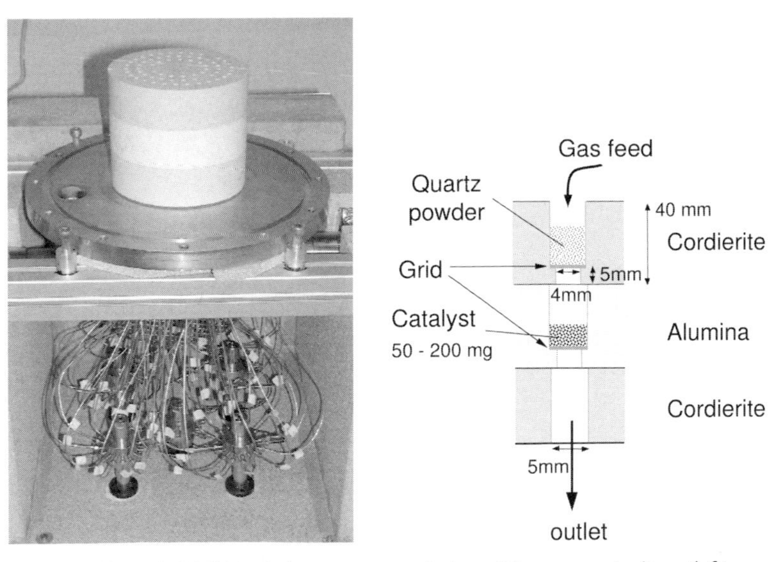

Fig. 4.11 Ceramic 64-fold tubular reactor consisting of three ceramic discs (left), and internal cross-sectional view of a single channel including pressure resistance (right).

The ceramic 64-channel reactor of Rodemerck et al. [28–30] bridges the gap between micro- and macroscopic fixed-bed reactors (Fig. 4.11). The catalyst containers consist of a massive tubular disc of alumina with borings for the catalyst powder clamped between two massive cordierite discs. This set-up can be operated up to 550 °C at pressures of up to 3 bar.

Using this set-up, two new catalysts were found for the oxidative dehydrogenation of ethane to ethene. With Cr/Mo-O_x and Co/Cr/Sn/W-O_x catalysts an industrially relevant product yield of more than 60% was reached.

4.2 Steady-state Reactor Set-ups

4.2.1 Methanol Steam Reforming

Cominos et al. [17, 55] have developed a single channel micro-reactor for testing catalysts for methanol steam reforming for PEM fuel cells. The stack-like stainless steel device took up a stack of 5 to 15 plates each consisting of micro-channels 500 μm in width and 350 μm in depth. Supported Cu/Zn catalysts were prepared by introducing γ-alumina wash-coats of an average thickness of 10 μm into the micro-channels. The average pore diameter amounted to 45 nm and the BET surface area of the catalysts was determined to be 45 $m^2 g^{-1}$. The residence times were adjusted from 100 to 200 ms at total flow rates of between 500 and 900 ml min^{-1} using only five coated plates. Increasing the temperature from 200 to 275 °C increased the conversion from 37% to 65%. Carbon monoxide formation started at temperatures exceeding 250 °C. When 15 plates were introduced into the reactor, 80% conversion was achieved at 290 °C, a residence time of 600 ms and a steam-to-carbon ratio of 2. The product stream contained more than 50% hydrogen and 0.25% CO. From these results a power output of the testing device of 43 W and a power density of 1.8 kW l^{-1} were calculated.

4.2.2 Propane Steam Reforming

A new modular screening reactor has been developed by Kolb et al. [36] to test up to 10 catalyst carrier plates in parallel (Fig. 4.5). The reactor is equipped with drawers into which the micro-structured plates are placed. The variation of the end-caps of the reactor enables two different operation modes: firstly, parallel screening of up to 10 plates and, secondly, serial operation of up to 10 identically coated catalyst plates. The latter gave the opportunity of modifying the residence time via the catalyst mass under identical flow conditions. Owing to the low stability of Ti at the reaction temperature (500 °C), a steel housing was placed around the titanium reactor. This additional housing increased the dimensions of the reactor (160×70×120 mm^3). The reactor had to be heated externally. The microstructured plates were 300 μm in depth, 500 μm wide and 100 mm long. Experiments performed with the reactor by Wörz at BASF on a proprietary reaction gave a 60% yield for the desired product at residence times as low as 40 ms in the micro-reactor. This performance was superior to the experiments performed in an aluminum capillary, which corresponds well to the reactor design of the industrial process. A 2000% gain in space–time yield was found for the porous coated microstructures compared with the aluminum capillaries. Later, a stainless steel version of the reactor was developed and applied to methanol steam reforming. This reactor was far more compact ($70 \times 55 \times 64\ mm^3$) than the former version. The channel length was reduced to 50 mm but the cross-sectional area of the channels was maintained. To heat the reactor, cartridge heaters were incorporated into the housing. The functionality of the reactor concerning cross-talk between individual channels in the parallel operation mode was verified. Ziogas et al. [56] have performed catalyst screening with the reactor with catalyst coatings made of various base aluminas, such as corundum, boehmite and γ-alumina. Testing of Cu/Cr and Cu/Mn catalysts based on the different coatings for methanol steam reforming revealed differences in activity that were ascribed to both the different surface area and morphology of the carrier material. The functionality of the serial operation mode of the reactor was also tested. There were indications for mass transfer limitations for higher flow rates for the same reaction system.

4.2.3
Catalytic Methane Combustion and Methods for Sample Preparation

Screening was originally developed in the pharmaceutical industry to examine samples positioned on so-called titer-plates with up to several thousand species. This approach was transferred to high-temperature catalyst screening by Müller et al. [37, 38]. Due to the danger of thermal cross-talk, this very large number of species cannot be expected for heterogeneous catalyst screening. Nevertheless, an increase of efficiency and flexibility was desired. The build-up of a library of catalysts with a very dense format for later reference is enabled by the standardized format of these plates. The standard format of the titer-plate is the 48-well format with a respective number of individual catalysts. Plates with other configurations and a different number of wells are possible.

The parallel reactor for the screening of the titer-plates consists of several modules, each of them is responsible for just a single operation (Fig. 4.12). The gas flow for example is preheated and evenly distributed within the distribution module and delivered to the wells on the titer-plate. The latter is clamped between the distribution module and the insulation module and also treated as a separate reaction module. The insulation module separates the heated section of the parallel reactor from the unheated section and is further cooled by the heat exchanger module on top of it. The last module, just above the heat exchanger module, is a multi-port valve that delivers the product gas to the gas-chromatograph.

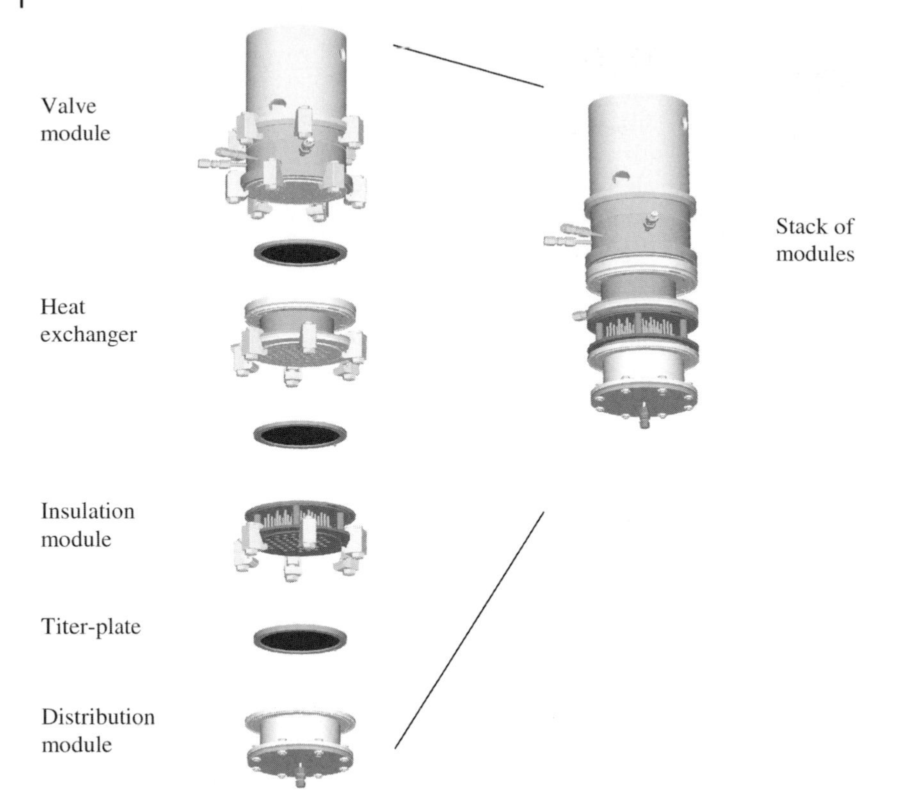

Fig. 4.12 Modular concept of the steady reactor configuration with a 48-fold titer-plate [37, 38].

4.2.3.1 Wet Chemical Procedure (Washcoating/Flow Impregnation)

Preparation of the catalysts by applying a wet-chemical method is a multi-step procedure (Fig. 4.13). Therefore, aqueous suspensions, consisting of γ-alumina, binder (PVA) and other additives are applied. Microstructured substrates are first cleaned in an ultrasonic bath with isopropanol (step 1). After positioning the substrates in the pocket of a holder, the area that should not become coated outside the microstructures is masked (step 2). Then, the γ-alumina slurry is washcoated into the microstructured channels of the titer-plate, filling the channels completely (step 3). Surplus slurry is removed by a razor blade (step 4). After drying the deposited slurry in the oven, the layer thickness in the channels shrinks and an open cross-section for the fluid develops (step 5). The final layer height is adjusted by the amount of alumina in the aqueous suspension. The wash-coats are then calcined at 600 °C (step 6), resulting in a porous layer with an internal surface area of ca. 70 $m^2 g^{-1}$, as measured by BET-surface analysis. Scanning electron microscope studies are used to show the porous surface of these wash-coats.

To improve the filling of the pores of the alumina layer with impregnation liquid, the samples are evacuated in an exsiccator and conditioned with carbon dioxide

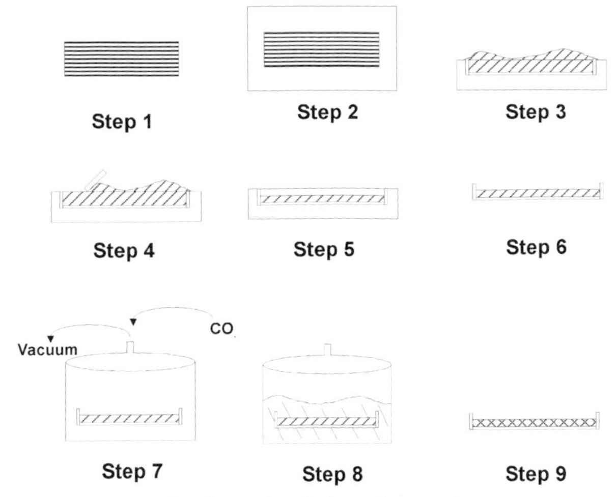

Fig. 4.13 Principle of wet-chemical catalyst coating.

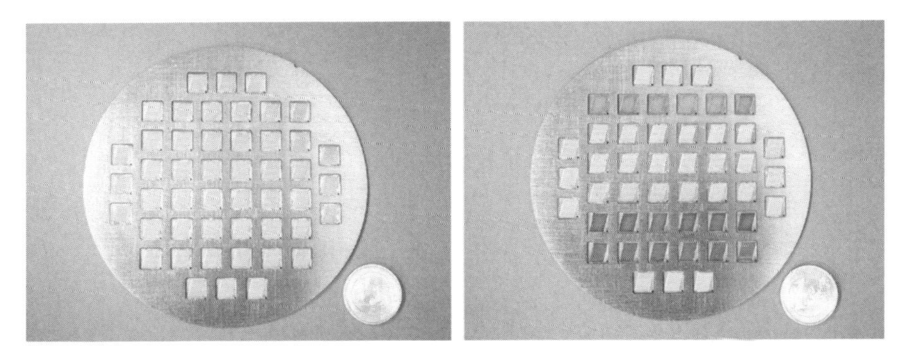

Fig. 4.14 A 48-fold titer-plate coated with γ-alumina after calcination (left) and after impregnation (right).

Tab. 4.1 Procedure for semi-automated sample preparation on a titer-plate.

Process steps	*Pretreatment of titer-plate*	*Wash-coating*	*Calcination*	*Flow impregnation*	*Calcination*
Expended time (h)	0.5	3	10	Variable [a)]	10
Temperature (K)	298	298	873	298	Variable [b)]
Amount of substance (ml)	–	48×0.5	–	48×0.5	–

a) Depends on the desired amount of catalyst in the alumina layer. One hour is often sufficient. Remark: only the time for the penetration of the liquid into the pores is considered. The time for the preparation of the impregnants is not included because this can be done by a laboratory sample robot.

b) Depends on the catalytic reaction. To hold the layer stable during the reaction, the calcination temperature should be higher than the reaction temperature.

(which is water-soluble, unlike air). In step 8, the γ-alumina precursors are dip-coated with the salt of transition metals. In addition to the manual impregnation by the dip-coating method, a semi-automated high-throughput method is used (here, the alumina layer is not evacuated). For this step, aqueous solutions of metal salts [e.g. H_2PtCl_6, $RhCl_3$, $Ni(NO_3)_2$] are used. After the impregnation, the samples are dried and calcined again (step 9). The procedure and further details are summarized in Tab. 4.1. A titer-plate coated with γ-alumina is shown in Fig. 4.14.

4.2.3.2 **Experimental and Discussion**

After coating, the titer-plates are inserted into the reactor to test their activity. The reaction conditions were held constant in the following experiments. The reactor was heated to 475 °C and held at a pressure of 0.2 bar (g). The heat exchanger was operated at 50 °C, and the throughput for a single well was adjusted to 1 ml min^{-1}, resulting in a total space velocity of 9000 h^{-1}. The residence time in the wells was 0.4 s.

An under-stoichiometric mixture of methane and oxygen (2:1) was diluted with 70% nitrogen, preheated and distributed within the distribution module and continuously delivered to the single wells. For the catalytic combustion of methane, a number of parallel and consecutive reactions occur [57]. During the tests mostly carbon dioxide was obtained according to the formula:

$$CH_4 + 2O_2 \rightleftarrows CO_2 + 2H_2O \quad \Delta H_r = -802.3 \text{ kJ mol}^{-1}$$

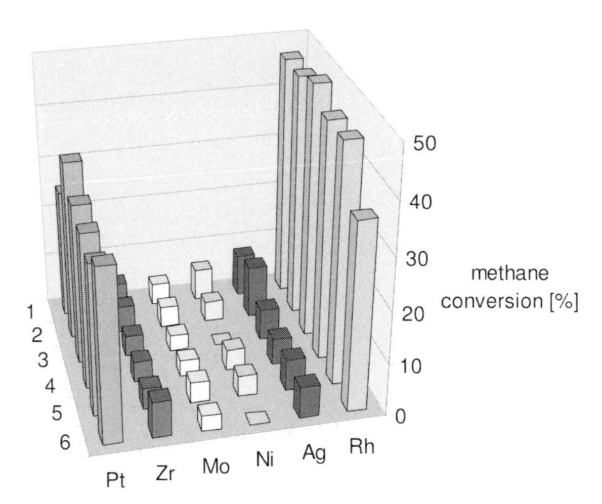

Fig. 4.15 Six metallic catalysts each deposited on six different wells onto a titer-plate (conversion is correlated into the correct geometric position of the wells).

Cullis and Keene [83] have also found that under low-temperature conditions the conversion of methane is nearly completely due to the formation of these total oxidation products.

Catalysts prepared by the wash-coating method were first used to check the reproduction of the measured values. For this reason, six elementary metal salts (platinum, zirconium, molybdenum, nickel, silver, and rhodium) were dissolved and impregnated onto a titer-plate. The catalysts were pre-reduced inside the reactor with 5% hydrogen in 95% nitrogen at 250 °C. The results were recorded first before the pre-reduction and then after the pre-reduction. The repeated measurements indicated good reproducibility in both cases. The conversion of methane with the rhodium catalyst is better after the pre-reduction. Methane conversion after 18 h runtime was still stable.

To check if fluidic or thermal cross-talk exists, the same six metal catalysts were coated in six rows on one titer-plate. Active catalysts were positioned on neighboring wells with less active catalysts (Fig. 4.15). The result was that the two active rows (platinum and rhodium) did not influence the neighbouring less-active rows (zirconium and silver). Thus, cross-talk was not observed. In the next test, elementary, binary and ternary mixtures of the same metal catalysts were examined. Every mixture was deposited on four different wells to check for reproducibility (Fig. 4.16). Except for some of the ternary mixtures, the results were well-reproduced and were consistent with the known well activity of rhodium (row 1) and the mixture of rhodium and platinum (row 5) that had the highest activity.

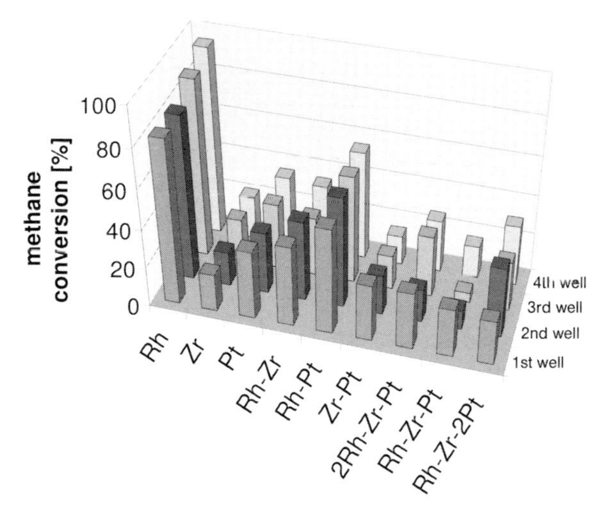

Fig. 4.16 Elementary (rows 1–3), binary (rows 4–6) and ternary (rows 7–9) catalysts respectively deposited on 4 different wells.

4.3 Transient-state Reactor Set-ups

4.3.1 Introduction

With the introduction of microreactors, transient reactor operations became interesting due to their low internal reactor volume and, thus, fast dynamic behavior. In 1999, Liauw et al. presented a periodically changing flow to prevent coke development on the catalyst and to remove inhibitory reactants in an IMM microchannel reactor [58]. This work was preceded by Emig in 1997, of the same group, who presented a fixed-bed reactor with periodically reversed flow [59]. In 2001, Rouge et al. [14] presented the catalytic dehydration of isopropanol in an IMM microreactor.

The use of transient reaction conditions to characterize catalysts began in 1950 with the single-pellet reactor [60]. The single-pellet reactor is basically a Wicke–Kallenbach-cell operated under reactive environments. It was extended by Scott in 1974 into the single pellet tube reactor, which is closer to industrial environments. Scott used a one-dimensional dispersion model to obtain the effective dispersion coefficient from pulsed input signals. Chromatographic techniques for the evaluation of kinetics in a micro-pulse reactor were already introduced in 1955 [61]. The model of Schneider and Smith [62] is often used to evaluate adsorption rates or effective diffusion coefficients from experimental data. From 1961 to 1969 Polanski and Naphtali [63] developed the frequency response experiments where the reactor volume is periodically changed and the resulting pressure fluctuation is used to describe adsorption rates on a catalyst surface. In 1998, Colin et al. [64] applied the frequency response method to rectangular microchannels close to and below one micrometer. They studied the slip effect at the channel wall with pulsed sinusoidal signals. In 1999, a highly miniaturized flow-through calorimeter on a silicon chip was presented [65] for the evaluation of kinetics using an integrated thermochemical detector.

4.3.2 Kinetics Derived from Tracer Signal Dispersion in a Channel Reactor

Screening in stationary mode will only give information about the activity of a single catalyst or a catalyst mixture. When a proper catalyst for a certain reaction is found, the next important information is the reaction kinetics. To obtain this information, several methods and reactors are recommended in the literature [66–73]. Most of them apply transient reactor operations to find detailed kinetic information. Microreactors are particularly suited for such an operation since their low internal reaction volumes enable a fast response to process parameter changes, e.g., concentration or temperature changes. This feature was already applied by some authors to increase the product yield in microreactors [70, 74, 75]. De Bellefon [76] reported a dynamic sequential method to screen liquid–liquid and liquid–

gas phase catalytic reactions, applying the method of the injection of different samples followed by barrier liquid respectively. While he applied the pulsed input to establish spatially separated samples, this method might well be applied to study dynamic behavior in gas-phase reactions.

Microreactors operated in pulsed mode were introduced by von Kokes in 1955 [68] but have became intensively used only in the last ten years. Such transient studies to obtain insight into reaction mechanisms were undertaken by Gleaves et al. in 1997 with the temporal analysis of products (TAP) reactor [72]. They observed rate coefficients of elementary reaction steps like adsorption and desorption by applying pulses of reactants to a catalytic microreactor combined with a quadrupole mass spectrometer. Not only concentration pulses were used as input signals. Wojciechowski used temperature ramps with his temperature scanning reactor [71, 77] while Kobayashi and Kobayashi [78] applied concentration step functions. Typical process parameters that can be changed are the pressure, the temperature or the composition of the gas mixture. Fast mixture or pressure pulses can be realized by the injection of reaction gas into the system by a micro-dispense valve. The transition into the next stationary mode is then recorded by an appropriate flow sensor. For the fast evaluation of such response signals a diffusion/dispersion model is necessary, which must also consider the heterogeneous wall reaction. A simplified dispersion model, assuming a first-order reaction, is presented below, which enables the concentration distribution to be predicted in rectangular microchannels. These results prepare the way to an extended type of secondary screening, where screening means the evaluation of catalysts by applying various reactor geometries and transient operations to collect kinetic information.

The limitation of such a model to first-order reaction rates is not as restricting as it seems. In fact, many reactions might at least be considered as "pseudo"-first order, which means that they behave macroscopically like first-order reactions. This is the case for diluted fluids, for non-catalytic gas–solid reactions such as the so-called shrinking core or shrinking particle model. Other examples are electrochemical reactions [79]. For the above applied oxidation of methane to carbon dioxide on some metal oxide catalysts, a first order reaction is also assumed [66, p. 182 and 193]. However, in combinatorial catalysis it may be sufficient to have a first rough idea of the underlying kinetics. Without prior information about the kinetics, the performance of a reactor is hugely uncertain. This is obvious if one considers the wide variation of reaction rates. Pre-exponential factors of reaction rate constants derived by the transition-state-theory vary widely, from ca. 10^1 to 10^{16} s^{-1} [66]. This first information might then be used to develop a pilot plant for the up-scaling and for further detailed kinetic examinations.

4.3.2.1
Experimental Aspects of Dispersion

A characteristic of micro-channel reactors is their narrow residence-time distribution. This is important, for example, to obtain clean products. This property is not imaginable without the influence of dispersion. Considering only the laminar flow would

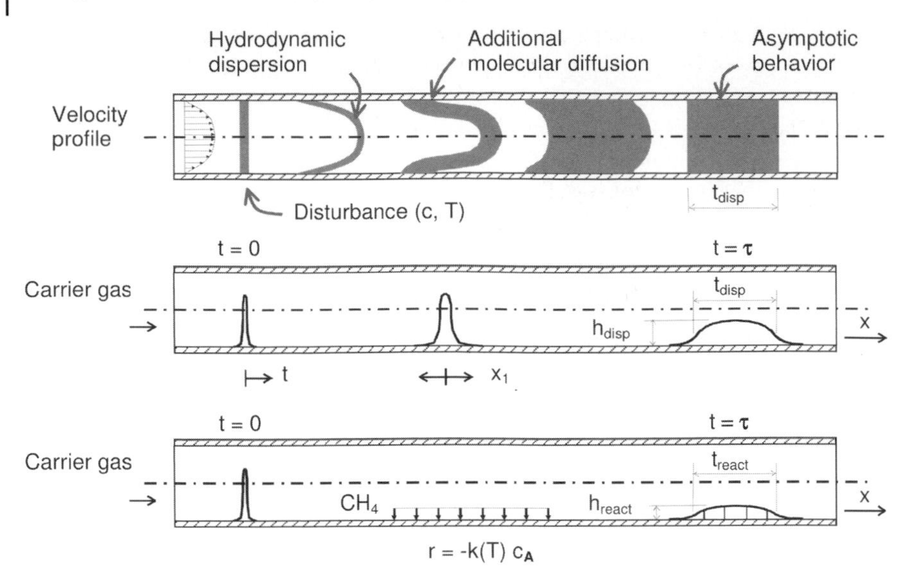

Fig. 4.17 Dispersion in a laminar flow channel reactor with inert walls and with catalytically active walls.

deliver an extremely wide residence-time distribution. The near-wall flow is close to stagnation because a fluid element at the wall of the channel is, by definition, fixed to the wall for an endlessly long time, in contrast to the fast core flow. The phenomenon that prevents such behavior is the dispersion effect, which is demonstrated in Fig. 4.17. The velocity profile in the channel is assumed to be fully developed and steady. Any disturbance, e.g., a concentration pulse or a local pressure change, is primarily distorted by the so-called hydrodynamic dispersion. Without considering diffusion, such a signal would be spread extremely long in an axial direction. Molecular diffusion counter-balances this effect and rapidly equilibrates the signal strength in the cross-section. This equilibration is due to the large radial concentration gradients originating from the laminar parabolic velocity profile that already delivers the disturbance in the center of the channel when the signal has not yet arrived at the walls. After a sufficiently long distance, the laminar flow profile develops into a plug flow profile. The average length of signal t_{disp} and the average signal height (h_{disp}), after the signal has traveled a time τ in the channel, is then a measure for the dispersion inside the channel, characterized by the dispersion coefficient k_r. If the walls are coated with a catalyst, this signal is lowered further by the wall reaction (h_{react}).

4.3.2.2
Reactor Configuration for Catalyst Screening in Transient Mode

To accommodate such measurements, the reactor set introduced in Section 4.2.3 was equipped with a second insulation module and a second rotary valve. The latter replaced the gas distribution module of the steady-reactor configuration

(Fig. 4.18). The input signal was injected by a valve into the reactor inlet valve, which mainly consisted of a short wound 1/16″ Teflon tube, which sent the signal through one of the tubes of the insulation module and a boring in the foil heater to the reaction plate. From here, the product gas was delivered via one of the tubes of the second insulation module, the Teflon tube of the second (exit) valve, directly to the sensor. Both the Teflon tubes and the steel tubes of the insulation modules have the internal diameter ($\phi = 1$ mm). The gas also passed etched channels in the distribution plate of the inlet and the exit valve, which is a concession to the application of these valves to steady screening. Due to the etching process, the cross-section of these channels in the distribution plate was U-shaped (0.5 mm wide, 0.3 mm deep and 50 mm long). The reactor core of this set-up consists of an exchangeable reaction plate, which can be structured with any desired micro-geometry. This feature results from the fact that gas inlet and outlet positions can be chosen from among 48 borings.

For the following tests, a wound spiral was manufactured by fine mechanical means (1.5 m long with a rectangular cross-section of 1.0×0.5 mm^2). The reaction plate could be heated by a 500 W foil heater that had a number of borings to allow gas to pass through the heater.

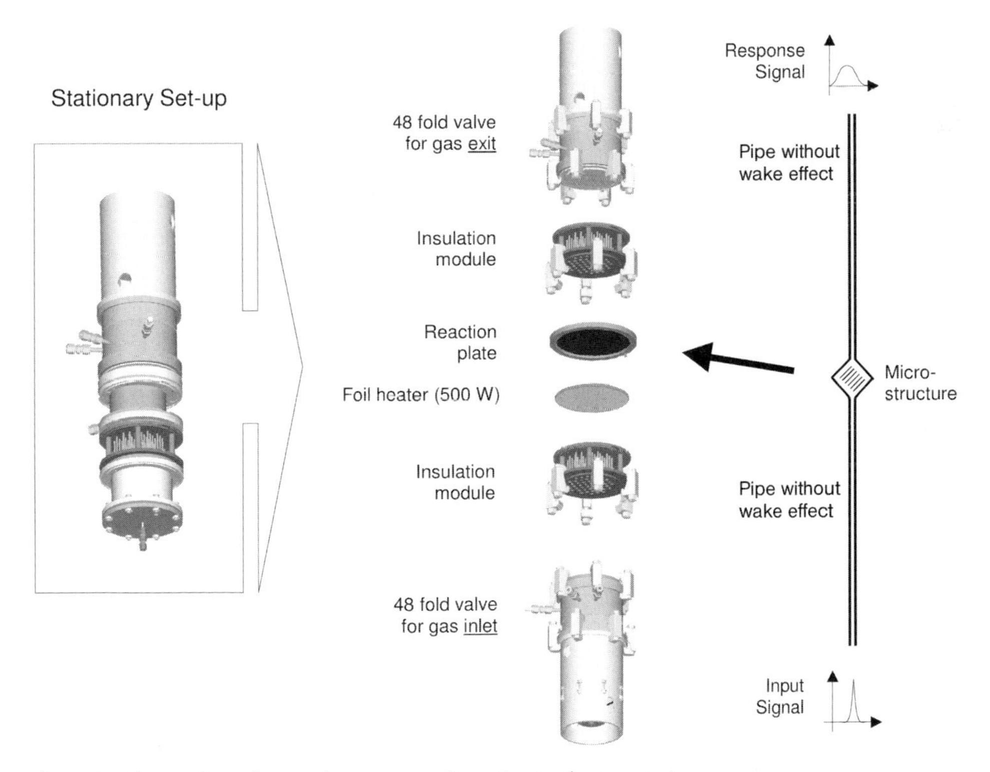

Fig. 4.18 Change from the steady reactor configuration to the transient configuration by the exchange of modules.

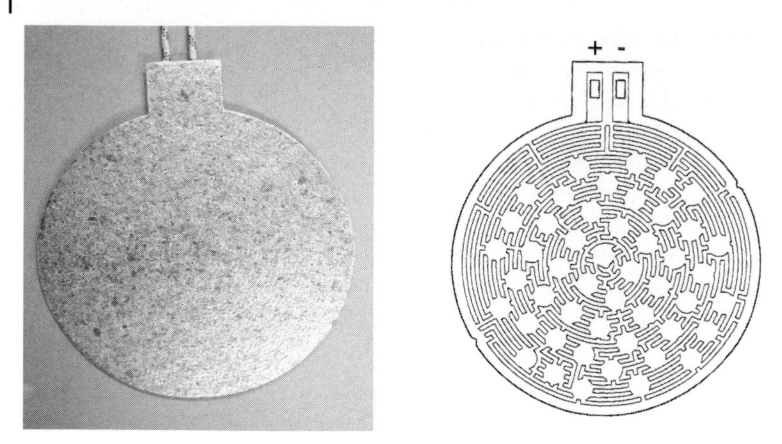

Fig. 4.19 Electrical foil heater and internal scheme of resistance wire (reproduced with kind permission of Watlow Electric GmbH, Kronau).

For the transient measurements, a special foil heater (0.7 mm thick) was designed, and manufactured by Watlow GmbH. Fig. 4.19 shows the internal wiring scheme of the electrical resistance wire. The 0.5 mm borings are in the center of the white spots of the wiring scheme. The wiring is the remaining part of an etched metal foil. A bottom and a top layer consisting of electrically insulating ceramic fiber material cover the metal foil. The foil heater thus has some flexibility and can stand the sealing force of the hydraulic press.

4.3.2.3 Methods for Gas Injection

Gas pulses can be realized by several methods. A gas sample might be trapped in a sample loop of an injection valve, as used for injecting gas samples in a gas chromatograph. However, for flexibility concerning the length and frequency of the pulse, a fast switching electromagnetic two-way valve has been used and connected to a pipe T-fitting (Fig. 4.20). These valves are commercially used for ink-jet printing and allow a shortest injection time of less than 1 ms. The valve is free from wake effects and the small internal volume here cannot influence the pulse shape.

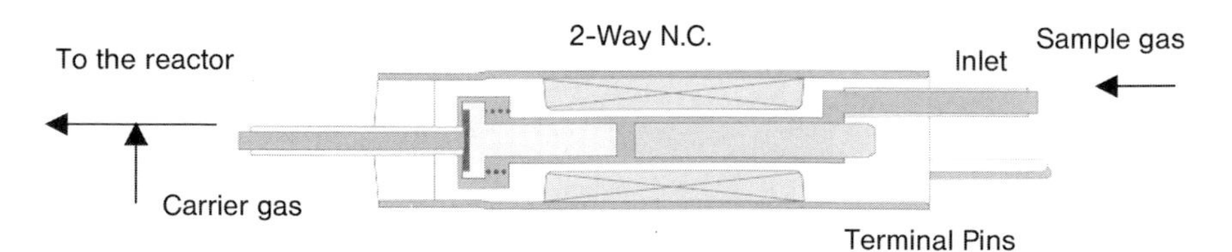

Fig. 4.20 Two-way injection valve (INKA type) combined with a pipe T-fitting (reproduced with kind permission of Lee Hydraulische Miniaturkomponenten GmbH, Frankfurt).

4.3.2.4 Methods for Pulse Signal Detection

Signals can be detected by the flame ionization detector (FID) of a conventional gas chromatograph. The FID measures the change of conductivity of the measure device, which is caused by organic radicals that develop during the combustion in an electric field. For these measurements, the built-in separation column in the gas chromatograph has to be bypassed, because a separation of the gas pulse into its single components is not intended here. The carrier gas flow normally supplied to the FID is replaced by the sample gas flow, consisting of nitrogen in excess and the sample gas pulse with a steady volume flow of 130 ml min^{-1}. These detectors are normally fed with hydrogen and air and are thus not sensitive to oxygen but they are sensitive to hydrocarbons such as methane.

4.3.2.5 Configuration of the Channel Reactor

Dispersion measurements and the coupled dispersion/reaction measurements used the same reaction plate. A 1.5 m long channel with a rectangular cross-section (1 mm wide and 0.5 mm deep) was manufactured out of a 1.4571 steel plate (Fig. 4.21). The plate was coated in a sputtering plant with a 300 nm layer of platinum. Reactants enter the channel through a 0.5 mm boring from below the plate and leave the channel through a similar boring in the cover plate. To enable long channels on the reaction plate, the channel was wound into a spiral. The curvature of the spiral is small because only the peripheral part of the reaction plate is structured, and thus secondary flow is assumed to be negligible. The gas channel is sealed by a graphite foil. The available reactor temperatures were thus limited to ca. 600 °C.

The mean velocity u_0/c of the carrier gas was determined by Taylor's restriction $D \ll k_r$ to be 3.24 m s^{-1}. This value resulted in a mean residence time of $\tau = cL/u_0 = 0.46$ s.

Before injecting the pulses into the reactor, the reactor is supplied with the carrier gas nitrogen and allowed to reach steady state. For the high-temperature experiments, the reactor was heated for at least one hour and the steady state observed with a temperature sensor on the reaction plate. The reactor temperature was mostly 450 °C, at close to atmospheric pressure.

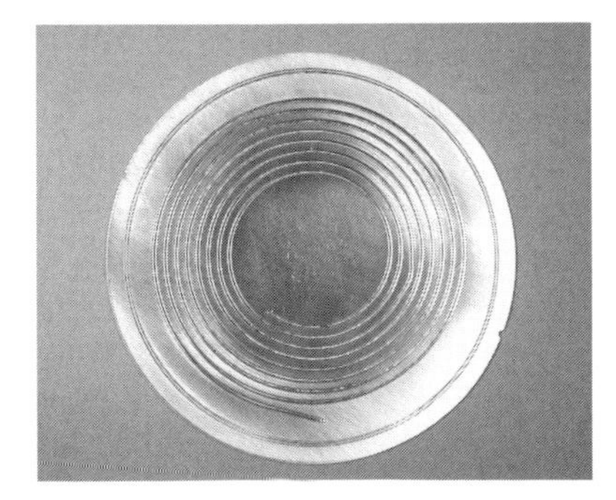

Fig. 4.21 Reaction plate coated with a dense layer of platinum.

Fig. 4.22 Infrared image of the central part of the reactor consisting of the reaction plate and two insulation modules (the temperature decrease of the pipes is only due to passive cooling by radiation and convection).

The increase in reactor temperature is observed with an infrared thermographic camera to detect possible inhomogeneously heated zones (Fig. 4.22). The temperature increase during the start up of the reactor is recorded at three different positions and, after ca. 30 min, a steady temperature distribution is reached; the tube end temperature did not reach a critical value for the O-ring seals in the rotary valves. This means that the desired thermal insulation can be reached by only passive cooling (active cooling with a liquid-cooling medium is excluded here because of the high temperatures). The temperature of the tube ends of the upper insulation module is higher than that of the tube ends in the lower insulation module, due to additional heat transfer from the reaction zone by natural convection in the upper module. In both cases, the heat loss is sufficient to prevent the rotary valves from damage by overheating.

4.3.2.6
Method of Signal Description

Similar to the sample delivery in a reactor operated in differential mode, the sample is injected into the steady laminar carrier flow in the channel which moves at the mean speed u_0/c.

The concentration of methane at room temperature at any channel position is described by the impulse response without reaction (Eq. 1) or by the extended impulse response (Eq. 2) [1].

1) Symbols of formulae see page 89.

$$c_A(x, t, T = 293\ \mathrm{K}) = \frac{M/4ab}{\sqrt{\pi}\sqrt{4k_r t}} \exp\left[-\frac{x_1^2}{4k_r t}\right] \quad (1)$$

$$s(x_1, t) = \frac{M}{2A}\left[erf\left(\frac{x_1 + X/2}{\sqrt{4k_r t}}\right) - erf\left(\frac{x_1 - X/2}{\sqrt{4k_r t}}\right)\right] \quad (2)$$

The dispersion coefficient for a rectangular channel with channel depth b and aspect ratio ε is given in Eq. (3).

$$k_r = \frac{2272}{8505}\frac{b^2 u_0^2}{D}\frac{(1+\varepsilon^4)^2}{(2+\varepsilon+3\varepsilon^2)^2(1+\varepsilon^6)} \quad (3)$$

At reaction temperature the concentration of methane obeys Eq. (4).

$$c_A(x_1, t, T = 723\ K) = \frac{M}{2A}\left[erf\left(\frac{x_1 + t k_v + X/2}{\sqrt{4k_r t}}\right) - erf\left(\frac{x_1 + t k_v - X/2}{\sqrt{4k_r t}}\right)\right] \exp\left[-D\phi^2\left(\frac{a+b}{ab}\right)^2 t\right] \quad (4)$$

(The apparent velocity increase k_ν is equal to $u_0\phi^2 f(\varepsilon)$).

4.3.2.7 Experimental Results

The influence of the reactor temperature on the conversion of methane was examined during pulsed operation. The heat performance of the foil heater was slowly increased while the reactor was continuously supplied with gas pulses consisting of pure oxygen at 129.5 ml min^{-1} together with a flow of methane (0.5 ml min^{-1}). The volume flow of the carrier gas nitrogen was adjusted to 130 ml min^{-1} at atmospheric pressure, delivering a residence time in the coated spiral of 0.4 s. The cat-

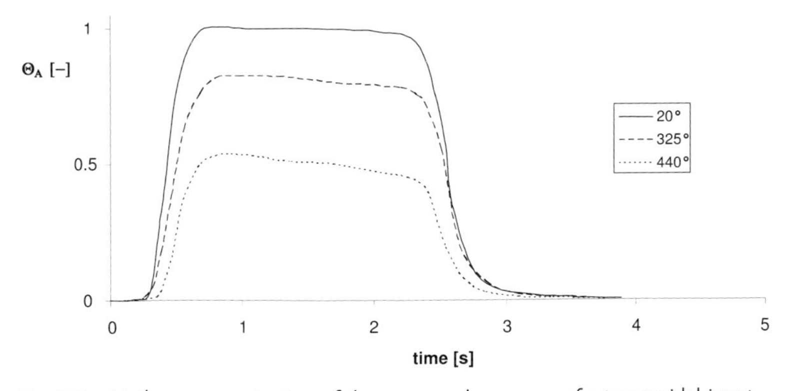

Fig. 4.23 Methane concentration of the measured response of a trapezoidal input signal for different reactor temperatures (two-way INKA valve, 2 s injection time, improved FID sensor).

alyst was not reduced. The result is given in Fig. 4.23 for gas pulses of 2 s followed by breaks of 2 s. No significant conversion was observed below 200 °C while at 325 °C the methane conversion already reached ca. 20%, which further increased to >50% at 440 °C.

This is consistent with the conversion values obtained earlier under steady operation with a platinum catalyst, ranging from 18 to 80%, depending on the pretreatment of the catalyst under similar residence times and temperatures.

Fig. 4.23 also indicates a slight decrease of the signal plateau which, at a first glance, was unexpected. In the following, a reactive dispersion model given in ref. [37] is applied to deduce rate constants for different reaction temperatures. A trapezoidal response function will be used. The temperature-dependent diffusion coefficient was calculated according to a prescription by Hirschfelder (e.g., [80], p. 68; or [79], p. 104] derived from the Chapman-Enskog theory. For the dimensionless formulation, the equation is divided by M/A (with M the injected mass and A the cross-section area). This analytical function is compared in Fig. 4.24 with the experimental values for three different temperatures. The qualitative behavior of the measured pulses is well met; especially the observed decrease of the plateau is reproduced. The overall fit is less accurate than for the non-reactive case but is sufficient to now evaluate the rate constant.

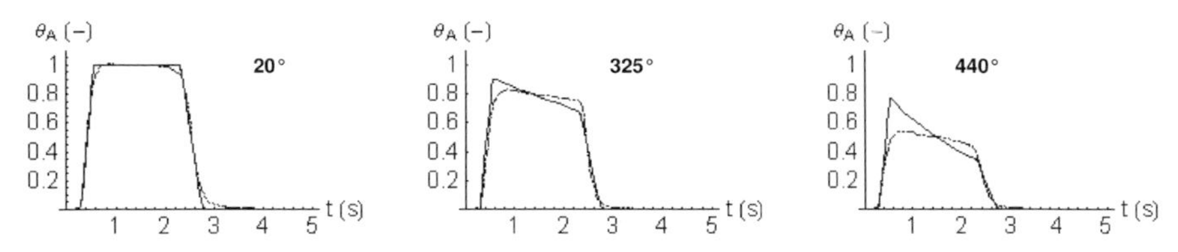

Fig. 4.24 Comparison of model-calculated and measured response of a trapezoidal input signal for three different reactor temperatures (solid line=model, broken line=experimental data; 2-way INKA valve, 2 s injection time, improved FID sensor).

For this purpose, the assumed reaction rate constants in the model were adjusted until the integral values of the calculated pulses were identical to the measured integral values and then drawn as a function of the $1/T$. This Arrhenius-type plot is given in Fig. 4.25 together with a first-order approximation. The straight line indicates a first-order reaction mechanism, which is confirmed by the results of Mezaki and Inoue [81] and Spencer and Pereira [82]. The slope of the function is, according to Eq. (5) (with k the rate constant, k_0 the pre-exponential term, E the activation energy), a function of the activation energy E (Eq. 6).

$$\ln\frac{k}{k_0} = -\frac{E}{\bar{R}}\frac{1}{T} \tag{5}$$

$$\text{slope} = \frac{E}{\bar{R}} = 2505\frac{1}{K}, \quad \bar{R} = 8.314 \text{ J mol}^{-1} \text{ K}^{-1}$$
$$\Rightarrow E = 20.8 \text{ kJ mol}^{-1} \tag{6}$$

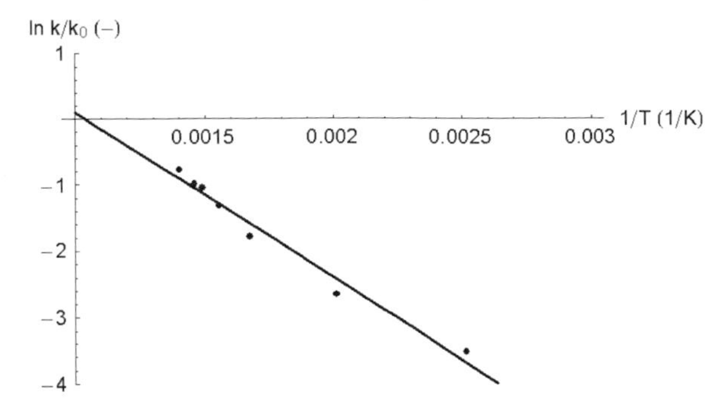

Fig. 4.25 Arrhenius plot of methane oxidation at a platinum surface.

This activation energy is close to the 24 kJ mol^{-1} reported by Mezaki and Inoue for the methane oxidation at a platinum catalyst on an alumina substrate at temperatures above 350 °C.

In the literature, higher values for the activation energy are also found [82, 83]. One reason for this could be the neglect of a pre-reduction of the platinum catalyst and also the low porosity of the sputtered catalyst. Another possibly important aspect is that here we actually measured intrinsic kinetic data compared with the diffusion-affected kinetic data in refs. [82] and [83].

4.3.3 Dynamic Sequential Method for Rapid Screening

Rapid liquid-phase screening in a single channel reactor is restricted by the large dispersion effects, which demand extremely long distances between two samples. A possible way of handling this problem is to use a multiphase system in which a barrier liquid, which is immiscible with the sample liquid, separates the fluids. In this context, a new concept for high-throughput screening has been developed [50] and applied to both gas–liquid (Fig. 4.26) and liquid–liquid systems (Fig. 4.27). This set-up consists in the core of a tubular reactor with an interdigital micro-mixer as dispersion unit (Fig. 4.28). The peripheral equipment consists of

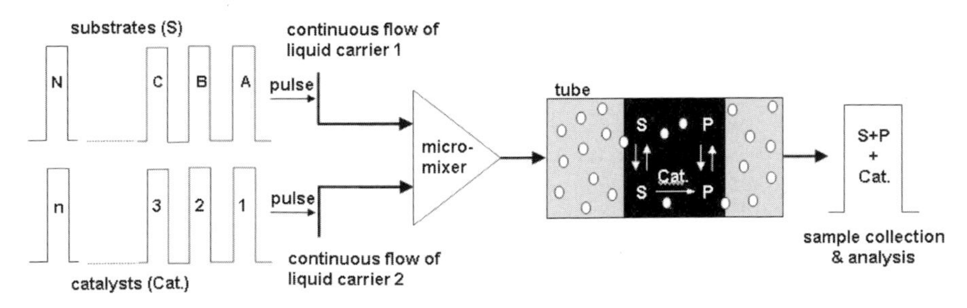

Fig. 4.26 Scheme of the screening device in liquid–liquid systems.

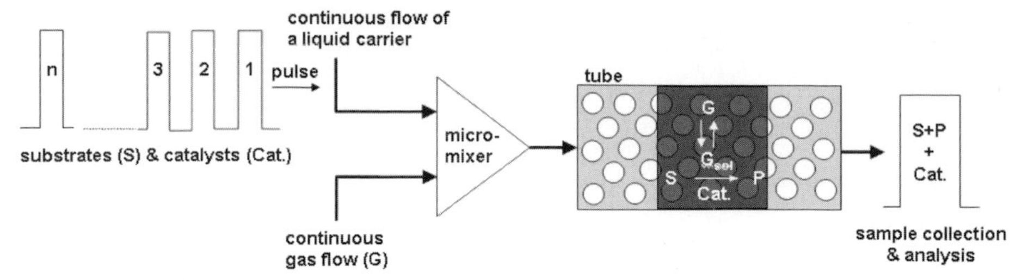

Fig. 4.27 Scheme of the screening device in gas–liquid systems.

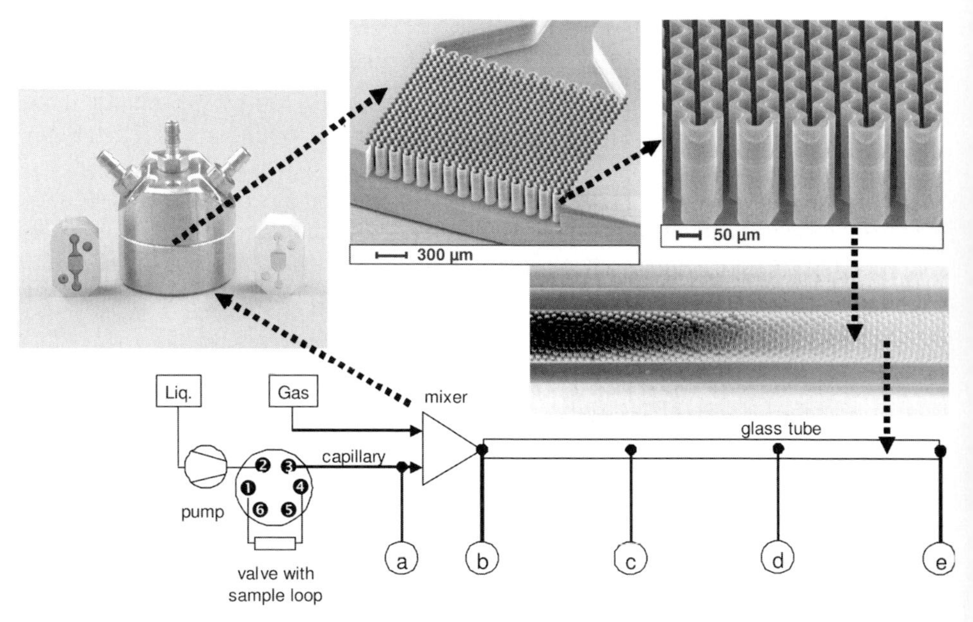

Fig. 4.28 Apparatus for liquid–liquid screening, set-up consisting of a micro mixer–tube reactor combination.

an automated pipetting robot, a fraction collector and a gas-chromatograph equipped with an automatic injector.

The micro-mixer is supplied continuously with two carrier fluids, either two immiscible liquids or a gas and a liquid. During screening, pulses of the dissolved catalyst to be screened and the substrate are injected simultaneously by a pipetting robot. The pulses are then mixed in a micro-mixer and form short defined reacting segments that move along the tubular reactor.

As a reaction in a gas–liquid multiphase system, the enantioselective hydrogenation of (*Z*)-(a)-acetamidocinnamic methyl ester has been investigated [50].

The micro-mixer then generates a continuous foam flow composed of small bubbles of hydrogen (ca. 200 μm in diameter) in the liquid (ethylene glycol–water

COOMe
Ph NAC H
Rh-diphosphane
H_2
water/glycol/SDS
H COOMe
Ph * NAc H

Reaction 1

OH
R R
Rh-cat.
water/heptane
OH
R R

Reaction 2

60:40 wt% with sodium dodecyl sulfate as surfactant). The reactor itself consists of a tubular glass reactor (156 cm long with an inner diameter of 2.85 mm). The reaction rate was proportional to the catalyst concentration and decreased with increasing surfactant concentration [84]. This screening concept has also been applied to liquid–liquid systems. As a test reaction, the isomerisation of allylic alcohols to carbonyls with water-soluble catalysts in a biphasic heptane–water system was chosen (Reaction 2)] [50, 85]. The catalysts (metals: Rh, Ru, Pd, Ni; and ligands: sulfonated phosphane or diphosphane ligands) were injected into the liquid carrier 2 (water). The substrates (different allylic alcohols) were injected into the liquid carrier 1 (heptane).

The reactor consisted of a glass tube (4 mm inner diameter and 80 cm long), resulting in a residence time of 100 s. The highest conversion of 53% was found with the rhodium-catalyst based on the ligand tris(m-sulfo-phenyl)phosphane. The results gained with the micro-mixer-based set-up were in good agreement with those found for a mini-batch reactor.

4.4 Future Prospects

4.4.1 Introduction

In current screening, data evaluation is ever more becoming the bottleneck as the number of samples increases. New sampling strategies need to be developed. An interesting approach was introduced by Rothenberg in 2002 [51]. He studied the effect of the distribution of samples along the time axis and found that using an equidistant distribution along the time axis is much less efficient than the use of an equidistant distribution along the ordinate. Another modern approach is the strategy of condensing information derived from test data. This leads directly to transient measurements, which “condense” stochastic conversion data either by numerical or analytical methods to a function (usually the reaction kinetics). Microstructures are increasingly used for catalyst screening despite the need to upscale the data derived from tests in microstructures to the size of fixed-bed reactors. The question could be raised: is it necessary to upscale microstructures or does an alternative exist? All these strategies would contribute to a reduction in valuable test time and are presented below.

4.4.2 Numerical Evaluation Methods

Numerical simulations have been conveniently used to describe complex fluid dynamic behavior in microstructures [21, 86]. Van der Linde et al. [87] solved the coupled diffusion equation for reacting species and compared the results with data from the oxidation of CO on alumina-supported Cr using the step-response method. Transient periodical concentration changes in microchannels have been numerically calculated by various authors [19, 58, 88].

In 2001, Mirodatos et al. [89] stressed the importance of transient studies as an alternative to steady continuous reactor operations. A combination of microkinetic analysis together with transient experiments should allow the determination of the global catalytic conversion from elementary reaction steps. Prerequisite for such analysis is the correlation of experimental data with the data of a model. Compliance between the data helps to derive the reaction mechanism.

In another recent publication [90], experimental impulse responses were compared with model data obtained by numerically solving the non-linear differential equations for an impulse of O_2 and Ne.

In the same publication, a method for the parallelization of TAP experiments was also indicated. It was stated that "...high-throughput transient kinetics carried out in addition to high-throughput catalyst synthesis and testing both accelerate the search for new catalytic materials and bring fundamental insights into reaction mechanisms."

One clear reason for this statement is that transient screening, in contrast to steady screening experiments, gives a denser information-content simply because during transient tests a complete function (the kinetics) is recorded instead of single stochastic data.

4.4.3 Taylor-type Evaluation Methods

Mass transport in laminar flow is dominated by diffusion and by the laminar velocity profile. This combined effect is known as dispersion and the underlying model for the theoretical derivation of a kinetic study had to be derived from the dispersion model, which Taylor [91] and Aris [92] developed. Taylor concluded that in laminar flow the speed of an inert tracer impulse initially given to a channel will have the same speed as the steady laminar carrier gas flow originally prevailing in this channel.

A reactive dispersion model proposed by Müller et al. [37] predicts a change of speed if the tracer impulse consists of reactants, which react at the walls of the channel (Figs. 4.29 and 4.30). Brenner found a quite fascinating explanation for

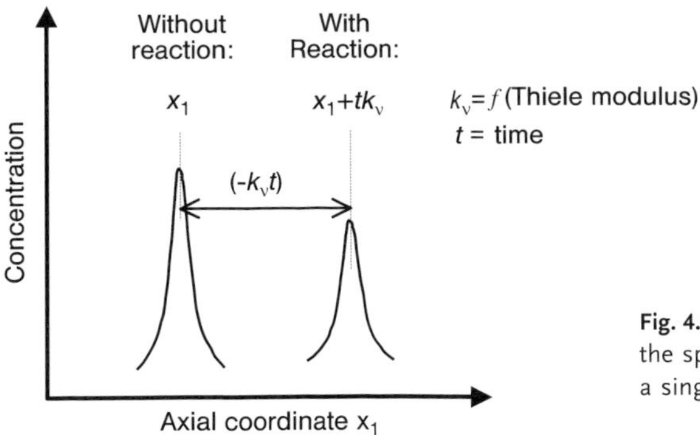

Fig. 4.29 A chemical reaction changes the speed of an impulse of reactants in a single channel reactor.

this phenomenon [93]. He regarded the stochastic Brownian motion of a tracer particle through a channel with reactive walls. Such a molecule will irreversibly react at the channel walls as soon as it touches the wall. An interesting side effect is that such a particle will also be delayed because it spends some time in the slow near-wall region, unlike a particle that does not approach the wall. Conversely, the latter particle will be accelerated compared with the mean flow velocity. Brenner states: "This phenomenon stems from the fact that only those solute molecules "smart enough" to stay away from the wall region survive their trip downstream, ..., those molecules "foolish" enough to meander (by lateral motion) over to the tube wall are destroyed by the reaction. As such, those solute mole-

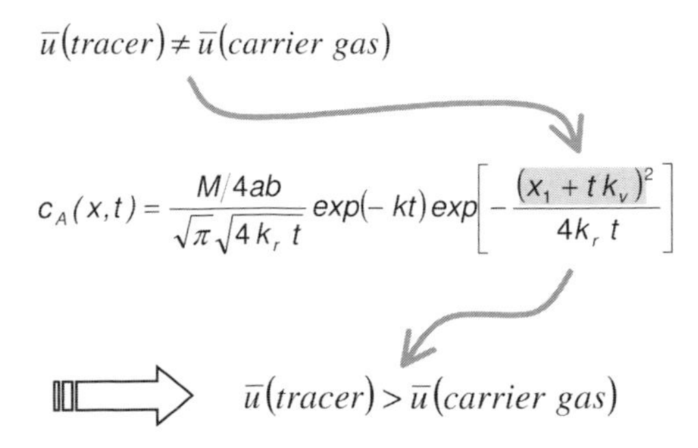

Fig. 4.30 In contrast to the (non-reactive) Taylor model, the tracer gas speed for a reacting gas is unequal to the carrier gas speed. This can be derived from the exponential term in the solution of the concentration of the tracer gas c_A at the channel exit.

cules that exit the tube (...) have not sampled the slower moving streamlines of the Poiseuille flow existing near to the wall."

The theoretical foundation for this kind of analysis was, as mentioned, originally laid by Taylor and Aris with their dispersion theory in circular tubes. Recent contributions in this area have transferred their approach to micro-reaction technology. Gobby et al. [94] studied, in 1999, a reaction in a catalytic wall micro-reactor, applying the eigenvalue method for a vertically averaged one-dimensional solution under isothermal and non-isothermal conditions. Dispersion in etched microchannels has been examined [95], and a comparison of electro-osmotic flow to pressure-driven flow in micro-channels given by Locascio et al. in 2001 [96].

A method was proposed to obtain the kinetic rate constant at a fixed temperature with a one-point measurement. This method is comparable to gas-chromatographic concentration measurements and can in principle be executed with a convenient gas-chromatograph equipped with a flame ionization detector (FID) [37]. The background of this method is introduced in the following.

In macroscopic reactors, knowledge of the velocity profile in the channel cross-section is a necessary and sufficient prerequisite to describe the material transport. In microscopic dimensions down to a few micrometers, diffusion also has to be considered. In fact, without the influence of diffusion, extremely broad residence time distributions would be found because of the laminar flow conditions. Superposition of convection and diffusion is called dispersion. Taylor [91] was among the first to notice this strong dominating effect in laminar flow. It is possible to transfer his deduction to rectangular channels. A complete fluid dynamic description has been given of the flow, including effects such as the influence of the wall, the aspect ratio and a chemical wall reaction on the concentration field in the cross-section [37].

Here, the response functions of the diffusion equation for a number of discrete input signals were calculated based upon the solution for a Dirac impulse input signal.

A set of transformation relations was used to obtain the solutions for a reacting gas directly from the solutions of a non-reacting gas:

$$M \Rightarrow M \exp\left[-D\phi\left(\frac{a+b}{ab}\right)^2 t\right]$$
$$x_1 \Rightarrow x_1 + t\,k_v \tag{7}$$

The response function for a reacting gas in Eq. (8) [the dispersion coefficients k_r and the coefficient k_v are given in Eq. (3) and by $u_0\phi^2 f(\varepsilon)$, respectively] is derived from the response function obtained for a Dirac impulse for a non-reacting gas (Eq. 9).

$$c_A(x,t) = \frac{M/4ab}{\sqrt{\pi}\sqrt{4k_r t}} \exp\left[-D\phi^2\left(\frac{a+b}{ab}\right)^2 t - \frac{(x_1+tk_v)^2}{4k_r t}\right] \tag{8}$$

$$c_A(x,t) = \frac{M/4ab}{\sqrt{\pi}\sqrt{4k_r t}} \exp\left[-\frac{x_1^2}{4k_r t}\right] \tag{9}$$

The response to the step-function for a non-reacting gas is given by integrating the response of the pulse function as Eq. (10).

$$h(x_1,t) = \frac{M}{4ab}\frac{1}{\sqrt{\pi}\sqrt{4k_r t}} \int_{x_1}^{\infty} \exp\left[-\frac{x_1^2}{4k_r t}\right] dx_1 = \frac{M}{8ab}\left[1 - erf\left(\frac{x_1}{\sqrt{4k_r t}}\right)\right] \tag{10}$$

The solution for the reactive case is then obtained without calculation directly from Eq. (10) by applying the transformations of Eq. (7):

$$h(x_1,t) = \frac{M}{8ab}\left[1 - erf\left(\frac{x_1 + tk_v}{\sqrt{4k_r t}}\right)\right] \exp\left[-D\phi^2\left(\frac{a+b}{ab}\right)^2 t\right] \tag{11}$$

A graphical representation is shown in Fig. 4.31. For the extended pulse function of width X, the solution given by Eq. (12) is obtained in the same manner.

$$s(x_1,t) = \frac{M}{8ab}\left[erf\left(\frac{x_1 + tk_v + X/2}{\sqrt{4k_r t}}\right) - erf\left(\frac{x_1 + tk_v - X/2}{\sqrt{4k_r t}}\right)\right] \exp\left[-D\phi^2\left(\frac{a+b}{ab}\right)^2\right] \tag{12}$$

The reaction introduces a spatial shift into the solution (tk_v), which leads to an asymmetric behavior of the response function in the moving frame of reference. The slight asymmetry of the lower solid line in Fig. 4.31 is only visible in a magnified view (Fig. 4.32).

This effect is more distinct in a very flat rectangular channel and even more so in a liquid system, as shown in Fig. 4.33 for the water–acetone system at 25 °C. Diffusion coefficients of liquids are close to 10^{-5} cm^2 s^{-1}, a value four orders of magnitude smaller than the diffusion coefficients of gases. In consequence, the dispersion

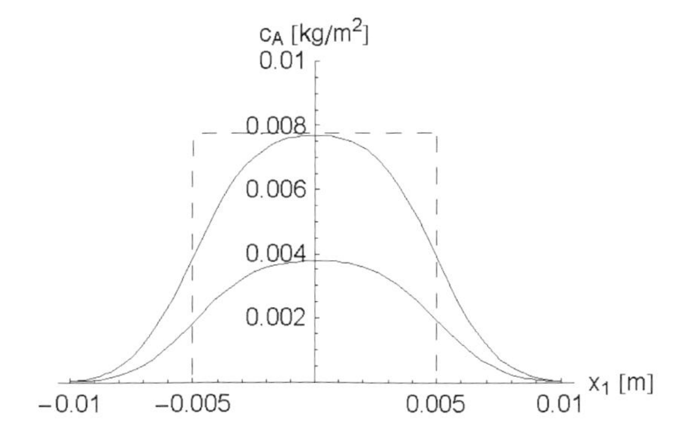

Fig. 4.31 Response to the extended pulse function for t=0 s (broken line) and t=0.01 s for a non-reacting gas (upper solid line) and a reacting gas (ϕ=0.3, lower solid line).

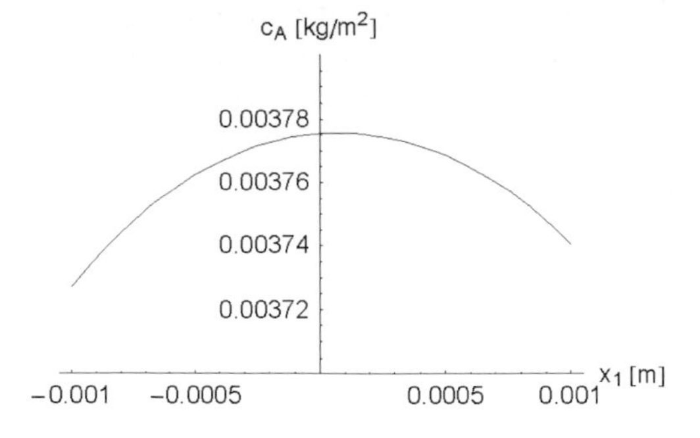

Fig. 4.32 Magnification of the lower solid line in Fig. 4.32 (ϕ=0.3).

and thus the pulse shift is also more distinct. Fig. 4.33 shows the concentration in a flat channel with such a small aspect ratio of 0.1 for different Thiele moduli (ϕ). Obviously, the maximum of the pulses has moved towards the channel exit (towards the right-hand side in Fig. 4.33). This ability of the heterogeneous wall reaction could be a means to measure the kinetic rate coefficient k directly from the arrival time of the pulse at the channel exit with a one-point measurement (see above). An experimental procedure for the fast determination of reaction kinetics can make use of the apparent velocity increase of the pulse movement due to the wall reaction to obtain the kinetics. Information about the intrinsic kinetics is obtainable by measuring the arrival time of the peak maximum of a pulse function at the reactor outlet. Only the position of the peak maximum has to be measured and not the reactor outlet concentration itself. As the effect is not strong, especially for gases, very flat channels and high-resolution equipment are certainly necessary. The examples above correspond to slow reactions such as the hydrogenation of benzene (ϕ>0.05) or the oxi-

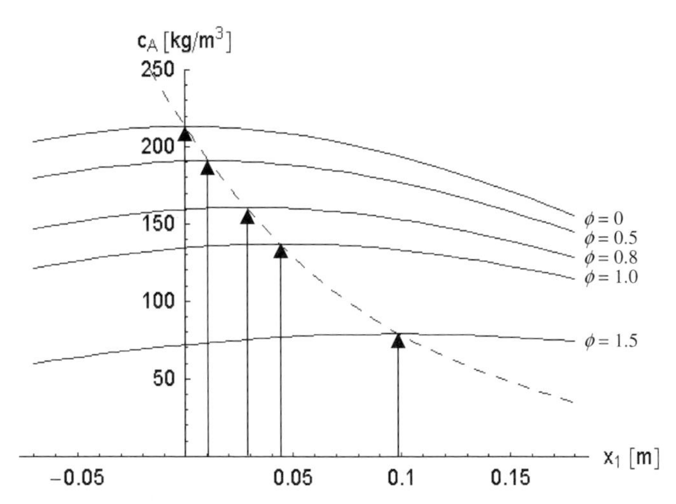

Fig. 4.33 Shift of concentration pulses (Dirac impulses) for an aspect ratio of 0.1 in the water–acetone system (arrows indicate the individual peak maximum; the hatched line connects the peak maxima).

dation of ethylene ($\phi = 0.08$). For fast reactions, such as the oxidation of methanol to formaldehyde ($\phi = 1.1$), the reaction time will reduce and thus the difficulties in observing the pulse shift will increase but, also, the observed asymmetry will be more distinct. Microreactors seem to be especially suited to measuring such pulse shifts because they exhibit short residence times and, thus, high time resolutions.

4.4.4
Evaluations Following Biological Means

A separate class of experimental evaluation methods uses biological mechanisms. An artificial neural net (ANN) copies the process in the brain, especially its layered structure and its network of synapses. On a very basic level such a network can learn rules, for example, the relations between activity and component ratio or process parameters. An evolutionary strategy has been proposed by Mirodatos et al. [97] (see also Chapter 10 for related work). They combined a genetic algorithm with a knowledge-based system and added descriptors such as the catalyst pore size, the atomic or crystal ionic radius and electronegativity. This strategy enabled a reduction of the number of materials necessary for a study.

Baerns et al. [98] (see also Chapter 6 for related work) stressed the problem of overtraining, which means that an ANN may well reproduce the screening data of the training data set but give bad results for an unknown set of data. A method to prevent this effect is, for example, the early stopping of training. The perceptron-type network was equipped with one hidden layer and network optimization was executed with the Levenberg-Marquardt approximation, a backpropagation method to minimize the error by the Gauss-Newton quadratic minimization method. With this method they could predict the yield of propene for the oxidative dehydrogenation of propane with an error of 5.4%.

Corma et al. [99] (see also Chapter 5 for related work) also used a neural net, which they adapted to the desired results by supervised learning. During training, the network adapts the weights between the neurons, thus acting as a kind of complex function approximator between training data as the input data and desired results as the output data. They also found a multilayer perceptron to have better performance than a radial-basis network or a self-organizing network using the Kohonen learning rule. As input data for the learning process, again the catalyst composition of 13 elements was fed into the net and adapted to the desired output data, the yield and the selectivity. Predicted and experimentally observed data for the test reaction of the oxidative dehydrogenation of ethane agreed sufficiently. They also reported problems with data singularities, a known disadvantage of neural nets, which are better at predicting smooth data relations. Rodemerck et al. [100] and Corma et al. prepared the catalyst compositions by genetic algorithms and combined this approach with an artificial neural network. This seems to be one of the most promising strategies for lean data mining. Rodemerck et al. trained a neural network to predict the composition of new catalysts. Together with a genetic algorithm new generations of catalysts were produced by mutation and recombination of catalysts from the old generation in an iterative process.

4.4.5
Is Up-scaling a Necessity?

Is there a way out of the up-scaling problem? The transfer of results obtained in small dimensions to large-scale reactors often fails as the experimental conditions change. Dimensionless descriptors like the Reynolds number or the Nusselt number would help to derive up-scaled geometries or process parameters but they are often hindered by the experimental realities such as diffusion limitations, temperature restrictions, etc. Hence, a method to avoid up-scaling would be highly interesting. This is certainly not a new approach as, for example, Heatric already offers large microstructured heat exchangers manufactured in a diffusion-bonding process. At the IMM a meso-scale counterflow heat exchanger has been developed using a laser welding technique. In a German public project a large vessel-type microstructured reactor has been developed for propylene oxide synthesis. These apparati consist of large microstructured metal plates. An unsolved problem is the coating of these plates with catalyst. So far only manual procedures exist, which deliver sufficient results with respect to homogeneity for small reactors but certainly not for large reactors. Another important aspect, besides that of reproducibility, is production costs. If coating has to be cost efficient, it must be executed in an automated continuous procedure. Such procedures exist, for example, in the non-woven industry. Fiber fleece products, for example, in use in household products are produced continuously and are also coated or impregnated continuously. The IMM currently modifies a coating device from the company Coatema in Dormagen to enable a new coating procedure for microstructured etched metal plates (Fig. 4.34). These metal plates are continuously unwound from a roll, which is to be microstructured by wet-chemical etching. If desired, the metal foil can be precut by laser-cutting at defined positions. This should enable either the later separation of the plates after the coating or the obtainment of small platelets of any desired shape for usage in small reactors, e.g., in catalyst screening. These etched and laser-cut foils are already commercially available products but they still must

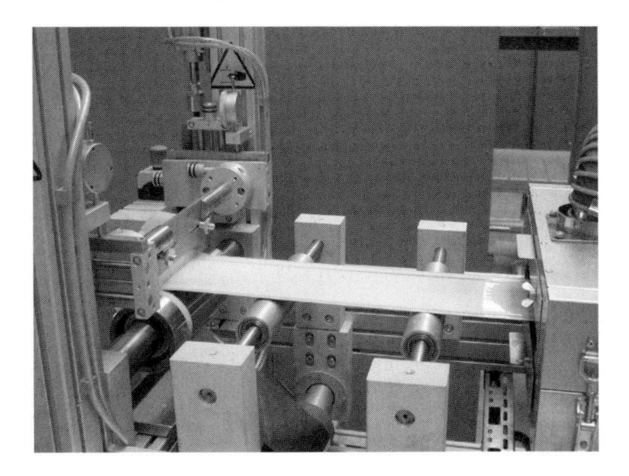

Fig. 4.34 Semi-automated continuous coating apparatus for catalyst and precursor preparation. The procedure is demonstrated here using an unstructured metal foil unwound from a roll. The wash-coat slurry is fed through a slot die onto the metal foil. After coating, the foil is sent through a drying compartment and wound up.

be integrated in the coating process. In such a way, small platelets and also large plates will be produced using the same coating procedure, by-passing an up-scaling process. However, these are so far only expected results. The complete process is still under development. Currently, the first tests have been executed with unstructured foils. For large reactors it is also intended to coat the catalyst and the catalyst precursor simultaneously. However, for the application in catalyst screening, the described method has to be combined with an impregnation facility such as a laboratory sample robot.

4.5 References

1 W. Ehrfeld, V. Hessel, V. Haverkamp, Microreactors, *Ullmann's Encyclopedia of Industrial Chemistry*, Wiley-VCH, Weinheim, 1999.

2 W. Ehrfeld, V. Hessel, H. Löwe, *Microreactors*, Wiley, Weinheim, 2000, p. 24.

3 A. Gavriilidis, P. Angeli, E. Cao, K.K. Yeong, Y.S.S. Wan, *Trans. IChemE.*, 80/A, 1 (2002), 3.

4 P.D. Fletcher, E. Pombo-Villar, B.H. Warrington, P. Watts, S.Y.F. Wong, X. Zhang, *Tetrahedron*, 58(24) (2002), 4735.

5 R.S. Wegeng, M.K. Drost, D.L. Brenchley, *Proc. 3rd International Conference on Microreaction Technology, 1999*, Springer, Berlin, 2000, p. 2.

6 K.F. Jensen, *Chem. Eng. Sci.*, 56 (2001), 293.

7 A. de Mello, R. Wootton, *Lab on a Chip*, 2 (2002), 7N.

8 R. Besser, X. Quyang, H. Surangalikar, *Proc. 6th International Conference on Microreaction Technology, 2002*, AIChE, New York, 2002, p. 254.

9 H. Kestenbaum, A. Lange de Olivera, W. Schmidt, F. Schüth, W. Ehrfeld, K. Gebauer, H. Löwe, T. Richter, *Stud. Surf. Sci. Catal.*, 130 (2000), 2741.

10 G. Wiessmeier, D. Hönicke, *Ind. Eng. Chem. Res.* 35 (1996), 4412.

11 R. Srinivasan, I.-M. Hsing, P.E. Berger, K.F. Jensen, *AIChE J.*, 43 (1997), 3059.

12 M. Fichtner, J. Mayer, D. Wolf, K. Schubert, *Ind. Eng. Chem. Res.*, 40 (2001), 3475.

13 E.V. Rebrov, M.H.J.M. de Croon, J.C. Schouten, *Catal. Today*, 69 (2001), 183.

14 A. Rouge, B. Spoetzl, K. Gebauer, R. Schenk, A. Renken, *Chem. Eng. Sci.*, 56, 2001, 1419.

15 O. Wörz, K.-P. Jäckel, T. Richter, A. Wolf, *Chem. Eng. Technol.*, 24 (2001), 138.

16 Y.S.S. Wan, J.L.H. Chau, A. Gavriilidis, K.L. Yeung, *Chem. Commun.*, (2002), 878.

17 V. Cominos, S. Hardt, V. Hessel, G. Kolb, H. Löwe, M. Wichert, R. Zapf, *Proc. 6th International Conference on Microreaction Technology, 2002*, AIChE, New York, 2002, p. 113.

18 H. Löwe, V. Hessel, *Chem. Eng. Technol.*, 25 (2003), 13.

19 J.M. Commenge, L. Falk, J.P. Corriou, M. Matlosz, *AIChE J.* 48 (2000), 345.

20 U. Hagendorf, M. Janicke, F. Schüth, K. Schubert, M. Fichtner, *Proc. 2nd International Conference on Microreaction Technology, 1998*, AIChE, New York, 1998, p. 81.

21 V.B. Makhijani, V.B. Raghavan, J. Przekwas, A. Przekwas, Simulation of Biochemical Reaction Kinetics in Microfluidic Systems, IMRET3: *Proceedings of the Third International Conference on Microreaction Technology*, Springer, Berlin Heidelberg, New York, 2000, p. 441–450.

22 F. Keil, *Diffusion und Chemische Reaktion in der Gas/Feststoffkatalyse*, Springer, Berlin, Heidelberg, 1999.

23 J. Dumesic, D. Rudd, L. Aparicio, J. Rekoske, A. Trevino, *The Microkinetics of Heterogeneous Catalysis*, American Chemical Society, Washington DC, 1993.

24 R. Adler, T. Hennig, M. Bauer, F. Schröder, M. Schreier, *Chem. Ingenieur Technik*, 74 (2002), 783.

25 S.M. Senkan, *Nature*, 394 (1998), 350.

26 S. Senkan, K. Krantz, S. Ozturk, V. Zengin and I. Onal, *Angew. Chem.*, 111 (1999).

27 S. Senkan, T. Miyazaki, K. Krantz, S. Ozturk, C. Leidholm, Discovery and Optimization of Heterogeneous Catalytic Materials Using Combinatorial Methodologies, *CombiCat2002 North America, Conference Proceedings*, 2002.

28 U. Rodemerck, O.V. Buyevskaya, P. Ignaszewski, M. Langpape, S. Kolf, P. Claus, M. Baerns, Parallel Synthesis and Fast Catalytic Testing of Heterogeneous Catalysis, Microreaction Technology: Parallel Preparation and Testing of Catalysts, 21st June 2001, Dechema, Frankfurt/Main, Germany, Book of Abstracts.

29 U. Rodemerck, P. Ignaszewski, M. Lucas, P. Claus, M. Baerns, *Proc. 3rd International Conference on Microreaction Technology, 1999*, Springer, Berlin, 2000, p. 287.

30 U. Rodemerck, P. Ignaszewski, M. Lucas, P. Claus, *Chem. Eng. Technol.*, 23 (2000), 5.

31 C. Hoffmann, A. Wolf, F. Schüth, *Angew. Chem.*, 18 (1999), 111.

32 F. Schüth, C. Hoffmann, A. Wolf, *Angew. Chem., Int. Ed.*, 38 (1999), 2800.

33 C. Hoffmann, S. Thomson, O. Busch, A. Wolf, C. Kiener, H.-W. Schmidt, F. Schüth, Book of Abstracts of "Microreaction Technology: Parallel Preparation and Testing of Catalysts", 21st June 2001, Dechema, Frankfurt/Main, Germany.

34 F. Rampf, W.A. Herrmann, *Chem. Ing. Technol.*, 73 (2001), 1.

35 T. Zech, D. Hönicke, A. Lohf, K. Golbig, T. Richter, *Proceedings of the Third International Conference on Microreaction Technology, 1998*, AIChE, New York, 1998, p. 261–266.

36 G. Kolb, V. Cominos, K. Drese, V. Hessel, C. Hoffmann, H. Löwe, O. Wörz, R. Zapf, *Proc. 6th International Conference on Microreaction Technology, 2002*, AIChE, New York, 2002, p. 61.

37 A. Müller, "A Modular Approach to Heterogeneous Catalyst Screening in the Laminar Flow Regime", Thesis, VDI-Verlag GmbH, Düsseldorf, 2004.

38 A. Müller, K. Drese, H. Gnaser, M. Hampe, V. Hessel, H. Löwe, S. Schmitt, R. Zapf, *Catal. Today*, 81 (2003), 377.

39 A. Schober, G. Schlingloff, A. Thamm, H.J. Kiel, D. Tomandl, M. Gebinoga, M. Döring, J.M. Köhler, G. Mayer, *Microsystem Technol.*, 4 (1997), 35.

40 T. Zech, D. Hönicke, Microreaction Technology: Parallel Preparation and Testing of Catalysts, 21st June 2001, Dechema, Frankfurt/Main, Germany, Book of Abstracts.

41 T. Zech, *Miniaturisierte Screening-Systeme für die kombinatorische heterogene Katalyse*, Fortschr.-Ber. VDI Reihe 3 Nr. 732. VDI, Düsseldorf, 2002.

42 K.F. Jensen, S.L. Firebaugh, A.J. Franz, D. Quiram, R. Srinivasan, M.A. Schmidt, Integrated gas phase micro-reactors, in: Harrison J., van den Berg, A. (eds.), *Micro Total Analysis Systems*, Kluwer Academic Publishers, Dordrecht, 1998, p. 463.

43 F. Jensen, M.A. Schmidt, *Proc. 2nd International Conference on Microreaction Technology, 1998*, AIChE, New York, 1998, p. 33.

44 S. Bergh, S. Guan, *Int. Pat.*, WO 00/51720, 2000.

45 M.T. Reetz, M.H. Becker, K.M. Kühling, A. Holzwarth, *Angew. Chem.*, 110 (1998).

46 J. Klein, C. Lehmann, H. Schmidt, W. Maier, *Angew. Chem.*, 110 (1998).

47 M. Orschel, J. Klein, H.W. Schmidt, W.F. Maier, *Angew. Chem.*, 111 (1999).

48 R. Potyrailo, R. Wroczynski, J. Lemmon, W. Flanagan, O. Siclovan, *J. Comb. Chem.*, 5 (2003), 8.

49 N. Morris, T. Mallouk, *J. Am. Chem. Soc.*, 124 (2002), 11114.

50 D. de Bellefon, N. Tanchoux, S. Caravieilhes, P. Grenouillet,

V. Hessel, Microreactors for dynamic high throughput screening of fluid-liquid molecular catalysis, *Angew. Chem., Int. Ed.*, 39(19) (2000), 3442.

51 G. Rothenberg, H. F. M. Boelens, D. Iron, J. A. Westerhuis, A high-throughput approach to kinetic studies, *CombiCat2002 North America, Conference Proceedings*, 2002. The catalyst group resources, Inc., Philadelphia, USA.

52 J. Lauterbach, *Proc. CombiCat2002 North America*, 2002. The catalyst group resources, Inc., Philadelphia, USA.

53 R. S. Besser, X. Quyang, H. Surangalikar, *Chem. Eng. Sci.*, 58 (2003), 19.

54 P. Cong, R. D. Doolen, Q. Fan, D. Giaquinta, S. Guan, E. McFarland, D. Poojary, K. Self, H. W. Turner, W. H. Weinberg, *Angew. Chem.*, 4 (1999), 111.

55 V. Cominos, S. Hardt, V. Hessel, G. Kolb, H. Löwe, M. Wichert, R. Zapf, *Chem. Eng. Commun.*, in press.

56 A. Ziogas, V. Hessel, G. Kolb, H. Löwe, R. Zapf, *Proc. of the 5th European Congress of Chemical Engineering (ECCE)*, 2003, Granada.

57 M. Fichtner, J. Mayer, Mikroreaktorsystem für die partielle Oxidation von Methan, Jahresbericht 1997 zu Teilprojekt 5: Entwicklung und Herstellung von Mikroreaktoren aus Metallen mit spanabhebender Materialbearbeitung für die Beispielreaktion Methanoxidation, Forschungszentrum Karlsruhe Technik und Umwelt, 1997.

58 M. Liauw, *Proc. 3rd International Conference on Microreaction Technology, 1999*, Springer, Berlin, 2000, p. 224.

59 G. Emig, H. Seiler, *Chem. Eng. Technol.*, 21 (1999), 6.

60 V. A. Roiter, G. P. Korneichuk, M. G. Leperson, N. A. Stukanowskaia, B. I. Tolchina, *Zh. Fiz. Chim.*, 24 (1950), 459.

61 R. Kokes, H. Tobin, P. Emmett, *J. Am. Chem. Soc.*, 77 (1955), 5860.

62 P. Schneider, J. M. Smith, *AIChE J.*, 14 (1968), 763.

63 L. Polanski, L. Naphtali, *AIChE J.*, Chemical engineering research and development, American Institute of Chemical Engineers, New York, ISSN: 0001-1541, 1969.

64 S. Colin, C. Aubert, R. Caen, *Eur. J. Mech., B/Fluids*, 17 (1998), 79.

65 M. Zieren, R. Willnauer, J. M. Köhler, N. Schwesinger, Detection of catalytic reactions in a miniaturized FIA set-up using a static micromixer and a flow-through chip calorimeter, Abstract collection '13. Ulm-Freiberger Kaloriemetrietage, Freiberg, Germany, 1999, p. 110.

66 J. Dumesic, D. Rudd, L. Aparicio, J. Rekoske, A. Trevino, *The Microkinetics of Heterogeneous Catalysis*, American Chemical Society, Washington DC, 1993.

67 C. Bennett, *Catal. Rev.-Sci. Eng.*, 13 (1976), 121.

68 R. Kokes, H. Tobin, P. Emmett, New microcatalytic-chromatographic technique for studying catalytic reactions, *J. Am. Chem. Soc.*, 77 (1955), 5860–5862.

69 C. Mirodatos, *Catal. Today*, 9 (1991), 83.

70 S. Walter, M. Liauw, Fast concentration cycling in microchannel reactors, *Proceedings of the Fourth International Conference on Microreaction Technology*, Atlanta, USA, March 2000, Ehrfeld, Eul, Wegeng, AIChE, Atlanta, USA.

71 B. Wojciechowski, *Catal. Today*, 36 (1997), 167.

72 J. Gleaves, G. Yablonski, P. Phanawadee, Y. Schuurman, *Appl. Catal. A: General*, 160 (1997), 55.

73 S. van der Linde, T. Nijhuis, F. Dekker, F. Kapteijn, J. Moulijn, Mathematical treatment of transient kinetic data: Combination of parameter estimation with solving the related partial differential equation, *Appl. Catal. A: General*, 151 (1997), 27–57.

74 R. Jahn, D. Snita, M. Kubícek, M. Marek, Evolution of spatio-temporal temperature patterns in monolithic catalytic reactor. *Catalysis Today*, 70 (2001) 393–409.

75 J. M. Commenge, L. Falk, J. P. Corriou, M. Matlosz, Optimal design for flow uniformity in microchannel reactors, *AIChE J.* 48(2), (2000), 345–358.

76 D. de Bellefon, N. Tanchoux, S. Caravieilhes, P. Grenouillet, V. Hessel, Microreactors for dynamic high throughput screening of fluid-liquid molecular catalysis, *Angew. Chem., Int. Ed.*, 39(19) (2000), 3442–3445.

77 B.W. Wojciechowski, *Catal. Today*, 36 (1997), 191.

78 H. Kobayashi, M. Kobayashi, *Catal. Rev.-Sci. Eng.*, 10 (1974), 139.

79 E.L. Cussler, *Diffusion Mass Transfer in Fluid Systems*, Cambridge University Press, Cambridge, UK, 1997.

80 M. Baerns, H. Hofmann, A. Renken, *Chemische Reaktionstechnik, Lehrbuch der Technischen Chemie*, Band 1, Georg Thieme, Stuttgart, New York, 1987.

81 R. Mezaki, H. Inoue, *Rate Equations of Solid-catalyzed Reactions*, University of Tokyo Press, Tokyo, 1991.

82 N.D. Spencer, C.J. Pereira, Partial oxidation of CH_4 to HCHO over a MoO_3-SiO_2 Catalyst: A kinetic study, *AIChE J.*, 33(11) (November 1987).

83 C.F. Cullis, D.E. Keene, D.L. Trimm, Studies of the partial oxidation of methane over heterogeneous catalysts, *J. Catal.*, 19 (1970), 378–385.

84 C. de Bellefon, N. Pestre, T. Lamouille, P. Grenouillet, V. Hessel, *Adv. Synth. Catal.* 345 (2003), 190.

85 C. de Bellefon, R. Abdallah, T. Lamouille, N. Pestre, S. Caravieilhes, *Chimia*, 56 (2002), 621.

86 D. Snita, J. Kosek, H. Sevcikova, J. Lindner, J. Havlica, M. Paces, M. Marek, *Proc. 3rd International Conference on Microreaction Technology, 1999*, Springer, Berlin, 2000, p. 336.

87 S. van der Linde, T. Nijhuis, F. Dekker, F. Kapteijn, J. Moulijn, *Appl. Cat. A: General*, 151 (1997), 27.

88 J.M. Commenge, L. Falk, J.P. Corriou, M. Matlosz, *Proc. 3rd International Conference on Microreaction Technology, 1999*, Springer, Berlin, 2000, p. 224.

89 A. Holzwarth, P. Denton, H. Zanthoff, C. Mirodatos, *Catal. Today*, 67 (2001), 309.

90 A.C. Van Veen, D. Farruseng, M. Rebeilleau, T. Decamp, A. Holzwarth, Y. Schuurman, C. Mirodatos, *J. Catal.*, 216 (2003), 135.

91 Sir G. Taylor, *Proc. Royal Soc. London, Ser. A*, Vol. 223, p. 186–203 (1954).

92 R. Aris, *Proc. Royal Soc. London* (1955), 67.

93 H. Brenner, *J. Surf. Colloids* (1990), 6(12).

94 D. Gobby, I. Eames, A. Gavriilidis, *Proc. 3rd International Conference on Microreaction Technology, 1999*, Springer, Berlin, 2000, p. 253.

95 D. Dutta, D.T. Leighton, *Anal. Chem.*, 73 (2001).

96 L.E. Locascio, D. Ross, T.J. Johnson, *Anal. Chem.*, 73 (2001), 2509.

97 C. Mirodatos, J.M. Serra, A. Corma, D. Farrusseng, L. Baumes, C. Flego, C. Perego, Styrene from toluene by combinatorial catalysis, *Catal. Today*, 81 (2003), 425–436.

98 M. Holena, M. Baerns, Feedforward neural networks in catalysis, a tool for the approximation of the dependency of yield on catalyst composition, and for knowledge extraction, *Catal. Today*, 81 (2003), 485–494.

99 A. Corma, J.M. Serra, E. Argente, V. Botti, S. Valero, Application of artificial neural networks to combinatorial catalysis: modeling and predicting ODHE catalysts, *CHEMPHYSCHEM*, 3 (2002), 939–945.

100 U. Rodemerck, Hochdurchsatz-Experimentation, Jahresbericht 2002, Institut für Angewandte Chemie Berlin-Adlershof e.V., May 2003.

5
Two Exemplified Combinatorial Approaches for Catalytic Liquid–Solid and Gas–Solid Processes in Oil Refining and Fine Chemicals

José M. Serra and Avelino Corma

5.1
Introduction

Drug development underwent drastic and successful change in the 1990s by means of the fast synthesis and screening of large libraries of diverse formulations on fully automated working stations and analytics. The so-called combinatorial approach rapidly extended to other research domains such as materials science and catalysis. The automated high-throughput screening of large libraries of solid catalysts is today entirely possible thanks to fast-growing technologies for automation, micro-mechanics and computation. Hence, fully automated robots specially designed for fast catalyst synthesis and multiple parallel reactors for catalytic testing are now available.

Combinatorial catalysis [1, 2] is understood as a methodology where a large number of new materials are prepared and tested in a parallel fashion. Combinatorial catalysis involves the co-ordination of high-throughput systems for preparation, characterisation and catalytic testing, large information data management, and rapid optimisation techniques for experimental design. With this methodology the number of variables examined can be increased, resulting in a potentially better performing final catalyst and shorter search times. Indeed, employing this methodology allows the study of promotion, synergy and other effects between different elements and supports, and their influence on activity, selectivity and the stability of each new material for a specific catalytic reaction. All this is reached by using not only high-throughput experimentation techniques for the preparation, characterisation and catalytic tests but also by computer-aided search strategies. Consequently, the global search/optimisation strategy is the main difference from the traditional catalyst research and should allow a reduction in the number of experiments needed to find an optimal catalyst composition and to understand the principles of the catalysis of the materials under exploration.

A global optimisation strategy rules the search process that decides which catalysts to synthesise, taking into account the catalytic performance shown by the previous set of catalysts. These search strategies can be categorised into two major groups: heuristic and stochastic. Heuristic techniques require a huge number of experiments to reach a good solution and are not adequate for high-dimensional

High-Throughput Screening in Chemical Catalysis
Edited by A. Hagemeyer, P. Strasser, A. F. Volpe, Jr.

ISBN: 3-527-30814-8

searches with several local maximums and, in many cases, can not be directly applied to solid catalysts. These techniques include simplex method, holographic search [3], split & pool [4], etc. Stochastic methods include genetic algorithms (GA) [5–7], neural networks [8], and simulated annealing [9]. These algorithms employ random search techniques and rules based on natural processes.

Genetic algorithms have been applied successfully to new catalytic materials discovery. GA optimisers are supposed to be particularly effective since: (a) the goal is to find an approximate maximum in a high-dimensional solution space and to reduce the number of preparations, in comparison with the traditional research methods [10]; (b) GA tolerate noisy experimental data; and (c) GA uses a population of points to conduct the search, which fits quite well with the parallel screening. Typically, this approach looks for combinations of elements and concentrations that would deal better experimental results. Through the search process, the optimisation procedure focuses on specific regions (exploitation routine) of the whole combinatorial space in which catalytic results are supposed to be better. Conversely, the combinatorial procedure tries to preserve the diversity up to a point, exploring different regions (exploration routine). All this is carried out by conveniently applying the operators: crossover for exploitation and mutation for exploration.

A new interesting approach applies *softcomputing* techniques in the combinatorial search of new materials. Softcomputing is a collection of methodologies that aims at tolerance for imprecision, uncertainty and partial truth to achieve tractability, robustness and low solution cost [11]. The principal constituents of softcomputing are fuzzy logic, neural computing, evolutionary computation, machine learning and probabilistic reasoning. What is particularly important about softcomputing is that it facilitates the use of different techniques in combination, leading to the concept of hybrid intelligent systems. Indeed, this new approach in catalyst optimisation uses a combination of neural networks and a genetic algorithm, improving significantly the optimisation performance and experimental error tolerance and providing a tool for modeling and analysing experimental data [12, 13].

Traditional processing and understanding of experimental outputs (characterisation and catalytic performances) is accomplished by researchers, who apply previous experience or fundamental knowledge to carry out the experimental design and to establish relationships between the different experimental results. With combinatorial catalysis, the large number of variables involved and the application of complex optimisation algorithms for the experimental design makes direct human interpretation of data derived from high-throughput experimentation difficult. Recently, data mining techniques have been applied [14, 15] to find relationships and patterns between the input and output data derived from accelerated experimentation. Hence, artificial intelligence (AI) techniques have an important potential in the modeling and prediction of complex high-dimensional data. Among these techniques, artificial neural networks (ANN) could be useful in catalysis. ANN have been successfully applied to conventional catalytic modeling and design of solid catalysts for ammoxidation of propylene [15], oxidative coupling of methane [16, 17], prediction of results of the composition of NO_x over zeolites [18], and the modeling and prediction of ODHE (oxidative dehydrogenation of ethylene) catalytic results derived from a combinatorial search [12].

5.1.1 Aim of the Work

The following examples illustrate the application of high-throughput screening tools together with heuristic search algorithms in the development of new enhanced catalyst for two fields of industrial interest, olefin epoxidation and the isomerization of light paraffins.

In both cases, the experimental space is based on fundamental knowledge. The definition of the search space was done taking into account the general principles stated in the literature, as well as in-house knowledge, thus constraining the number of possible combinations of the different parameters/elements and their range of variation. Alternatively, it can be considered that in both cases the primary screening has been substituted by a profound state-of-the-art analysis and the experimental optimisation process constitutes a secondary screening.

Catalyst preparation and activation conditions are included as parameters of importance in the optimisation algorithm. The preparation and activation procedures are very relevant aspects since minor variations in such conditions would cause major changes to the final phase of the solid and, consequently, to its catalytic properties. Typical preparation variables are promoter precursors, type of impregnation, calcination atmosphere, time and temperature, time and temperature for metal reduction and so forth.

Experimental work done in the present work illustrates the integrated employment of high-throughput tools: (1) automated robotic systems for catalyst preparation, including hydrothermal synthesis, post-synthesis surface modifications and wet impregnation, and (2) multiple parallel reactors operating under realistic reaction conditions (pressure, temperature, reagents nature) but utilising small amounts of solid catalyst (5–500 mg).

5.2 Search for New Catalytic Materials for the Epoxidation of Olefins

5.2.1 Epoxidation on Ti-based Materials

An important breakthrough in the use of Lewis acids as heterogeneous catalyst was the discovery of the activity of Ti-silicalite (TS-1) as catalysts for a series of useful oxidation reactions using different oxidants [19]. TS-1 is a zeolite with an MFI structure whose topology is formed by a two-channel system with diameters of 0.53×0.56 and 0.51×0.51 nm. Titanium was incorporated in the framework of the silicalite as isolated tetrahedral Ti(IV), exhibiting interesting catalytic properties as isolated Lewis acid site. The synthesis of other Ti-silicate materials such as Ti-beta [20] and Ti-MCM-41 [21, 22] has broadened the scope of these interesting catalysts to include the oxidation of large hydrocarbon molecules. With these catalytic systems the hydrophobic/hydrophilic properties of the surface are just as important as the number of active sites. Control of the hydrophobicity of the molecular sieve allows the optimisation of the adsorption of reactants and products. This is especially true for the epoxidation of olefins where the epoxide has a higher polarity than the olefinic substrate. As a result, the epoxide competes more favourably for adsorption on the hydroxylated surfaces on the Ti-silicates. In addition, the presence of both silanol and =Ti–OH groups allows adsorption of the epoxide and the formation of diols, which tend to strongly adsorb on the Ti sites and lead to partial deactivation of the catalyst.

5.2.2 Experimental Design using Softcomputing Techniques

Genetic algorithms are effective optimization techniques. They rely on random elements in parts of their operations rather than being determined by specific rules. A genetic algorithm tries to find the optimal solution to the problem by investigating many possible solutions simultaneously. The possible solution population is evaluated and the best solutions are selected to form next generations. Over a number of generations, good traits dominate the population, resulting in an increase of the quality of the solutions. An effective genetic algorithm requires a quick feedback of the fitness values of the samples. Hence the combination of GA with a NN seems suitable [23], and we have shown that it works well for catalytic problems [12, 13].

Genetic algorithm convergence can be mistaken if the problem codification is not appropriate. If the selected codification for the problem were wrong, it would be possible that the algorithm would solve a different optimization problem from the one under study. In the problem explained here, each variable belongs to a continuous domain and so it has been decided to adopt real-coded GAs to optimize the catalyst parameters (composition, synthesis and post-synthesis conditions). Therefore, each chromosome will be composed of a set of genes that represent a variable of the problem. Several studies guarantee this kind of real codification [24].

In this proposal (Fig. 5.1), each generation is tested by a neural network to obtain its fitness evaluation. This fitness evaluation depends on the predictions of the epoxide field values. Therefore, the GA obtains a predicted quality rank of samples ordered by this fitness function from maximum to minimum. Afterwards, crossover and mutation operators are applied to the whole generation with a certain probability. Mutation operator would be able to modify each gene with a new value, jumping randomly anywhere within the allowed domain. Thus, the mutation is an explorer operator, which looks for new solutions, and prevents system converging quickly on a local maximum, thus avoiding loss of genetic diversity. This hybrid algorithm can design the next generation in a more efficient way,

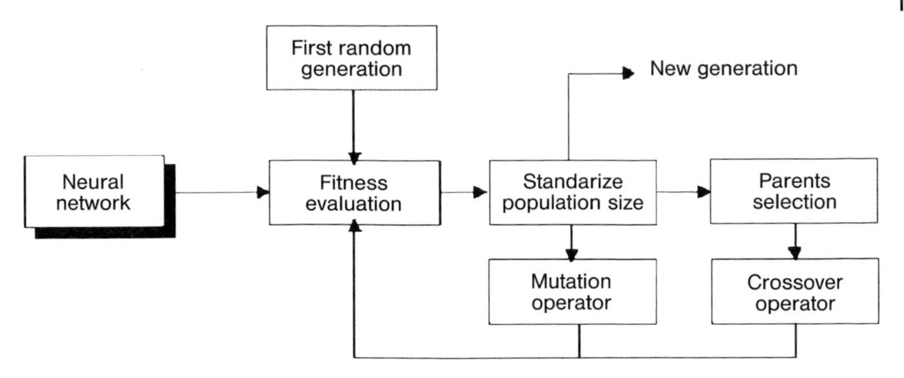

Fig. 5.1 Operating process of the NN-GA hybrid algorithm.

employing the parameters space "knowledge" extracted from the NN-modeling of all the previous generations. Fig. 5.1 shows a simple representation of the operation process of the NN-GA hybrid algorithm.

Conversely, a crossover operator is used based on confidence intervals. This operator uses information from the best individuals in the population. Moreover, the crossover operator is associated with the capacity of interpolation (exploration). This capacity is related to the belonging of a population parameter to a confidence interval. The crossover operator is also associated with the capacity of extrapolation (exploitation). To select the suitable parents for the next generation, the roulette wheel selection method is used. This method consists of a random selection in which the best quality individuals have more possibilities to be selected. In this way, the explained operators create new individuals that are added to the population. To produce the next generation, that extended population is reduced to its original size using the rank-space method. This selection procedure links fitness to both quality rank and diversity rank. Thus, it promotes not only the survival of individuals, which are extremely fit from the perspective of quality, but also the survival of those that are both quite fit and different from others.

Different analyses were performed employing an hypothetical model to optimise the performance of the algorithm: (1) analysis of the different parameters of the genetic algorithm, (2) analysis of the impact of the initial random generation goodness on GA performance and (3) softcomputing model applied to different catalytic reactions, GA was combined with a NN (GA-NN hybrid). The main conclusion of those analyses is that the initial random generation should be carefully designed, to improve the behavior of this soft computing technique, since GA-NN optimisation performance is poor when the initial generation has a very low quality.

5.2.2.1 **Definition of the Catalytic Space to be Explored**

Composition and preparation conditions of mesoporous materials containing titanium are screened to optimise the catalytic activity and selectivity in the epoxidation of cyclohexene using tert-butyl hydroperoxide as oxidant. Important parameters to be varied include synthesis temperature and time, nature of Si precursor, nature of Ti precursor, Ti content in gel, surfactants content in gel, nature of surfactant, method for removing the surfactant, silylating molecule, silylation conditions, etc. We have selected five variables for the present study: (1) pH of the gel, (2) surfactant I content in gel, (3) surfactant II content in gel, (4) titanium content, and (5) type of surfactant removing procedure (calcinations and/or extraction). Other experimental parameters were fixed at values established by preliminary studies.

5.2.3 Experimental

5.2.3.1 Catalyst Preparation

Ti-silicate materials were prepared utilising automated robotic equipment. The preparation required three steps: (1) gel synthesis, (2) material crystallization under hydrothermal conditions, and (3) post-synthesis treatment that includes at least one of calcination, organic extraction and silylation.

The gel synthesis operation was carried out using an in-house built robotic system (Fig. 5.2), which can perform automatically the following routines:

1. Dosing of liquid reagents using syringes (range 10–5000 mg).
2. Dosing of solid reagents (powders) employing a novel weighting procedure (20–1000 mg).
3. Evaporation of solvents (water, ethanol) including IR heating and air/N_2 sweep flow.
4. Gel homogenisation by vigorous stirring.

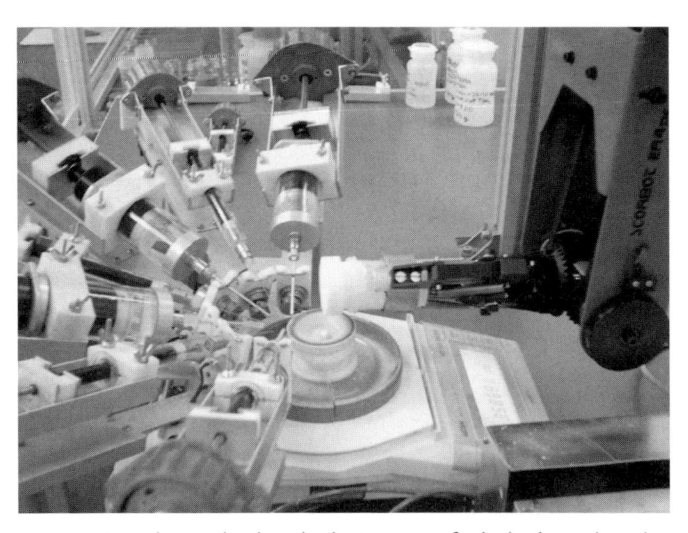

Fig. 5.2 An in-house-developed robotic system for hydrothermal synthesis: detail of solid dispensing operation.

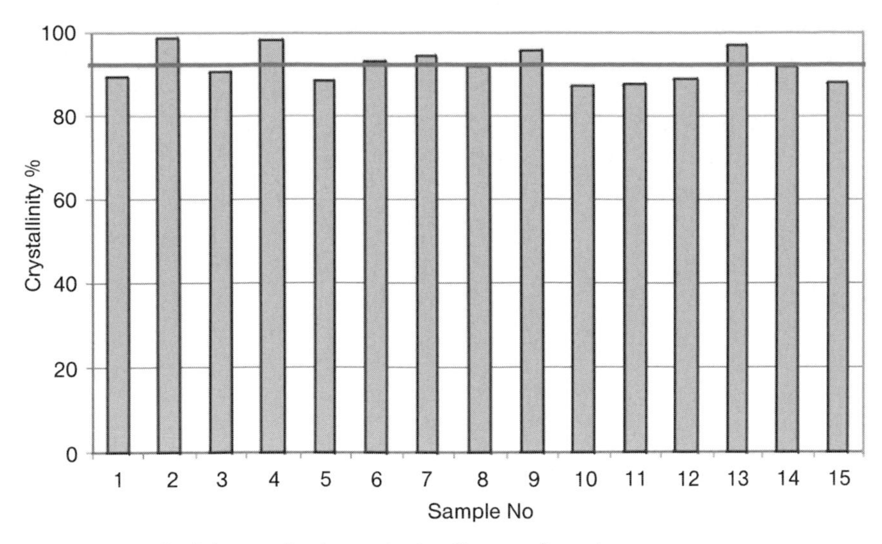

Fig. 5.3 Reproducibility test for the synthesis of beta zeolite using a robotic system for the gel synthesis and a multi-autoclave rack for the zeolite crystallization.

This automated system can handle simultaneously 15 to 30 syntheses. The procedure employed for the synthesis is practically the same as that used in the manual operation but on a lower scale. Thus, problems regarding the scalability of the synthesis should be strongly reduced. The composition of the gel is perfectly known, since the correct dosing of liquids and solids is always checked by the weighting of each vial. Gel synthesis is carried out in Teflon vials that at the end of the process are placed automatically in a 15-well stainless steel rack, which is closed manually in a very simple and quick manner. The Ti-silicates were synthesised at 140 °C during one day. The maximum gel volume per individual synthesis was 3.5 ml and the typical solid mass obtained after washing and drying ranged from 0.1–0.5 g, depending on the molar gel composition and the crystallization yield. Reproducibility of the system was very good with a standard deviation lower than 4% (Fig. 5.3).

Surfactant extraction and silylation process is performed with a Konik-type liquid-handling robot endowed with a stirring and heating station.

5.2.3.2 **Catalytic Testing**

The different solid materials are catalytically tested in the liquid phase at 60 °C, employing a fully automated system. This enables multiple parallel batch reactions in up to 21 vials under the following conditions:

- Reactor volume: 2–25 ml.
- Temperature range: 25–150 °C.
- Magnetic stirring with individual control for each reactor.
- Typical reactor material is glass with a silicon septum tip.

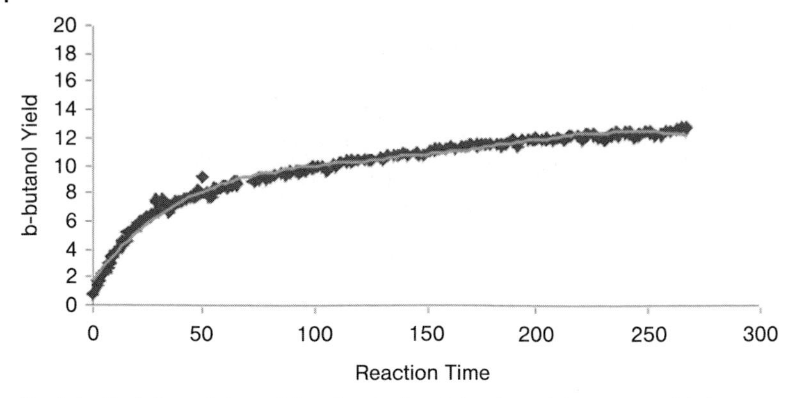

Fig. 5.4 Evolution of the activity of a single catalyst in cyclohexene epoxidation: 300 fast GC analyses.

The robotic system can perform the operations such as (1) automatic dispensing of a plurality of liquid chemicals and (2) reaction and analysis (fast GC). The software application permits the easy and automatic configuration of reaction settings: initial composition of each batch reactor, reaction conditions, sampling configuration, post-analysis calculations. It is also possible to carry out post-analysis calculations, scheduling the obtained data and exporting it to Microsoft Excel or other compatible software applications. This allows the continuous monitoring of the reaction evolution on each catalyst and does not require data handling after the reaction process.

Fig. 5.4 shows the evolution of the concentration of one reaction product with time. For this individual testing, high-speed systems can take a sample and analyse the product every 35 s; the experimental error deviation being below 2%. The system allows the volume of the reactor to be changed. We currently use reactors of 2, 4 and 25 ml that are independently stirred at up to 1000 rpm.

5.2.4
Results of the Optimisation Process

Eight catalyst generations of 21 samples have been synthesised and tested, and a highly active and selective catalyst for the epoxidation of cyclohexene with tert-butyl hydroperoxide found, which can be applied to the epoxidation of other olefins, especially propylene, using organic peroxides as oxidants [25].

Through the NN-GA optimisation process, an important improvement in the activity and selectivity of the starting materials has been achieved (Fig. 5.5). The figure shows the cyclohexane epoxide yields for the eight evolved generations (8×21 samples). The best materials have low titanium contents, and were extracted and silylated. These materials have a Ti-MCM-41 structure and a very hydrophobic surface.

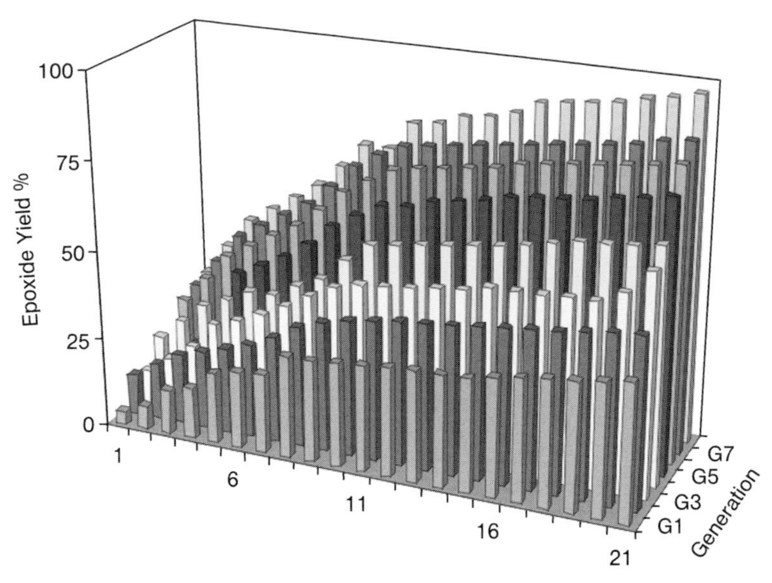

Fig. 5.5 Evolution of cyclohexene epoxide yield through the optimisation process.

5.3
Search for New Catalytic Materials for Isomerization of Light Paraffin

5.3.1
Isomerization of Light Paraffins

The hydroisomerization of light paraffins (C_5–C_6–C_7) to produce their branched isomers is an important industrial process aimed at improving the octane number of the light straight rum stream (LSR). Reformulated gasolines with their impact on olefins and aromatics reduction [26, 27] have increased the number of LSR isomerization units. Table 5.1 gives the octane number of the different C_5–C_7 linear and branched paraffins.

Hydroisomerization of light paraffins is a thermodynamically controlled acid-catalyzed process. As Fig. 5.6 shows, the equilibrium shifts to high octane branched paraffins when the reaction temperature decreases. Thus, it is convenient to reduce as much as possible the isomerization temperature to maximize the yield of high-octane branched isomers. At present, the isomerization is industrially carried out on catalysts such as halogen-treated alumina or mordenite zeolite. A chlorinated alumina catalyst works at low temperature but involves corrosion problems, and is very sensitive to sulfur and water. Zeolite-based catalysts, especially mordenite, are more resistant to sulfur and water but require higher reaction temperatures, giving a product with a lower octane number, and have a significant activity for cracking reactions. Nanocrystalline zeolites have shown a significant improvement in isomerization selectivity by reducing this cracking

Tab. 5.1 Research and motor octane number (RON and MON) of C_5–C_7 alkanes.

Hydrocarbon	*RON*	*MON*
i-Pentane	92	90
n-Pentane	62	62
2,2-Dimethylbutane	92	93
2,3-Dimethylbutane	101	94
2-Methylpentane	73	74
3-Methylpentane	75	74
n-Hexane	25	26
2,2-Dimethylpentane	93	93
2,4-Dimethylpentane	83	82
2-Methylhexane	72	45
3-Methylhexane	52	56
n-Heptane	0	0

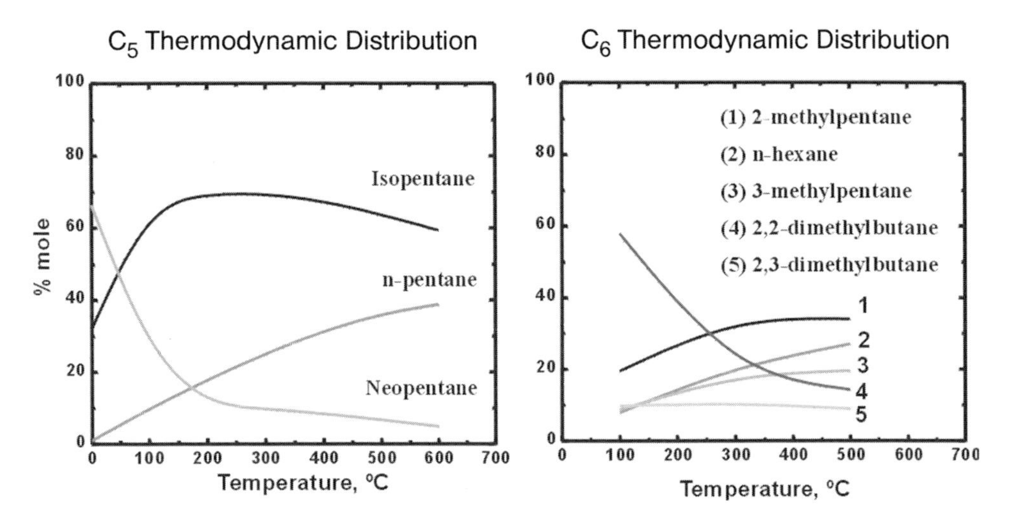

Fig. 5.6 Thermodynamic distribution of the pentane and hexane isomers.

yield. This effect is due to the rapid diffusion out of the zeolite pore system, decreasing the possibility of secondary reactions of the generated iso-paraffins. Oxides such as ZrO_2, TiO_2, SnO_2, Fe_2O_3, HfO_2 have, after sulfation, shown a remarkable increase in the surface acidity and in the catalytic activity for carbenium ion reactions [28–33]. Especially, platinum on sulfated zirconia and tungsten oxide on zirconia are very active and selective in the isomerization of light paraffin at low temperatures. Unfortunately, these catalysts are extremely sensitive to sulfur and water in the feed. Fig. 5.7 summarizes different types of isomerization catalysts and the positive and negative aspects of their catalytic performance.

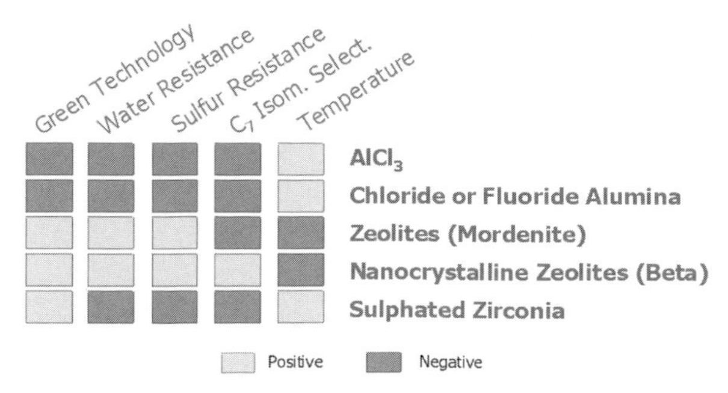

Fig. 5.7 Positive and negative aspects of the different paraffin isomerization catalysts.

Considering the available technologies for isomerization, it would be interesting to develop a new generation of environmentally friendly catalysts with high activity and selectivity to branched isomers that can work at lower reaction temperatures than zeolites and show an increased resistance to sulfur and water.

5.3.2 Search Methodology

A conventional genetic algorithm has been used for the experimental design, allowing optimisation of the catalytic parameters defined, considering the catalytic requirements, i.e., acidity, hydrogenating–dehydrogenating activity and thioresistance.

5.3.2.1 Definition of Variables

The possible catalyst components have been classified into three groups: metal oxide support, acidity enhancers, and promoters. As metal oxide supports, γ-Al_2O_3, ZrO_2, TiO_2 have been considered. Acidity enhancers SO_4^{2-}, BO_3^{3-}, PO_4^{3-} and WO_x were used, for which the content of SO_4^{2-}, BO_3^{3-}, PO_4^3 ranged from 0.5–6 wt% and of WO_x from 0.5–36%. For promoters (water and thioresistance) the effect of Pt, Ce, Pd, Sn, Ni, Mn and Nb was explored (0.5–6 wt% for each metal). Each material contains one support, one acidity promoter and at least two metallic promoters, of which one is Pt (0.5 wt%).

5.3.2.2 Structure of the Optimisation Algorithm

The genetic algorithm (GA) is initialised by generating 24 catalyst compositions, with random combinations of the catalytic parameters established following the patterns explained above. Succeeding generations were introduced by the genetic algorithm, taking into account the catalytic performances of the previous generation, applying crossover and mutation operators. Fig. 5.8 shows a simple representation of the optimisation process performed by the GA. The objective function is derived from experimental results of selectivity to iso-pentane and n-pentane con-

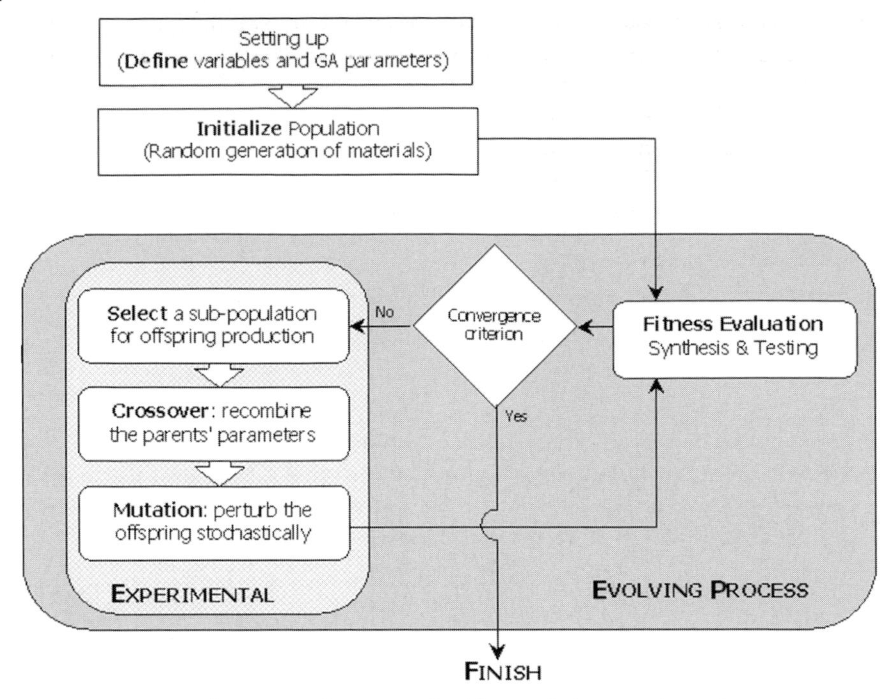

Fig. 5.8 Conventional structure of a genetic algorithm applied to experimental design in catalysis.

version. The population size is 24 catalysts, i.e. 0.167% of the total solution space. An elitism approach is employed, so that the new generation includes the genetic recombination of the selected catalysts and the outstanding catalysts of the last generation.

- *Coding strategy*: Five catalytic parameters are considered as genes: type of support, type of enhancer, content of enhancer, type of promoter, and content of promoter. Each gene is coded as an integer. A coding example of a catalytic material with five parameters is: $/Al_2O_3/SO_4^=/2\%/Sn/4\%/ \rightarrow /1/1/2/3/4/$.
- *Recombination of genes (crossover)*: In the exploration process a two-point crossover mechanism is performed such that new combinations contain the same loading of the corresponding element. Therefore, the chromosome (catalyst representation) can be divided into three groups of parameters: support type, acidity enhancer element and proportion, and promoter element and proportion. Exploitation process: When the problem is converging, the algorithm has focused the research in the maximum area, and, therefore, the diversity of the population has been decreased. To avoid a deadlock of the evolving process, we have considered some additional measures: (1) not to cross materials with very similar composition and (2) if parents only differ in a loading parameter their values are averaged.

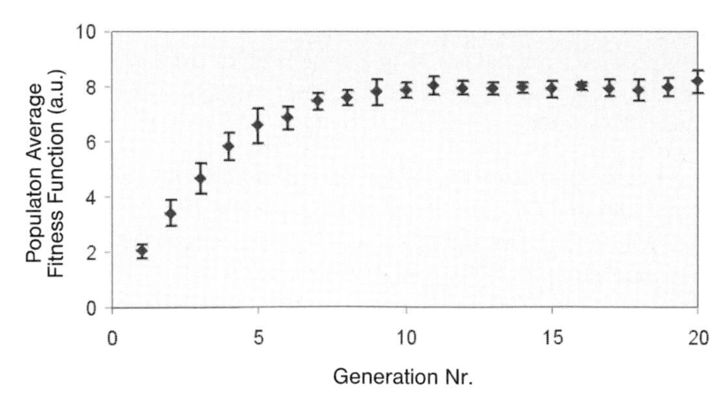

Fig. 5.9 Simulation of evolution process using a hypothetical objective function.

- *Mutation*: A mutation probability equal to 1.67% was chosen.

Fig. 5.9 shows a simulation of the evolution process employing the present algorithm, using a hypothetical objective function (average results of 15 different simulations). Population-averaged fitness function is represented as a function of generation number. Steady state for this average value is reached after six evolving cycles. A quasi-global maximum is usually observed after three or four generations.

5.3.3 Experimental

5.3.3.1 Catalyst Preparation

The different promoting elements were incorporated to the corresponding supports by incipient wetness impregnation following a sequential impregnation procedure. A commercial robotic system (Zymate II from Zymark®) was employed, which enables dosing and weighting different solids and liquids, handling different types of vials and vortex stirring of viscous slurries. The preparation was carried out in parallel for each generation, applying a sequential impregnation procedure. The following steps carried out by the automated system included: (1) impregnation of the different supports with an aqueous solution of the corresponding acidity enhancer and subsequent drying at 160 °C for 16 h; (2) impregnation with an aqueous solution of the corresponding promoter precursor (except Pt) and drying at 160 °C for 16 h; and (3) final impregnation with a H_2PtCl_6 aqueous solution, to reach 0.5% Pt, and drying at 160 °C for 16 h. The final materials are then composed of one oxide support impregnated with one acidity enhancer, one promoter and platinum.

5.3.3.2 Catalytic Testing

High-throughput testing was accomplished using a system of 16 continuous fixed-bed parallel microreactors, able to work up to 80 bar. The principal characteristics of this reactor are summarized as:

- Realistic testing conditions (pressure, temperature and fluid dynamics).
- Temperature, pressure and flows are measured in each catalyst bed.
- Individual controlled feed of one gas and one liquid for each microreactor.
- Rapid analysis of reaction products.
- Possible to perform any previous process over the catalysts such as reduction or calcination.
- Contact time and partial pressures changed independently in each reactor.
- It is possible to study catalyst decay.
- High-throughput:
  - Number of reactors: 16
  - Pressure range: 1.5–80 bar
  - Temperature range: 35–700 °C
  - Liquid flow range: 0–225 μl min^{-1}
  - Gas flow range: 0–100 ml min^{-1}

Two different software applications have been developed for this complex reaction system: (1) *Hardware control and automation*: this application enables one to set and control the pressure, liquids and gas flow and pressure, as well as the position of the mechanical parts of the system. It also allows one to program the variation of the different reaction conditions (64 variables in each reaction step); (2) *Analysis and reaction monitoring*: this application enables the *on-line* monitoring of the GC analysis results and reporting, which facilitates the *off-line* data analysis and leads to no-human data manipulations in the transfer to the genetic algorithm application.

Catalysts pre-treatment (calcination and reduction) was performed in the same testing system or in a parallel automatic activation system prior to reaction test. Calcination is carried out at 600 °C under airflow for 8 h and reduction at 250 °C for 2 h under hydrogen flow. Catalytic tests were carried out at 30 bar total pressure, temperature range 200–240 °C, and 2.26 h^{-1} WHSV, H_2/hydrocarbons molar ratio of 2.93. Each fixed bed microreactor contained 500 mg of catalyst (particle size 0.4–0.6 mm, for which there are no internal diffusion limitations). Reaction products distribution are analysed using a gas chromatograph (Varian 3380GC) equipped with a Plot Alumina capillary column.

5.3.4 Results of the Optimization Process

The search was conducted by a genetic algorithm, which designed the compositions of the new set of catalysts to be screened. Each catalytic material consisted of three components (one support + one acidity enhancer + promoters) having for each catalyst set 24 new materials (Tab. 5.2 shows the compositions of the most active catalysts of each generation). Each catalyst set was synthesized and tested

Tab. 5.2 Composition of the seven best-ranked catalytic materials of the three generations.

	Support	Acid (%)	Enhancer	Promoter I (%)		Promoter II (%)	
1st Generation							
1	ZrO_2	S	1.75	Ce	0.97	Pt	0.51
2	ZrO_2	S	2.18	Pd	1.00	Pt	0.46
3	TiO_2	W	17.38	Mn	0.87	Pt	0.48
4	ZrO_2	W	5.58	Ni	0.55	Pt	0.51
5	Al_2O_3	W	45.57	Nb	4.60	Pt	0.52
6	Al_2O_3	W	11.27	Sn	5.52	Pt	0.47
7	ZrO_2	B	0.54	Ni	1.06	Pt	0.50
2nd Generation							
1	ZrO_2	S	2.22	Pd	0.49	Pt	0.47
2	ZrO_2	S	1.76	Ce	0.97	Pt	0.51
3	ZrO_2	S	2.57	Nb	4.05	Pt	0.47
4	ZrO_2	S	2.05	Mn	1.74	Pt	0.48
5	ZrO_2	B	2.33	Mn	1.86	Pt	0.51
6	Al_2O_3	P	1.42	Mn	7.87	Pt	0.52
7	TiO_2	P	3.72	Ce	0.94	Pt	0.49
3rd Generation							
1	ZrO_2	S	2.21	Nb	0.48	Pt	0.56
2	ZrO_2	S	2.15	Ce	1.18	Pt	0.48
3	ZrO_2	S	0.59	Pd	0.52	Pt	0.56
4	ZrO_2	S	1.81	Nb	5.11	Pt	0.42
5	ZrO_2	S	2.44	Nb	3.96	Pt	0.52
6	ZrO_2	S	2.63	Ce	1.10	Pt	0.47
7	ZrO_2	S	2.55	Nb	0.88	Pt	0.48

for the n-pentane isomerization according to the methods described above. After each catalytic evaluation, the optimization algorithm designed the next set of catalysts, taking into account the compositions and catalytic performances of the previous generation. The objective function was iso-pentane yield, which corresponds to n-pentane conversion due to the very low yield of cracking products. Fig. 5.10 shows the abundance in each generation of the different catalytic components: supports (γ-Al_2O_3, ZrO_2, TiO_2), acidity enhancer (S, B, P, W) and metallic promoters. The convergence toward ZrO_2-S- and ZrO_2-W-based materials can be seen.

Three evolving cycles have been examined; the total number of catalysts processed being 72. Fig. 5.11 shows the catalytic performances of the ten most active catalysts for the three successive generations at three different reaction temperatures. An important enhancement in the activity is seen through the *evolution*. After three generations, the search process is focused on the exploration of the ZrO_2-S area, where the general catalyst formulation based on ZrO_2-S-Nb-Pt shows the highest activity. However, it cannot be warranted that this formulation is the absolute maximum of this well-defined solution space. By means of this optimization process, a substantial improvement in the isomerization activity of catalytic materials has been achieved.

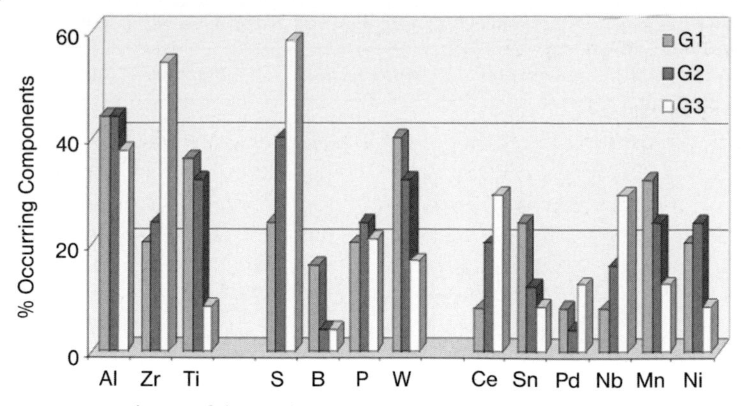

Fig. 5.10 Evolution of the catalyst composition in the three evolved generations.

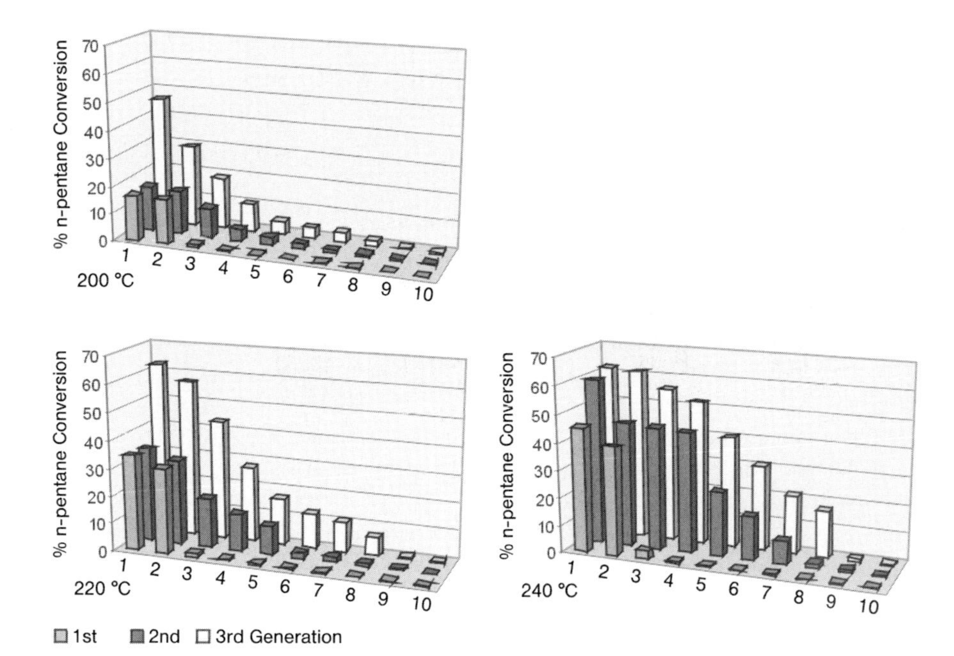

Fig. 5.11 Catalytic performance of the ten most active catalysts for the three successive generations at three different reaction temperatures.

5.3.5
Validation of the Best Catalytic System

5.3.5.1 Poisoning Resistance

The best catalytic formulation found in the GA exploration was studied for sulfur and water tolerance. Catalyst decay experiments were carried out in a conventional reaction system using as feed a mixture of n-pentane and n-hexane (60–40 wt%) at 2.26 h^{-1} WHSV, with a H_2/hydrocarbon molar ratio of 2.93 and H_2O (20 ppm) or H_2S (20 ppm). The amount of catalyst (SZNbPt) was 2 g and reactor temperature was 160 °C. The catalytic results presented in Fig. 5.12 show that this new formulation has a higher sulfur and water resistance than a platinum sulfated zirconia (SZPt) catalyst. Clearly, the isomerization activity of conventional sulfated zirconia (ZS) is strongly reduced after 400 min on stream while the new catalyst formulation maintains a high conversion. Analogously to the test with water, the formulation SZNbPt exhibits an improved sulfur resistance, having a higher activity than conventional sulfated zirconia after 400 min on stream. The new catalyst formulation found by the GA approach has an improved isomerization activity and also an improved water and sulfur resistance.

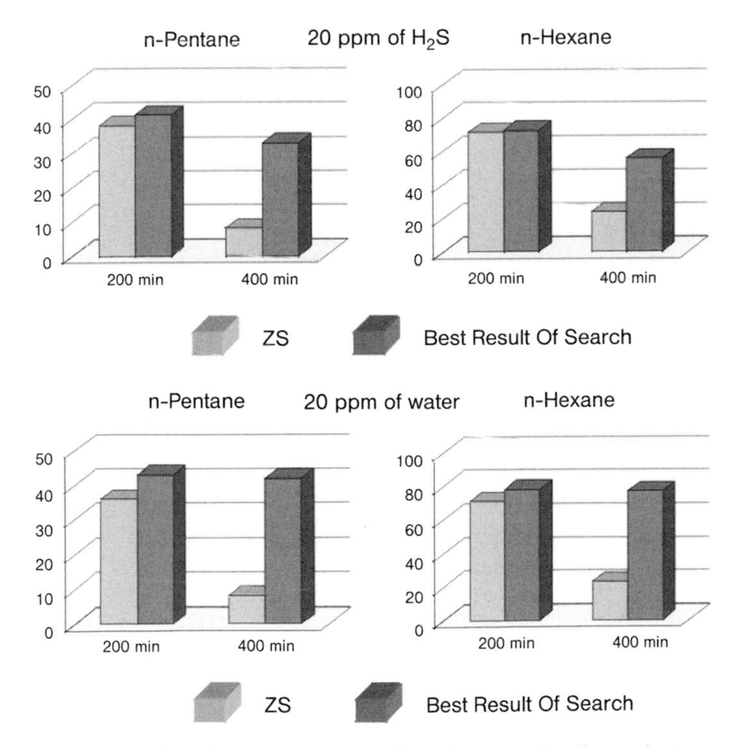

Fig. 5.12 Catalyst decay experiments: effect of water and sulfur on the isomerization activity of the new sulfated zirconia formulation (SZNbPt) and conventional sulfated zirconia (ZS).

5.3.5.2 Isomerization of a Simulated Industrial Stream

The most active formulation (ZSNbPt) was tested in a conventional reactor using as feedstream a mixture of light n-alkanes [n-pentane (20 wt%), n-hexane (60 wt.%) and n-heptane (20 wt%)] to simulate an industrial stream. Experiments were carried out in a conventional reaction system using a fixed-bed continuous-flow reactor. Reaction was carried out under the same conditions as the poisoning resistance experiments. The activity and selectivity of this catalyst (Fig. 5.13) have been compared with those obtained with sulfated zirconia impregnated with platinum (ZS). Fig. 5.13 represents the evolution of the conversion with reaction temperature. Clearly, the reactivity of the n-paraffin follows the order n-heptane > n-hexane > n-pentane for both catalysts, as expected when taking into account the adsorption heats of the different hydrocarbons [34].

The SZNbPt catalyst shows a higher isomerization selectivity and activity for all hydrocarbons. While n-pentane and n-hexane isomerization selectivity is very similar on both catalysts, important differences occur with n-heptane. When comparing both catalysts at the same conversion level (60%), SZNbPt shows a higher selectivity ($\sim$50%) than the conventional sulfated zirconia ($\sim$30%). However, the isomerization selectivity of n-heptane SZNbPt is still too low when the temperature is increased to achieve high C_5–C_6 isomerization yields.

5.3.6
Study of WO_x/ZrO_2 System Employing a Factorial Design

Analysis of the catalytic results achieved by the GA methodology shows that catalysts based on the $WO_x/ZrO_2/Pt$ system exhibit a significant activity, but one that is still lower than those reported previously [35]. The WO_x/ZrO_2 system is a well-known high activity material able to isomerize light alkanes at low temperatures [36]. For this reason, a systematic study of this system (Zr-W-Ce-Pt) was performed to achieve a better understanding of the determining variables in the final isomerization activity. The variables studied included composition but also preparation conditions, i.e., calcination temperature, tungsten content and nature of cerium precursor salt. The variation level of each variable was chosen by considering previous works: (1) cerium precursor nature: sulfate-nitrate-chloride; (2) calcination temperature: 500–650–800 °C; and (3) tungsten content: 8–16–24 wt%.

The systematic study was carried out using a fractional factorial design [37] (Fig. 5.14). If a 3-level factorial design is considered for the study of the influence of these three variables the test of 3^3 samples would be required. We have selected two building blocks of the whole 3^3 design, corresponding with two 3^2 factorial designs. In this case, the total number of samples to be processed is reduced to 15. These materials were tested for isomerization activity in the multiple reactor (calcinations ex situ) using as reactor feed a mixture of n-pentane and n-hexane (60–40 wt%).

Conversion of n-pentane and n-hexane for the samples of the factorial design are shown in Figs. 5.15 and 5.16. Important conclusions can be extracted from the results:

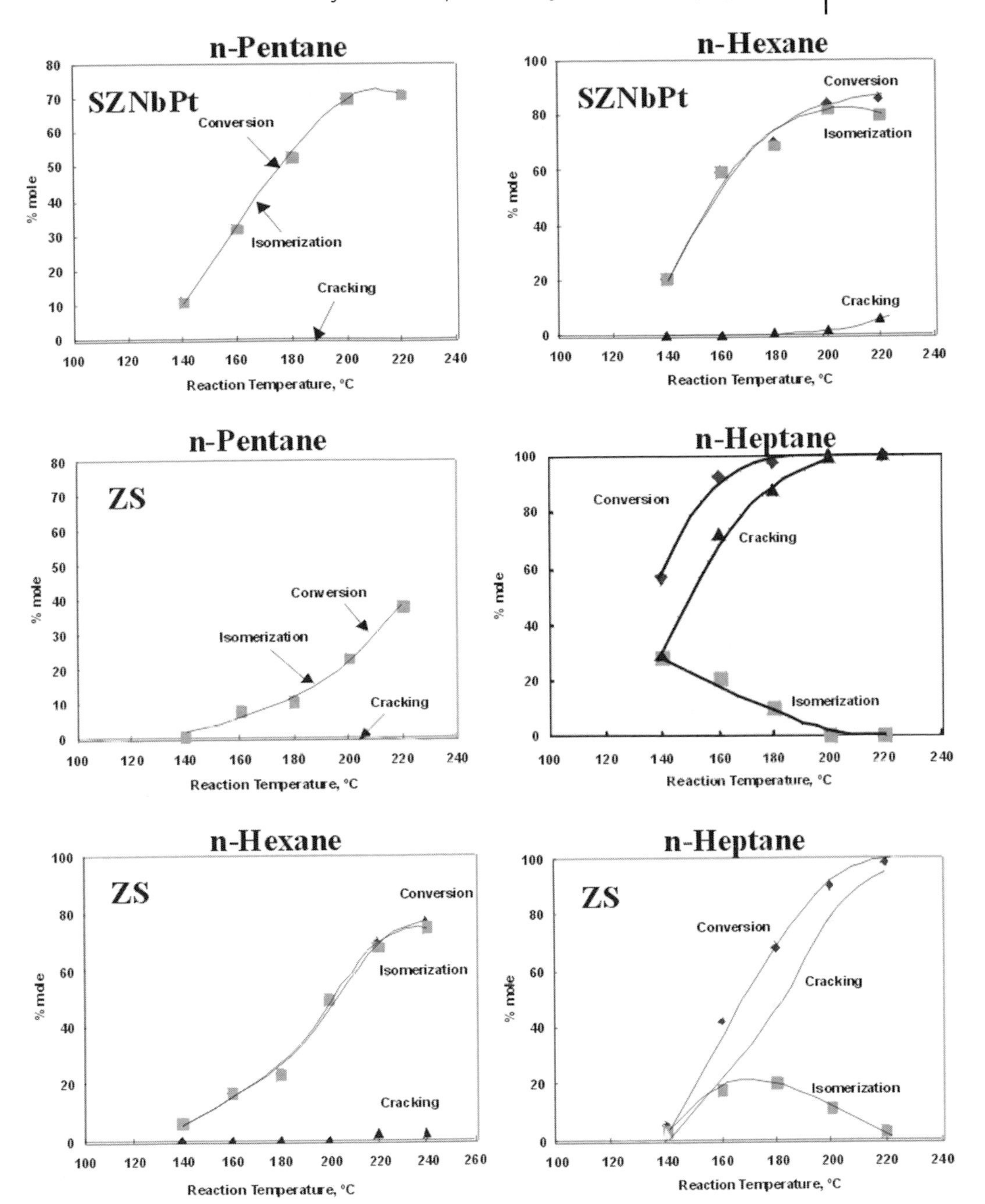

Fig. 5.13 Evolution of C_5–C_7 isomerization performance with reaction temperature for the new sulfated zirconia formulation (SZNbPt) and conventional sulfated zirconia (ZS).

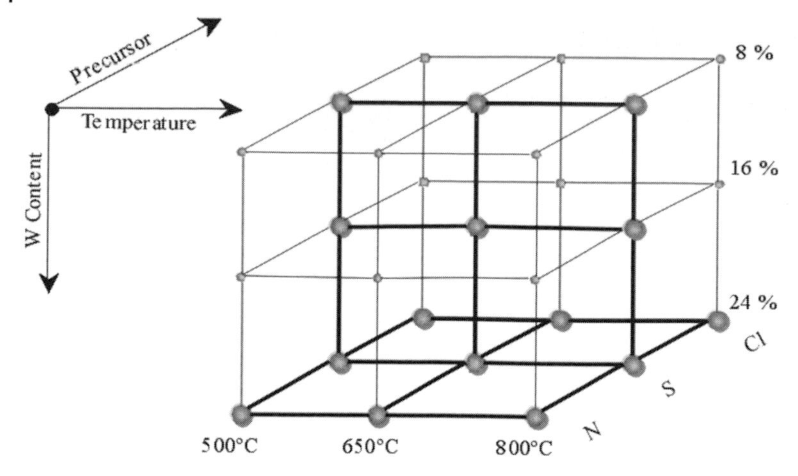

Fig. 5.14 Fractional factorial design employed for the detailed study of catalysts based on ZrO_2-W-Ce-Pt.

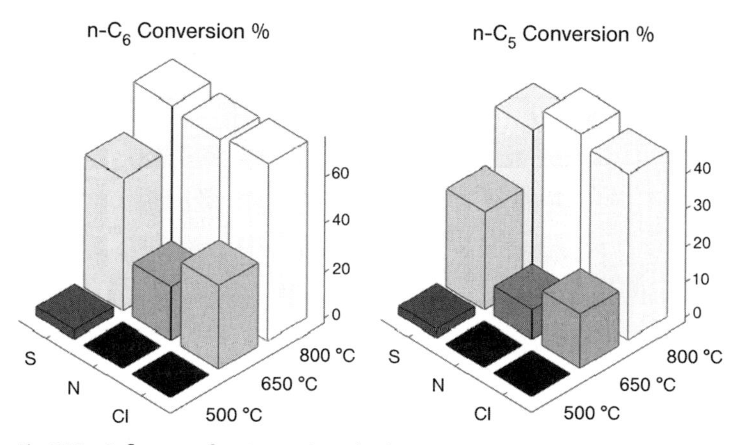

Fig. 5.15 Influence of cerium salt and calcination temperature on the isomerization performance of the ZrO_2-W-Ce-Pt catalyst.

- *Precursor salt and calcination temperature:* The best performing materials are those calcined at 800 °C, the cerium precursor salt being of minor relevance. However, when comparing samples calcined at 650 °C, the sample prepared with sulfate salt shows a much higher activity, which can be explained by considering that the sulfate interacts with the zirconia, thereby improving its acidity and, consequently, its isomerization activity. By means of calcination at 800 °C, however, the sulfate is totally removed from the catalyst surface – as determined by analysis – and the acidity is only due to the tungsten effect. For low calcination temperatures (500 °C) the samples obtained are inactive, indicating that these temperatures are too low to produce active sites for alkane iso-

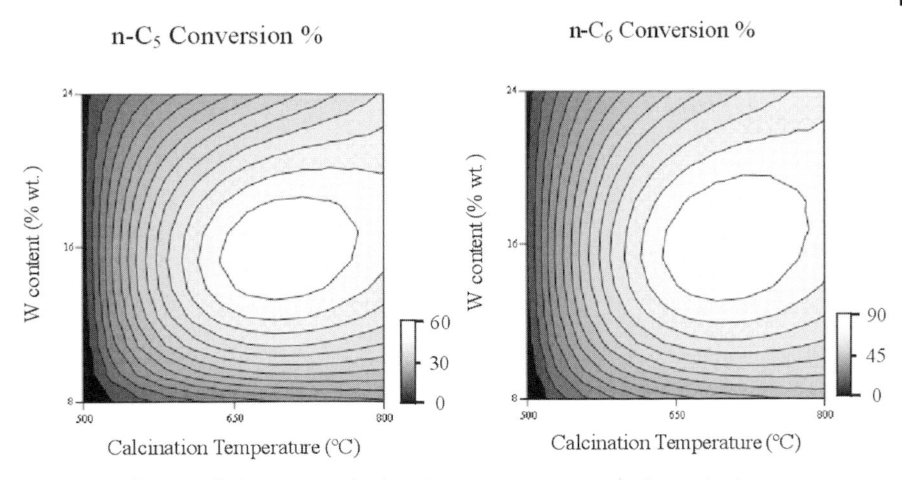

Fig. 5.16 Influence of W content and calcination temperature on the isomerization performance of the ZrO_2-W-Ce-Pt catalyst.

merization. Therefore, calcination temperature and precursors salt are important factors that can explain the low activity shown by the WO_x/ZrO_2 materials in the GA process and should be considered in further studies.

- *Tungsten content and calcination temperature*: The best performing material is obtained for 16 wt% W with calcination at 650 °C. However, the activity of the same composition calcined at 800 °C is also high, proving a hypothetical maximum located in the area between both calcination temperatures (white area of Fig. 5.16). The presence of a maximum can be explained considering that, with increasing temperature, the formation of W-based acid sites is promoted whereas the S-based acid sites disappear.

5.4 Conclusions

Two examples of the integrated application of high-throughput experimentation tools with advanced optimisation algorithms for experimental design have been described. The high-speed experimentation was guided by artificial intelligent techniques, i.e., a conventional genetic algorithm (GA) and a hybrid algorithm consisting of a GA coupled with a neural network. Especially, this hybrid intelligent system seems to be a useful new tool for the intelligent discovery of new catalytic materials, since it has appropriate tools for high-dimensional optimisation but maintains *in memory* the whole "history" of the search, reducing the screening of statistically-poor active materials.

Through both experimental processes, an important improvement in the catalyst activity has been found. Conversely, both compositional and processing variables are of paramount importance in the final activity/selectivity of the solid catalytic materials. Thus, the study of materials preparation conditions, activation conditions and characterization of the catalysts during the search process has to be considered to enlarge the possibilities of discovery in heterogeneous catalysis.

5.5 Acknowledgments

The authors thank CICYT (MAT2003-07945-C02-01) for financial support.

5.6 References

1 S. Senkan, *Angew. Chem., Int. Ed.*, 40 (2001) 312–329.
2 X.D. Xiang, X.D. Sun, G. Briceño, Y.L. Lou, K.A. Wang, H.Y. Chang, W.G. Wallacefreedman, S.W. Chen, P.G. Schultz, *Science*, 265(5218) (1995) 1738–1740.
3 L. Végvári, A. Tompos, S. Gbölös, J. Margitfalvi, *Catal. Today*, 81 (2003) 517–527.
4 J. Klein, T. Zehn, J.M. Newsan, S.A. Schunk, *App. Catal.* 254(1) (2003) 121–131.
5 D. Wolf, O.V. Buyevskaya, M. Baerns, *Appl. Catal.*, 200 (2000) 63–77.
6 L. Weber, *Drug Discovery Today*, 3 (1998) 379.
7 D.E. Goldberg, *Genetic Algorithms in Search, Optimization and Machine Learning* (1989) Addison/Wesley, Boston.
8 A.B. Bulsari, *Neural Networks for Chemical Engineers*, Elsevier, Amsterdam, 1995.
9 S. Kirkpatrick, C.D. Gerlatt, Jr, M.P. Vechi, *Science*, 220 (1983) 671.
10 D.L. Trimm, *Design of Industrial Catalysts*, Elsevier, Amsterdam, 1980.
11 L. Zadeh, Fuzzy logic, neural networks and soft computing, in *Communications of the ACM*, 37 (1994) 77–84.
12 A. Corma, J.M. Serra, E. Argente, S. Valero, V. Botti, *Chem. Phys. Chem.*, 3 (2002) 939–945.
13 S. Valero, E. Argente, V. Botti, J.M. Serra, A. Corma, *SoftComputing Techniques Applied to Catalytic Reactions*, Proceedings of CAEPIA, San Sebastian (Spain) 2003, Vol. 1, 213–222, Ed: CAEPIA-TTIA Organising Committee, 2003 (ISBN: 84-8373-564-4).
14 M. Negnevitsky, *Artificial Intelligence: A Guide to Intelligent Systems*, Addison/Wesley, Harlow, 2002.
15 P. Gedeck, P. Willett, *Curr. Opin. Chem. Biol.*, 5(4) (2001) 389–395.
16 T. Hattori, S. Kito, *Catal. Today*, 23 (1995) 347–355.
17 K. Huang, F.Q. Chen, D.W. Lü, *Appl. Catal. A*, 219 (2001) 61–68.
18 Z.Y. Hou, Q.L. Dai, X.Q. Wu, G.T. Chen, *Appl. Catal. A*, 161 (1997) 183–190.
19 M. Taramasso, G. Perego, B. Notari, US 4410501.
20 A. Corma, M.A. Camblor, P. Esteve, A. Martinez, J. Pérez-Pariente, *J. Catal.* 151 (1994) 145.
21 A. Corma, M. Domine, J.A. Gaona, J.L. Jorda, M.T. Navarro, F. Rey, J. Pérez-Pariente, J. Tsuji, B. McCulloch, L.T. Nemeth, *Chem. Commun.* (1998) 2211.
22 A. Corma, J.L. Jorda, M.T. Navarro, F. Rey, *Chem. Commun.* (1998) 1899–1900.
23 R.C.Rowe, R.J.Roberts, *Intelligent Software for Product Formulation*, Taylor and Francis Ltd, London, 1998.
24 D.E. Goldberg, *Complex Systems*, 5 (1991) 139–167.

25 J. Pérez-Pariente, A. Corma, M.T. Navarro, EP 0655278, WO9429022 to UPV-CSIC, 1994.

26 A. Chica, A. Corma, *J. Catal.*, 187 (1999) 167–176.

27 A. Chica, A. Corma, P.J. Miguel, *Catal. Today*, 65(2–4) (2001) 101–110.

28 M. Hino, K. Arata, *Chem. Lett.*, 1259 (1979).

29 M. Hino, K. Arata, *J. Chem. Soc., Chem. Commun.* (1979) 1148.

30 M. Hino, K. Arata, *J. Chem. Soc., Chem. Commun.* (1980) 851.

31 K. Tanabe, M. Itoh, K. Morishige, H. Hattori, in: *Preparation of Catalysts*, B. Delmon, P.A. Jacobs, G. Poncelet (eds.), Elsevier, Amsterdam, 1976, p. 65.

32 T. Jin, M. Machida, T. Yamaguchi, K. Tanabe, *Inorg. Chem.*, 23 (1984) 4396.

33 T. Jin, T. Yamaguchi, K. Tanabe, *J. Phys. Chem.*, 90 (1986) 4974.

34 W.O. Haag, in: *Abstracts of Papers of the American Chemical Society*, Vol. 211, Iss. MAR, pp 6-PTR, 1977 (UA48J).

35 K. Arata, M. Hino, in: *Proceeding, 9th International Congress on Catalysis*, Calgary, 1988 (M.J. Phillips, M. Tenan, eds.) 1727. Chemical Institute of Canada, Otawa.

36 S.J. Soled, N. Dispeziere, R. Saleh, in: *Progress in Catalysis*, Elsevier, Amsterdam, 1992, p. 77.

37 D.C. Montgomery, Design and analysis of experiments, 5th Ed. John Wiley, New York, 2001.

6
Present Trends in the Application of Genetic Algorithms to Heterogeneous Catalysis

Martin Holeña

6.1
Introduction

The challenge of finding an optimal catalyst composition and preparation for a specific reaction has, during the last 10–15 years, lead to preparing catalysts with active elements, support and dopants selected from increasingly large pools and prepared in an increasing number of different ways. Consequently, the optimum has to be searched for in a high-dimensional descriptor space (nowadays, 20–30 descriptors are not exceptional). For high-dimensional spaces, the traditional DOE (design of experiments) methods are no longer applicable. As an alternative, since the late 1990s, function optimization methods have been employed to searching for optimal catalysts in the descriptor space. The optimized functions are functions describing the dependence of some performance indicator of the catalyst (e.g., yield or conversion) on its composition and on other descriptors. However, standard mathematical optimization can not be directly used to this end because values of the optimized function can be obtained only from experimental measurements, but there is insufficient theoretical knowledge of the reaction mechanism to describe that function analytically. Analytical expressibility is a prerequisite for all standard kinds of efficient optimization methods, such as gradient methods, conjugate-gradient methods, or various second-order methods (e.g., Gauss-Newton, Levenberg-Marquardt, or variable-metric methods), to get sufficiently precise estimates of the gradient or second-order derivatives of the optimized function. To tackle this difficulty, two different approaches are in principle possible, and both have been used in heterogeneous catalysis:

(1) To employ an optimization method not requiring the gradient or second-order derivatives of the optimized function, either a deterministic one, such as the simplex method [1], or a stochastic one, such as simulated annealing [2–4] or genetic algorithms [5–14]. Especially genetic algorithms have become very attractive in heterogeneous catalysis, mainly due to the possibility of establishing a straightforward correspondence between optimization paths followed by the algorithm and channels of the high-throughput reactor in which the catalysts proposed by that algorithm are subsequently tested.

High-Throughput Screening in Chemical Catalysis
Edited by A. Hagemeyer, P. Strasser, A. F. Volpe, Jr.

ISBN: 3-527-30814-8

(2) To construct, using the available data, an analytically expressible approximation of the optimized function, for which the gradient and second-order derivatives can be precisely enough estimated, thus making standard optimization methods applicable. Since the function describing the dependence of catalyst performance on descriptors can be quite complicated, and may even contain discontinuities, a method enabling the approximation of very general functions is needed. Most promising in this direction are approximation methods relying on various classes of artificial neural networks. Hence, it is not surprising that all of the heterogeneous catalysis applications in which this approach has been used have employed approximation methods based on a particular kind of neural networks, i.e. the so-called multilayer perceptron [15–22].

This chapter focuses on present trends in the optimization of catalyst performance with genetic algorithms, but also pays attention to the possible synergy resulting from its integration with the second approach. In the next section, the theoretical and methodological principles of genetic algorithms-based optimization and neural network-based approximation are recalled. In Section 3, the shift from predefined genetic algorithms to genetic algorithm generators, i.e., metasystems generating a particular algorithm at run-time according to user requirements is explained and illustrated on a metasystem that is currently under development at the Institute for Applied Chemistry Berlin-Adlershof (ACA). Finally, Section 4 reviews different ways in which the integration of genetic algorithms and artificial neural networks has been attempted in catalysis, concentrating on the most recent among them – the use of neural networks to simulate the experimental feedback, yielding values of the optimized function.

6.2 Theoretical and Methodological Principles

6.2.1 Genetic Algorithms

Genetic algorithms (GAs) are a *stochastic* optimization method. This means that, when searching for maxima or minima of the optimized function, the available information about its response surface is complemented with random influences. In the context of catalysis, the optimized function is typically some performance measure of the catalyst, such as yield, activity, conversion, selectivity. Together with simulated annealing, genetic algorithms now belong to the most popular representatives of such optimization methods. The term "genetic algorithms" refers to the fact that their particular way of incorporating random influences into the optimization process has been inspired by the biological evolution of a *genotype*. Basically, this consists of:

- Randomly exchanging coordinates of points between two particular points in the input space of the optimized function (*recombination, crossover*).

- Randomly modifying coordinates of a particular point in the input space of the optimized function (*mutation*).
- *Selecting* the points for crossover and mutation according to a probability distribution, either uniform or skewed towards points at which the optimized function takes high values (the latter being a probabilistic expression of the survival-of-the-fittest principle).

The particular way in which selection, crossover and mutation are accomplished is strongly problem-dependent. Therefore, no universally applicable genetic algorithm exists. Each implementation is designed with a particular class of problems in mind, and its design determines the spectrum of problems for which it can be later employed. With catalysis, it is useful to differentiate between *qualitative mutation*, which consists in adding a new component (element or support material) to the catalyst or in removing one of its constituting components, and *quantitative mutation*, which consists in merely changing the proportions of constituting components (Fig. 6.1).

A detailed description of various kinds of genetic algorithms, as well as of other kinds of the broader class of evolutionary algorithms, can be found in specific monographs [23–34]. Here, only those particular features of genetic algorithms

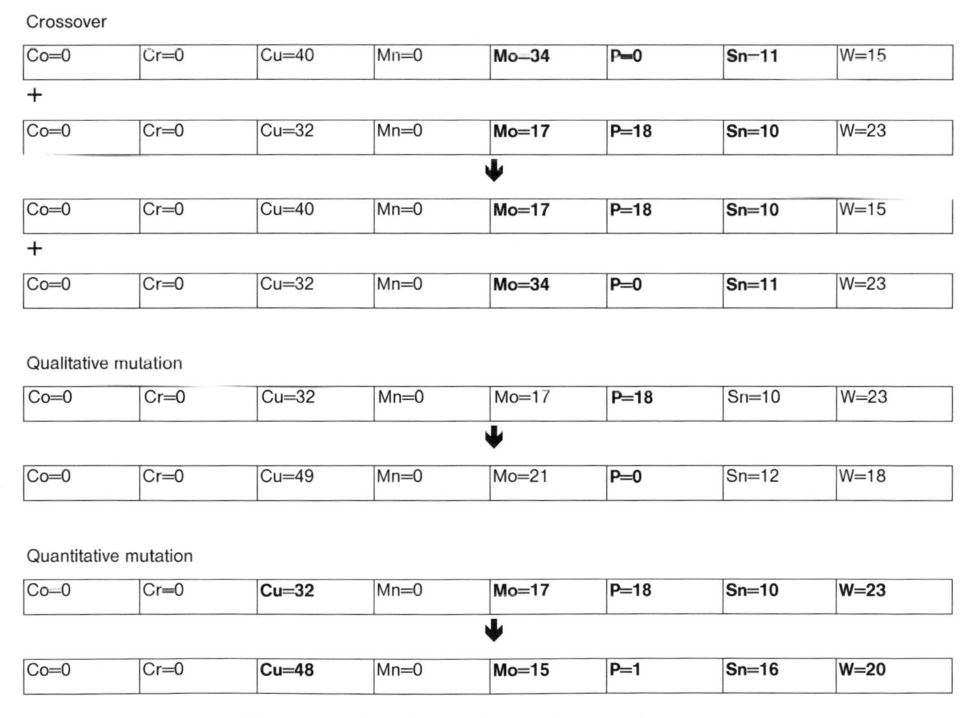

Fig. 6.1 Operations used in genetic algorithms (values in the examples are proportions of elements in the active component of the catalyst expressed in mol%).

that can be considered the most important in the context of catalysis will be pointed out.

1. Genetic algorithms do not require knowledge of the gradient (first derivatives) of the optimized function (like various kinds of gradient methods), or its second derivatives (like the Gauss-Newton method, Levenberg-Marquardt method, or variable metric methods). Moreover, the optimized function even does *not* have to be sufficiently smooth, i.e., to have derivatives in the whole area where the optimization is performed. Since the functions optimized in catalysis are not analytically expressible, their derivatives can not be numerically estimated with sufficient precision. If, in addition, the catalyzed reaction includes a phase transition, then the optimized function even may contain discontinuities.

2. The crossover and mutation operations provide an excellent possibility for the genetic algorithm to be applied to more individuals at each iteration, and thus to follow *many* optimization paths in parallel, a synchronization among which is needed only after each iteration. Because of the biological inspiration of genetic algorithms, individual iterations of a genetic algorithm are called *generations,* and all points in which the optimized function is evaluated in a particular generation (e.g., all catalysts whose performance has been measured at that generation) are denoted as *population* added in that generation. The fact that genetic algorithms follow many optimization paths in parallel is actually the main reason for their attractiveness in high-throughput catalysis, because it is then possible to establish a straightforward correspondence between those optimization paths and channels of the high-throughput reactor in which the catalysts are experimentally tested.

3. Genetic algorithms heavily rely on heuristics. Therefore, they usually contain a large number of *adjustable* control parameters that need to be *tuned* to achieve the required performance of each incorporated heuristics and, ultimately, the required performance of the whole algorithm. As any other optimization algorithm, even a genetic algorithm is required to get sufficiently close to the searched optimum (maximum or minimum) of the optimized function, and to get there within a sufficiently low number of iterations. In mathematical terms, the algorithm is required to *sufficiently quickly converge* to the searched optimum. It is important to realize that if the parameters are tuned so that many or even all of the individual optimization paths quickly approach the searched optimum, then they also quickly approach each other. Consequently, the *diversity* among the individuals *decreases* at the same speed as the algorithm converges. In the worst case, we may end up with a population of identical individuals. Needless to say, such a quick decrease of diversification is highly undesirable because then the advantage of multiple optimization paths may get completely lost. Moreover, the correspondence between multiple optimization paths and high-throughput testing in catalysis would lead to the situation that the same catalyst would be tested in many or even all channels of

the reactor, which is not acceptable from the experimentation point of view. Therefore, control parameters of a genetic algorithm should be not only tunable for a high speed of convergence, but they should also allow to slow-down the decrease of diversity among the individuals.

In heterogeneous catalysis, genetic algorithms have been used for approximately five years [5–14]. At ACA, a simple genetic algorithm has been implemented following the proposal in [5]. It provides the possibility to divide all components of the considered catalysts into several types, each with an individual preparation mode or a set of preparation modes, and with an individual probability of modification through a heuristics-based crossover and mutation at the second and following generations. At the first generation, compositions are proposed randomly, with proportions of individual components chosen optionally from uniform, log-uniform or exponential distribution. On the compositions resulting from crossover or mutation in the second and further generations, no constraints can be imposed, with the exception of the most fundamental constraint that the proportions of all catalyst components have to sum to 100%. Conversely, simple additional constraints may be imposed at the first generation, namely constraints requiring that proportion ratios of particular pairs of components have to lie within prescribed intervals. Meanwhile, a much more sophisticated system is under development, which allows the dynamic generation of problem-tailored genetic algorithms following the principles explained in Section 6.3.

6.2.2
Artificial Neural Networks

Artificial neural networks (ANNs) are distributed computing systems that attempt to implement some part of the functionality of biological neural systems. They are based on the concepts of a *neuron*, whose biologically inspired meaning is some elementary signal processing unit, and of a *connection* between neurons or between a neuron and the environment, which enables the transmission of signals. Neurons that can receive signals from the environment are the input neurons of the network, while those that can send signals to the environment are its output neurons. The state of the remaining neurons has no one-to-one correspondence with signals observable in the environment and, therefore, they are called hidden neurons. The partition of the set of neurons into input, output and hidden neurons, and the structure of connections between neurons form the *architecture* of the artificial neural network. Most networks encountered in applications are feedforward networks, in which the set of neurons is partitioned into a sequence of disjointed layers (layered architecture) such that signal transmission between neurons is possible only from a lower layer to a higher layer (Fig. 6.2). The most common kind of feedforward networks are *multilayer perceptrons*; all neural networks employed so far in heterogeneous catalysis are of this type [15–22]. Detailed treatment of multilayer perceptrons can be found in the monograph [35] – in the context of catalysis they have been described in some detail in [22].

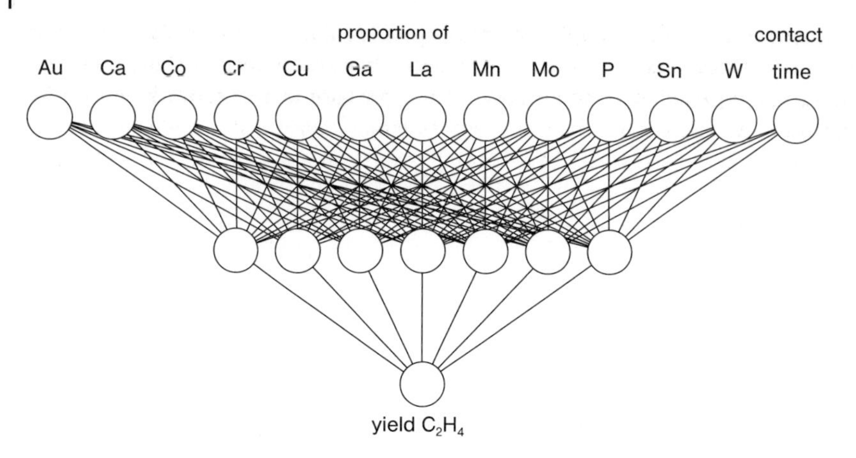

Fig. 6.2 Example architecture of a neural network used to study the dependence of ethylene yield on catalyst composition and contact time in high-throughput experiments for the oxidative dehydrogenation of ethane.

The network architecture determines possible paths through which signals can propagate from input to output neurons, thereby restricting the possibilities for mapping input signals to output signals. Nevertheless, the architecture itself does not yet determine that mapping. To this end, we need to know all transformations that the signals undergo on the paths through which they propagate, i.e., all transformations they undergo in any hidden or output neuron or on any connection between neurons (conventionally, input neurons only receive signals from the environment, and no transformations are applied to them). Inspired by biological terminology, transformations that are applied to signals directly in neurons are called *somatic* operations, whereas transformations applied to signals in connections are called *synaptic* operations. A somatic operation applied in a hidden neuron mostly includes a one-dimensional non-linear transformation, called an *activation function*, which is typically the same in all hidden neurons.

One of the most remarkable features of feedforward networks is the possibility to approximate, with an arbitrarily prescribed precision, even extremely complicated and extremely general dependencies [36–43]. In catalysis we are primarily interested in dependencies of catalyst performance, expressed as products yields, catalyst activity, conversion of feed molecules and products selectivity, on composition of the catalysts, their physical properties, and on reaction conditions. It is for the approximation of such dependencies that artificial neural networks have been used in catalysis so far [15–22].

Despite restrictions imposed by the architecture, each artificial neural network allows infinitely many mappings of received signals to sent signals. To be able to choose among them we need to quantify the agreement or disagreement of each mapping with the particular purpose for which we are using the network, such as approximation of unknown dependencies, classification of objects, clustering of objects according to their similarity, forming associations between concepts, and extraction of logical rules from data. To this end, various agreement/disagreement functions can be used, depending both on the purpose of the network and on its architecture. A number of disagreement functions used with feedforward networks have been surveyed [22]. In a network with n_I input neurons and n_O output neurons that computes a mapping F, each disagreement function in some way compares the sequence $F(x_1),\ldots, F(x_p)$ of n_O-dimensional points to which F maps a given sequence $x_1,\ldots, x_p$ of n_I-dimensional points with sequence $y_1,\ldots, y_p$ of n_O-dimensional points such that for $j=1,\ldots, p$, y_j is empirically known to correspond to x_j in the dependence approximated with F. For example, if we want to approximate the dependence of a product yield on catalyst composition, then $x_1,\ldots, x_p$ are compositions of p catalysts that were used in the performed experiments, and $y_1,\ldots, y_p$ are the yields of the considered product. The most common disagreement function, called mean squared error (MSE), is simply the mean squared deviation between the empirical points $y_1,\ldots, y_p$ and the points $F(x_1),\ldots, F(x_p)$ obtained by means of the mapping F, i.e.,

$$\mathrm{MSE}(F,(x_1,y_1),\ldots,(x_p,y_p)) = \frac{1}{p}\sum_{j=1}^{p}\left\|F(x_j)-y_j\right\|^2 = \frac{1}{p}\sum_{i=1}^{n_O}\sum_{j=1}^{p}\{[F(x_j)]_i-(y_j)_i\}^2$$

If a neural network architecture and a disagreement function are fixed, then the task to find the mapping whose disagreement with the purpose of the network is minimal turns into a standard optimization task. Since all methods for optimization of general functions are iterative, the process of improving the agreement of a network with its purpose is also always iterative. Inspired again by biological terminology, this iterative process is called *learning* or *training*, and the pairs $(x_1, y_1),\ldots, (x_p, y_p)$ are denoted as *training pairs* or *training data*. Principally, any optimization method may be used as neural network learning method, even in the particular context of feedforward networks and the disagreement function MSE. In the early decades of artificial neural networks, a variant of gradient descent was very popular in that context, called back-propagation, due to the way the elements of the MSE gradient are calculated [35, 44]. Nowadays, back-propagation increasingly competes with the Levenberg-Marquardt method, a hybrid method combining gradient descent and the Gauss-Newton quadratic optimization with Jacobian-based estimation of second derivatives [35, 45, 46]. Compared to back-propagation, the Levenberg-Marquardt method has much higher storage requirements and is substantially more difficult to implement, but its convergence speed increases from linear to quadratic near the minimum, whereas back-propagation converges always only linearly.

A crucial problem pertaining to neural-network learning is that the information available for learning does not concern the complete unknown dependence to be approximated, but only the finite sequence $(x_1, y_1), \ldots, (x_p, y_p)$ of training data. This entails the danger that the resulting approximation will very precisely fit the training data, including all noise present in them, but values outside the points

$x_1, \ldots, x_p$ will be quite irrelevant to the approximated dependence. Such a phenomenon is called "overtraining", "overlearning" or "overfitting". To recognize whether overtraining occurs, and to what extent, another sequence of data $(x'_1, y'_1), \ldots, (x'_q, y'_q)$ is needed, called test data, which are also governed by the approximated dependence but have not been involved in the learning process. The lower the value of a disagreement function for the test data is, the better the trained network approximates the unknown dependence, whereas the higher that value, the more the network has been overtrained. Using the value of a disagreement function for the test data, it is possible to compare the quality of approximation both between different architectures, e.g. between feedforward networks with different numbers of hidden neurons, and between different iterations of the learning process. This suggests a simple method to reduce overtraining, called early stopping – to stop network learning at the iteration at which the disagreement function for the test data starts to decrease. However, if a sequence of test data is used in the early stopping method to stop network learning, then it gets involved in the learning process in this way, and can not be further used as test data. Therefore, we need a second, separate set of test data $(x''_1, y''_1), \ldots, (x''_s, y''_s)$ for early stopping, which is usually denoted validation data. Another method for overtraining reduction relies on Bayesian statistics and is called Bayesian regularization [47, 48]. This method views parameters parametrizing the mapping of received signals to sent signals (such as weights and biases of a multilayer perceptron) as realizations of random variables with a particular distribution, and overtraining as realizations that are not likely for that distribution. The values of parameters corresponding to minimal overtraining can then be obtained by adding to MSE a term penalizing improbable realizations.

For a detailed treatment of artificial neural networks, readers are again referred to specific monographs [35, 49–51], for a survey of their applications in chemistry to overview books [52, 53], reviews [54–56], and relevant sections of publications [57–59]. For heterogeneous catalysis, a recent overview has explained the applicability of feedforward networks to the approximation of unknown dependencies and to the extraction of logical rules from experimental data [22].

6.3
Automatically Generated Problem-tailored Genetic Algorithms

As mentioned in Section 6.2.1, genetic algorithms are strongly problem-dependent, and any design decision made when such an algorithm is implemented restricts the class of problems for which the implementation can be subsequently employed. In catalysis, this can pose serious problems since the change of focus to other reactions, the increased variability of catalyst composition, and the emergence of new catalyst preparation methods may substantially decrease the usefulness of an existing genetic algorithm implementation after several years. To keep the usefulness of the implementation high for a long time would require the previewing, during its design, of all reactions, catalysts and their preparation methods for which the implemented algorithm might be employed in the future. Nevertheless, there is no guarantee that all of them really can be previewed at the design time. In addition, the more possibilities the implementation of a genetic algorithm attempts to cover, the higher becomes the number of parameters that have to be set or even costly tuned whenever the implementation is used.

In this section, a conceptually different approach to the problem-specificity of genetic algorithms is described. Its basic idea is to postpone the implementation of the algorithm as much as possible, i.e., to implement the algorithm immediately before using it to solve a particular problem. Since, at that time, all requirements specific to the problem are known this approach provides the possibility of implementing precisely problem-tailored genetic algorithms. However, traditional implementation by humans is not possible in such a situation: it is too slow, too expensive and too error-prone. Hence, we need genetic algorithm implementations that are generated automatically according to actual user requirements by some *software metasystem*. Needless to say, such a metasystem has to be first implemented by humans, and its implementation will necessarily entail restrictions, both on the user requirements that it accepts and on how those requirements are reflected in the generated genetic algorithm implementations. But since those restrictions are much more abstract than restrictions entailed by genetic algorithm implementations themselves, they allow for a much broader spectrum of problems than any particular single implementation could ever cover. In addition, tuned parameters are included among the user requirements in this approach – hence the user has full control over the extent of parameter tuning.

At ACA, a prototype metasystem GENACAT implementing the outlined approach is under development. In the remainder of this section, the structure and key functionality of that metasystem are briefly sketched.

The above-stated purpose of the metasystem implies that its key module is a genetic algorithm generator, i.e., software capable of generating implementations of genetic algorithms according to user requirements. To ensure that the requirements are precisely accomplished in the resulting implementation, they have to be specified in a rigorously formal way. At ACA, a special catalyst description language has been developed to this end. It provides the possibility of specifying all details of the algorithm to be generated and of the catalysts that the algorithm should propose, in particular:

- All components constituting the pool from which the catalysts should be composed, and optionally the division of all those components into an arbitrarily complex *hierarchy of component types* (such as support, active components, dopants), component subtypes, subsubtypes, etc.

- The choice, for a component type or any other class of the component types hierarchy, from a final set of possibilities (for example, support is one of α-Al_2O_3, γ-Al_2O_3, ZrO_2), each of which may be again chosen from a final set of possibilities, etc., hence completing the component types hierarchy with an arbitrary complex *selection hierarchy* ("OneOf" hierarchy).

- The *proportion* (either molar, or the weight-proportion) of any component within the catalyst or within any class of the component types hierarchy, optionally also the proportion of any class of the component types hierarchy within any higher class.
- The *ratio* between particular proportions, no matter the proportions of which and within which they are (e.g., proportion of active components within the catalyst to the proportion of support within the catalyst = 1 : 5; proportion of Ga within the active components to the proportion of Mo within the active components = 1 : 1).
- *Equality and inequality constraints* on any proportion or any group of proportions (e.g., 80% < proportion of support within the catalyst < 90%, the proportion of Mg within the active components + the proportion of Mn within the active components = 50%).
- Equality and inequality constraints on any ratio or any group of ratios between proportions (e.g., 1 : 2 < proportion of Ga within the active components to the proportion of Mo within the active components < 2 : 1).
- The *probability distribution* of the values of any proportion or any ratio between proportions on any prescribed interval (such as the uniform distribution, exponential distribution, or log-uniform distribution).
- The *number* of different components, component types, subtypes etc. that should be simultaneously present in any proposed catalyst, also, optionally, the number of different components, component subtypes etc. among those present in the catalyst that should belong to a given component type or any other given class of the component types hierarchy (e.g., 5 components altogether, including the support, 3 active components and 1 dopant).
- Equality and inequality constraints on the number of simultaneously present components or on the number of those among them that belong to a given class of the component types hierarchy (e.g., the number of all component lies between 3 and 6, the number of active components is at least twice that of dopants).
- The choice, for the number of components or for any other quantity, from a final set of possibilities, hence extending the selection hierarchy also to quantities.
- Identification of the way in which any component, component type or any other class of the component types hierarchy should be prepared, the *hierarchy* of employed ways of preparation may be arbitrarily complex.
- Which particular implementation or implementations of *genetic operators* should be used.
- Which particular parts of the algorithm output should be stored in the *database*, and in which tables and fields should they be stored.

```
ExperimentInformation GlobalParameter Test Experiment
PopulationSize GlobalParameter 60
ODHE-ht-cat ComposedOf support-a-Al2O3 InProportion fraction-support,
&active-components InProportion fraction-actives PreparedUsing seq-prep,
fraction-support FromInterval 0.6,0.9 WithPrecision 0.01
fraction-actives FromInterval 0.1,0.4 WithPrecision 0.01
fraction-support + fraction-actives = 1
active-components ComposedOf alkali-earth InProportion fraction-alkali-earth,
&rare-earth InProportion fraction-rare-earth, trans-metal InProportion
&fraction-trans-metal, alkali InProportion fraction-alkali, P InProportion
&fraction-P,
fraction-alkali-earth OneOf 0,positive-alkali-earth DistributedAs 1,3
fraction-rare-earth OneOf 0,positive-rare-earth DistributedAs 1,5
fraction-trans-metal OneOf 0,positive-trans-metal DistributedAs 1,3
fraction-alkali OneOf 0,positive-alkali DistributedAs 1,3
fraction-P OneOf 0,positive-P DistributedAs 3,1
positive-alkali-earth FromInterval 0.001,fraction-actives WithPrecision 0.001
positive-rare-earth FromInterval 0.001,fraction-actives WithPrecision 0.001
positive-trans-metal FromInterval 0.001,fraction-actives WithPrecision 0.001
positive-alkali FromInterval 0.00001*fraction-actives,0.001*fraction-actives
&WithPrecision 0.000001
positive-P FromInterval 0.00001*fraction-actives,0.001*fraction-actives
&WithPrecision 0.000001
fraction-alkali-earth + fraction-rare-earth + fraction-trans-metal +
&fraction-alkali + fraction-P = fraction-actives
alkali-earth OneOf Mg,Ca,Sr,Ba
rare-earh ComposedOf count-rare-earth FromAmong La InProportion fraction-La,
Nd &InProportion fraction-Nd, Sm InProportion fraction-Sm, Ce InProportion
&fraction-Ce, Y InProportion fraction-Y, Ho InProportion fraction-Ho, Yb
&InProportion fraction-Yb
count-rare-earth OneOf 1,2 DistributedAs 3,2
fraction-La,fraction-Nd,fraction-Sm,fraction-Ce,fraction-Y,fraction-Ho,
```

Fig. 6.3 Fragment of an example description in the catalyst description language.

- *Precision* with which any part of the algorithm output should be stored in the database; the generated algorithm then ensures, within the given precision, that any catalyst already present in the database is not proposed again.
- Whether *simulation* should be used to obtain the values of the empirical objective function, in which generations should it be used, and which particular simulation program should be employed to this end.
- How the actual values of the empirical objective function, which has been obtained either from experimental testing or from simulation, can be got from the database.
- Which other information should be taken from the database, and in which tables and fields is it stored.

A fragment of an example description in the catalyst description language is shown in Fig. 6.3. However, the user is not required to write such descriptions and to understand the description language in which they are written. The meta-

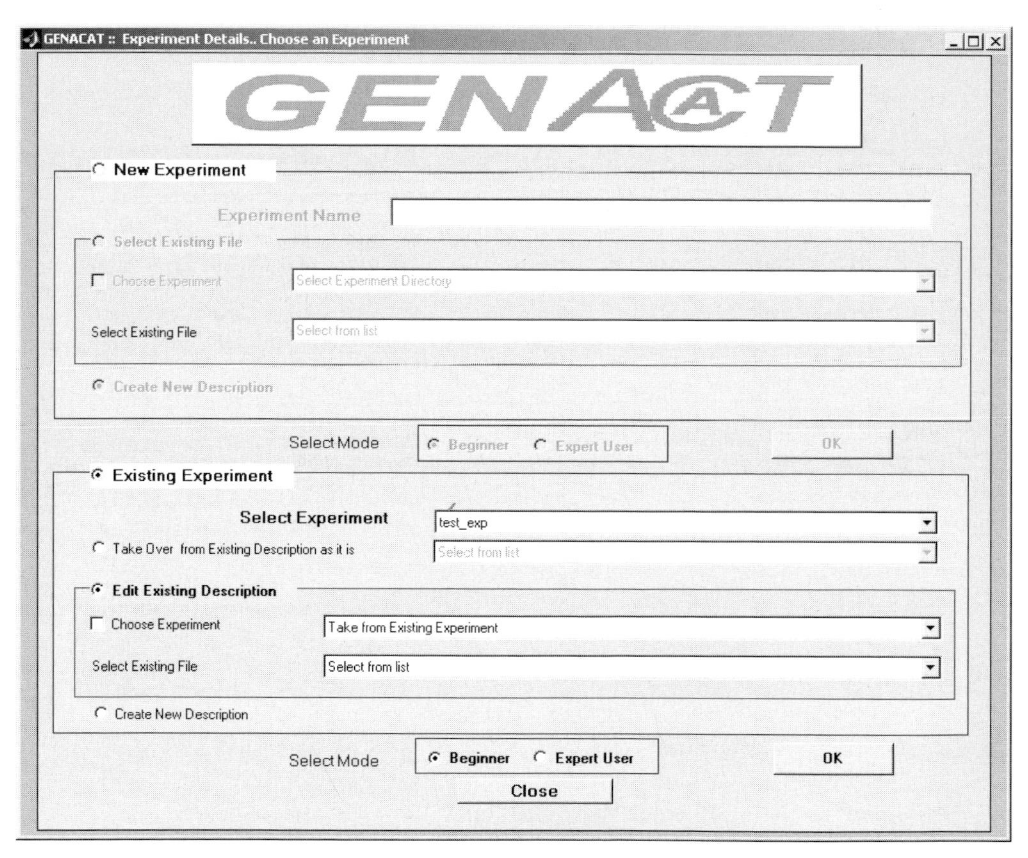

Fig. 6.4 Introductory screen to the graphical user interface of GENACAT.

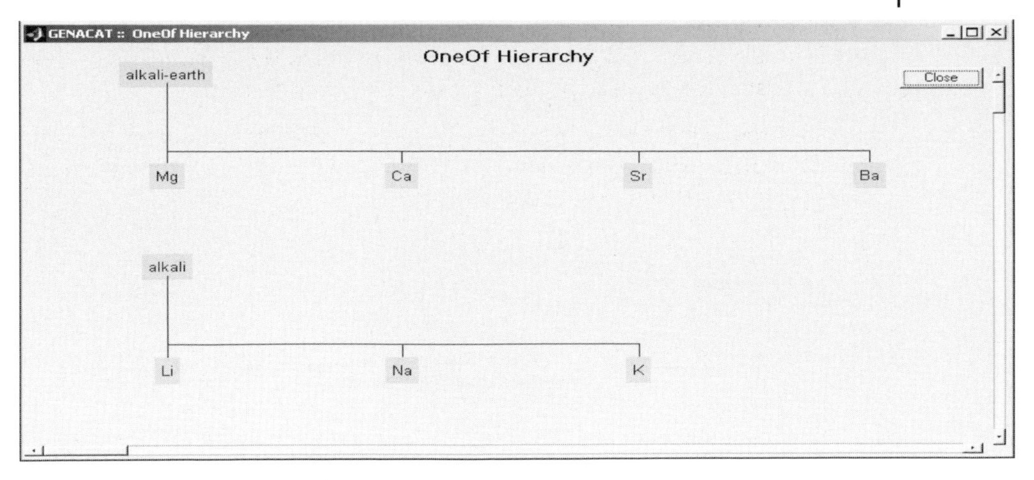

Fig. 6.5 Example of the visualized selection hierarchy.

system provides a user-friendly graphical user interface (GUI) that asks for all the information needed (Fig. 6.4). Moreover, it is possible to take over an existing description and to introduce only necessary changes in it, which may be arbitrarily small, as well as to store fragmental descriptions for later reopening and completing. These features have two-fold importance. First, much time in preparing the description can be saved if the intended experiment is similar to an earlier one. Second, the user is not bored with the necessity of dealing with information that lies outside his or her area of competence and that can be provided by a colleague. This is intended as the first place for information concerning database and simulations, which is typically provided by a database administrator or a data analyst, but it also allows information about catalyst composition and preparation methods to be provided by two independent experimenters.

In addition, the GUI also provides the possibility to visualize the component types' hierarchy and the selection hierarchy defined so far (Fig. 6.5).

The overall schema of the metasystem under development at ACA is given in Fig. 6.6.

6.4 Integrating Genetic Algorithms with Neural Networks

Because genetic algorithms and artificial neural networks have different objectives, they tackle the problem of searching high-performance catalysts from different directions. However, they do not need to be applied to that problem separately and independently. Integrating both approaches can have a synergistic effect, allowing one to find catalysts with high performance more quickly than their independent application. In catalysis, three different ways have already been followed to achieving such a synergistic effect of integrating genetic algorithms and artificial neural networks:

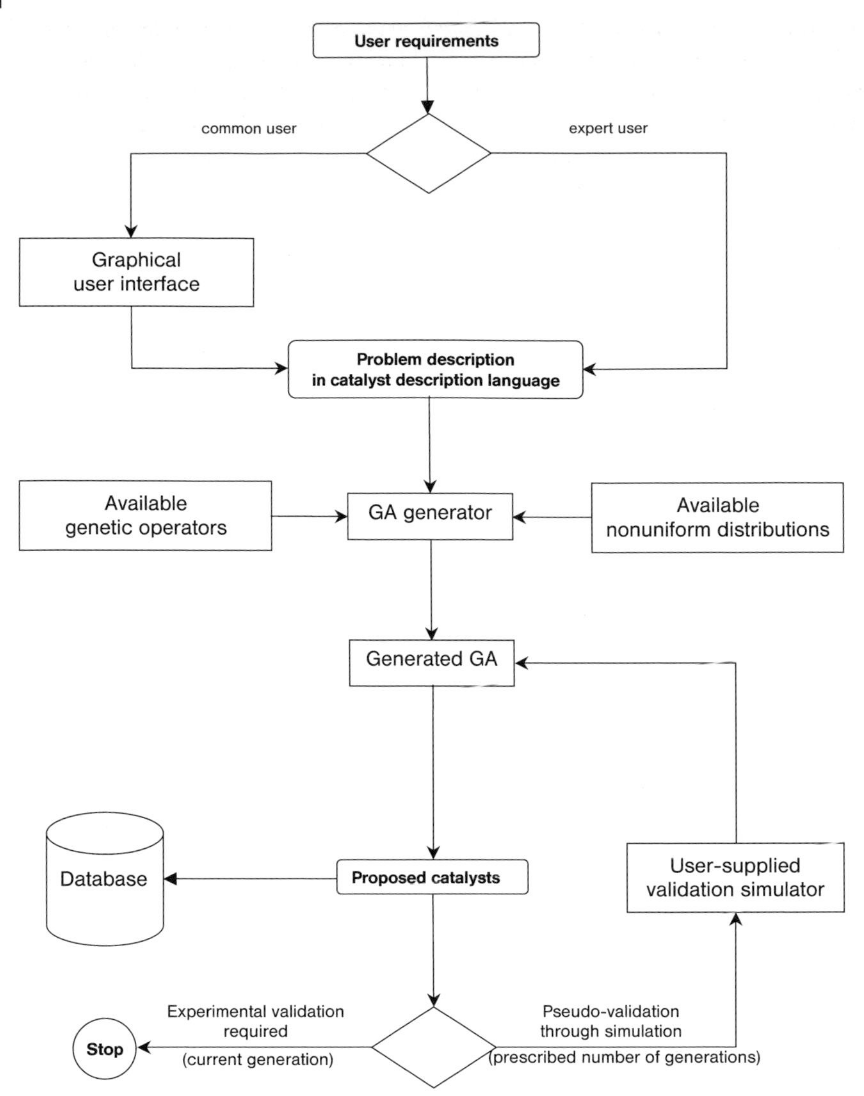

Fig. 6.6 Schema of the metasystem automatically generating problem-tailored genetic algorithms being developed at ACA.

In the simplest strategy, both approaches can be straightforwardly integrated by employing genetic algorithms for the optimization task that must be performed with neural networks, i.e., for network training. Genetic algorithms are actually used to this end, though they are far less important in this respect than two other optimization methods, i.e., back-propagation and the Levenberg-Marquardt method (the role of genetic algorithms in evolving neural networks has been surveyed in [60]). In catalysis, a genetic algorithm has been used [61] to train an artificial neural network for tuning parameters of a three-way catalytic converter.

In connection with neural network-based approximations of the dependence of catalyst performance on various descriptors, such as catalyst composition, physical properties and reaction conditions, another integration strategy is much more important – to use genetic algorithms as an optimization method to find the global maximum of such an approximation. In other words, genetic algorithms are used to find a vector of globally optimal composition, properties and conditions, i.e., a vector describing a catalyst with the best theoretically achievable performance. The global maximum being searched for is actually always a constrained maximum, subject to some natural restrictions, e.g., that the proportion of any component must be non-negative, and that the sum of all proportions must not exceed 100%.

This strategy of integrating neural networks with genetic algorithms has been used to search for the optimal composition of a catalyst for the ammoxidation of propane [62]. In that case, no experiments were performed; the network was trained with data published earlier by other authors [63]. However, those data were for only 26 catalysts, thus forming a quite small training set. Even more importantly, the predicted performance of the optimal catalyst, expressed by means of acrylonitrile yield, was not experimentally verified.

The same integration strategy has also been used [64] to find the optimal Cu:Zn:Al ratio in mixed oxide catalysts for methanol synthesis from Syngas.

A similar strategy has been employed for optimizing the process variables of the TS-1-catalyzed hydroxylation of benzene [65]. First, a neural network-based process model was developed, and then five different suggestions for process conditions leading to high benzene conversion, efficient H_2O_2 utilization and high values of phenol selectivity were found by means of a genetic algorithm. The process conditions were reaction time, catalyst concentration, reaction temperature, benzene-to-H_2O_2 ratio, benzene-to-H_2O ratio. Also in this case, the training set was very small, consisting of only 24 experiments. Nevertheless, the five suggested vectors of process conditions were experimentally verified, validating the predictions of the process model with good accuracy.

A crucial feature of this integration strategy is that no new experimental results need to be provided to the genetic algorithm during its search for a global maximum of catalyst performance. Indeed, even when new experiments were performed in the above applications, their purpose was solely to verify an already found optimal catalyst, not to search for it. On the one hand, this decreases the importance of using genetic algorithms to this end. Indeed, their main advantage in high-throughput catalysis, i.e., a straightforward correspondence between the optimization method and the experiment (between multiple optimization paths followed by the algorithm and multiple channels of an experimental reactor), becomes immaterial. In addition, the analytic form of the neural-network approximation makes it very suitable for optimization using gradient-based methods, which are much faster and readily available in common software systems for function optimization. Gradient-based methods are local optimization methods, i.e., their optimization paths converge in general not to the global maximum of the optimized function but to the local maximum in the attraction area of which the starting point of the path has been chosen. Nevertheless, they still may be superior to a GA in searching for the global maximum, in spite of the tendency of genetic algorithms to find a global maximum instead of a local one. The reason is that the attraction area of the global maximum is usually large; thus if we repeat the optimization many times with different starting points, hence obtaining many optimization paths such as in the case of a GA, there is a high probability that at least one of those starting points will lie in the attraction area of the global maximum. Since the form of a function in a close neighbourhood of a maximum or minimum is very well described by its gradient (and even better by its second derivatives), using the information about gradient and/or second derivatives will lead the optimization path much closer to the maximum than in the case of a genetic algorithm, which does not use that information.

To illustrate this phenomenon, Fig. 6.7 presents the results of optimizing a neural-network approximation of the dependence of ethylene yield on catalyst composition, obtained using data on 1412 catalysts for the oxidative dehydrogenation of ethane (ODHE). The simple genetic algorithm described in Section 2.1 has been run 10 times with a population size 60, searching for the maximum of that ethylene yield approximation. Fig. 6.7 shows that the maximal ethylene yield among the catalysts proposed by the GA is on average more than 2% lower than the global maximum of the approximation, which has been reached by a gradient-based sequential quadratic programming method [66–68].

Conversely, the population size and the number of generations, which are critical for the applicability of a genetic algorithm as a stand-alone optimization meth-

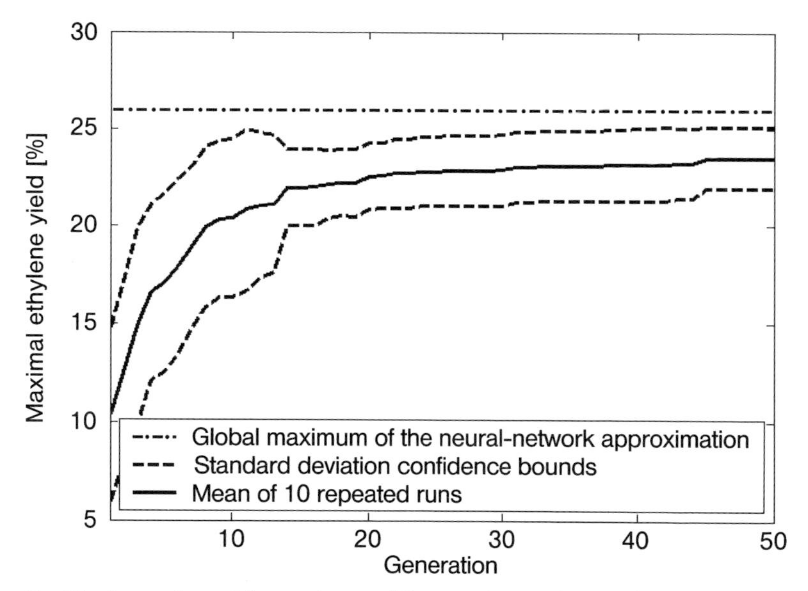

Fig. 6.7 Comparison of the maximum of the neural network approximation of the ODHE ethylene yield obtained in 10 runs of the genetic algorithm with a population size 60, and the global maximum obtained with a sequential quadratic programming method run for 15 different starting points.

od in given experimental conditions, do not matter at all if the algorithm is integrated in the above way with a trained neural network. This motivated the third, most recent integration strategy – using neural networks to *simulate experimental feedback* concerning the performance of the proposed catalysts; to simulate it for as many generations and as many simultaneously followed optimization paths as desired. Even more importantly, the genetic algorithm can be run several times for different population sizes, and it can be seen how quickly it, for each population size, converges to the global maximum, as well as how quickly the diversification of catalysts kept for further evolution decreases. In the same way, the influence of other adjustable parameters of the genetic algorithm on the convergence speed and diversification decrease can also be studied.

The necessity to study the dependence of the behavior of genetic algorithms on their adjustable parameters was recognized already earlier in combinatorial chemistry [69]. At ACA, the convergence speed and diversification decrease of the simple GA described in Section 2.1 has been investigated in the context of the oxidative dehydrogenation of propane (ODHP). In the investigation, all adjustable parameters of the employed genetic algorithm have been taken into account. These were the following parameters:

- population size;
- component type-specific probability of any modification through crossover and mutation;
- ratio between the probabilities of qualitative mutation and crossover;
- asymptotic ratio between the probability of quantitative mutation and the component type-specific probability of any modification;
- coefficient of quantitative mutation, used to multiply or divide the quantitatively mutated values.

The results of that investigation can be summarized as follows:

1. How close the algorithm gets to the best achievable performance of proposed catalyst is typically recognizable already during early generations. In the study performed at ACA, three increasingly strong milestones of convergence have been employed, all of them based on the difference between the best simulated performance among the catalysts proposed up to the present generation and the global maximum of the simulated performance. The weakest milestone, which was reached in 90% of the performed simulations, was reached on average already in the 4th generation. The second milestone, reached in 50% of the simulations, was reached on average in the 7th generation. Finally the strongest milestone, reached in 18% of the simulations, was reached on average in the 8th generation.

2. The convergence speed of the genetic algorithm tends to increase with increasing population size. However, this is merely a general tendency, which interferes with the influence of the remaining adjustable parameters, and only for particular combinations of them becomes really apparent.

3. Diversification of catalysts proposed by the algorithm tends to decrease more quickly with increasing asymptotic probability ratio of quantitative mutation to any modification, and with increasing coefficient of quantitative mutation. Again, these are only general tendencies, really apparent only for particular combinations of the remaining parameters.
4. Diversification of the proposed catalysts decreases more quickly if crossover and qualitative mutation occur with equal probability than if one of them substantially prevails.
5. Finally, also in the case of those results, the phenomenon shown in Fig. 6.7 was observed – the genetic algorithm was less successful in searching for the global maximum of the neural-network approximation than a gradient-based optimization method. To this end, the sequential quadratic programming method has again been used [66–68], but this time only seven different starting points were employed.

6.5 Conclusions

In heterogeneous catalysis, genetic algorithms have been used for approximately five years and are the most promising means to combine high-throughput experimentation with mathematical optimization. In this chapter, two recent innovations in the application of genetic algorithms to this area have been reviewed. One of them involves genetic algorithm generators, generating a particular algorithm at run-time according to user requirements, thus covering a much broader spectrum of different reactions and task than the traditional predefined genetic algorithms. The second is the synergistic effect resulting from the integration of genetic algorithms with artificial neural networks, another sophisticated mathematical method that came into use in heterogeneous catalysis for the same reason as genetic algorithms, i.e., to tackle the challenge of finding optimal catalysts for specific reactions.

The innovations presented here are the results of research in the area of mathematical optimization of catalyst performance. As this research is ongoing, similarly important innovations can be expected in the future. Concerning the integration of genetic algorithms and artificial neural networks, it has been, and will continue to be, additionally influenced by ongoing research into the approximation of the dependence of catalyst performance on its composition and other properties, as well as into the closely related area of knowledge extraction from catalytic data. In those areas, especially the application of other kinds of neural networks than multilayer perceptron, and of new rule extraction methods [22], may be of great importance in the near future. The Institute of Applied Chemistry Berlin-Adlershof is actively participating in all these research areas.

6.6 References

1 Holzwarth, A., Denton, P., Zanthoff, H., Mirodatos, C. *Catal. Today*, 2001, 67: 309–318.

2 Li, B., Sun, P., Jin, Q., Wang, J., Ding, D., *J. Mol. Catal. A: Chem.*, 1999, 148: 189–195.

3 McLeod, A. S., Gladden, L. F. *J. Chem. Information Comput. Sci.*, 2000, 40: 981–987.

4 Eftaxias, A., Font, J., Fortuny, A., Giralt, J., Fabregat, A., Stüber, F. *Appl. Catal. B: Environ.*, 2001, 33: 175–190.

5 Wolf, D., Buyevskaya, O. V., Baerns, M. *Appl. Catal. A: General*, 2000, 200: 63–77.

6 Baerns, M., Buyevskaya, O., Grubert, G., Rodemerck, U. in *Principles and Methods for Accelerated Catalyst Design and Testing*, Derouane, E. G., Parmon, V., Lemos, F., Ribeiro, F. R., Eds., Kluwer, Dordrecht, 2002, 85–100.

7 Buyevskaya, O. V., Wolf, D., Baerns, M. *Catal. Today*, 2000, 62: 91–99.

8 Buyevskaya, O. V., Brückner, A., Kondratenko, E. V., Wolf, D., Baerns, M. *Catal. Today*, 2001, 67: 369–378.

9 Grubert, G., Wolf, D., Dropka, N., Kolf, S., Baerns, M. in *Proceedings of 4th World Congress on Oxidation Catalysis, Book of Extended Abstracts, Volume I*, Dechema, Berlin/Potsdam, 2001, 113–119.

10 Langpape, M., Grubert, G., Wolf, D., Baerns, M. In: Proceedings of DGMK-Conference *Creating Value from Light Olefins – Production and Conversion*, DGMK, Hamburg, 2001, 227–234.

11 Rodemerck, U., Wolf, D., Buyevskaya, O.V., Claus, P., Senkan, S., Baerns, M. *Chem. Eng. J.*, 2001, 82: 3–11.

12 Corma, A., Serra, J. M., Chica, A. in *Principles and Methods for Accelerated Catalyst Design and Testing*, Derouane, E. G., Parmon, V., Lemos, F., Ribeiro, F. R., Eds., Kluwer, Dordrecht, 2002, 153–172.

13 Vauthey, I., Baumes, L., Hayaud, C., Farrusseng, D., Mirodatos, C., Grubert, G., Kolf, S., Cholinska, L., Baerns, M., Pels, J. R. in *EuroCombiCat 2002, European Workshop on Combinatorial Catalysis, Book of Abstracts*, Ischia, 2002, 44–46.

14 Wolf, D., Baerns, M. in *Experimental Design for Combinatorial and High Throughput Materials Development*, Cawse J. N., Ed., John Wiley & Sons, New York, 2003, 147–161.

15 Kito, S., Hattori, T., Murakami, Y. *Appl. Catal. A: General*, 1994, 114: L173–L178.

16 Hattori, T., Kito, S. *Catal. Today*, 1995, 23: 347–355.

17 Sasaki, M., et al., *Appl. Catal. A: General*, 1995, 132: 261–270.

18 Hou, Z.-Y., et al., *Appl. Catal. A: General*, 1997, 161: 183–190.

19 Sharma, B. K., et al., *Fuel*, 1998, 77: 1763–1768.

20 Huang, K., Feng-Qiu, C., Lü, D. W. *Appl. Catal. A: General*, 2001, 219: 61–68.

21 Corma, A., Serra, J. M., Argente, E., Botti, V., Valero, S. *ChemPhysChem*, 2002, 3: 939–945.

22 Holeňa, M, Baerns, M. in *Experimental Design for Combinatorial and High Throughput Materials Development*, Cawse, J. N., Ed., Wiley, New York, 2003, 163–202.

23 Bäck, T. *Evolutionary Algorithms in Theory and Practice: Evolution Strategies, Evolutionary Programming, Genetic Algorithms.* Oxford University Press, New York, 1996.

24 Banzhaf, W., Nordin, P., Keller, R. E., Francone, F. D. *Genetic Programming: An Introduction on the Automatic Evolution of Computer Programs and its Applications.* Morgan Kaufmann, San Francisco, 1998.

25 Fogel, D. B. *Evolutionary Computation: Toward a New Philosophy of Machine Intelligence*, 2nd Edition, IEEE Press, New York, 1999.

26 Freitas, A. A. *Data Mining and Knowledge Discovery with Evolutionary Algorithms.* Springer, Berlin, 2002.

27 Goldberg, D. *Genetic Algorithms in Search, Optimization, and Machine Learning.* Addison-Wesley, Reading, 1989.

28 Holland, J. *Adaptation in Natural and Artificial Systems.* MIT Press, Cambridge, 1992.

29 Genetic Programming: *On the Programming of Computers by Means of Natural Selection*. MIT Press, Cambridge, 1992.

30 Koza, J. R. *Genetic Programming II: Automatic Discovery of Reusable Programs*. MIT Press, Cambridge, 1994.

31 Koza, J. R., Bennett, F. H., Andre, D., Keane, M. A. *Genetic Programming III: Darwinian Invention and Problem Solving*. Morgan Kaufmann, San Francisco, 1999.

32 Langdon, W. B. *Genetic Programming and Data Structures: Genetic Programming + Data Structures=Automatic Programming*. Kluwer Academic Publishers, Boston, 1998.

33 Michalewicz, Z. *Genetic Algorithms + Data Structures= Evolutionary Programs* 3rd Edition. Springer, New York, 1996.

34 Mitchell, M. *An Introduction to Genetic Algorithms*. MIT Press, Cambridge, 1996.

35 Hagan, M.T., Demuth, H., Beale, M. *Neural Network Design*. PWS Publishing, Boston, 1996.

36 Barron, A. R. *Machine Learning*, 1994, 14: 115–133.

37 Chui, C. K., Li, X., Mhaskar, H. N. *Adv. Computat. Mathematics*, 1996, 5: 233–343.

38 De Vore, R. A., Oskolkov, K. I., Petrushev, P. P. *Ann. Numer. Mathematics*, 1997, 4: 261–278.

39 Hornik, K. *Neural Networks*, 1991, 4: 251–257.

40 Hornik, K., Stinchcombe, M., White, H., Auer, P. *Neural Comput.*, 1994, 6: 1262–1275.

41 Kůrková, V. *Neural Networks*, 1992, 5: 501–506.

42 Kůrková, V. *Neural Networks*, 1995, 8: 745–750.

43 Leshno, M., Lin, V. Y., Pinkus, A., Shocken, S. *Neural Networks*, 1993, 6: 861–867.

44 Rumelhart, D. E., Hinton, G. E., Williams, R. J. in *Parallel Data Processing*, Rumelhart, D. E., McClelland, J. L., Eds., MIT Press, Cambridge, 1986, Vol. 1, 318–362.

45 Hagan, M. T., Menhaj, M. *IEEE Trans. Neural Networks*, 1994, 5: 989–993.

46 Marquardt, D. *SIAM J. Appl. Mathematics*, 1963, 11: 431–441.

47 Foresee, F. D., Hagan, M. T. in *Proceedings of the 1997 International Joint Conference on Neural Networks*, IEEE, Houston, 1997, pp. 1930–1935.

48 MacKay, D. J. C. *Neural Computation*, 1992, 4: 415–447.

49 Haykin, S. *Neural Networks. A Comprehensive Foundation*. IEEE, New York, 1999.

50 Mehrota, K., Mohan, C. K., Ranka, S. *Elements of Artificial Neural Networks*. MIT Press, Cambridge, 1997.

51 White, H., *Artificial Neural Networks: Approximation and Learning Theory*. Blackwell Publishers, Cambridge, 1992.

52 Zupan, J., Gasteiger, J. *Neural Networks for Chemists*. Wiley-VCH, Weinheim, 1993.

53 Zupan, J., Gasteiger, J. *Neural Networks in Chemistry and Drug Design, An Introduction*. Wiley-VCH, Weinheim, 1999.

54 Henson, M. A. *Comput. Chem. Eng.*, 1998, 23: 187–202.

55 Melssen, W. J., Smits, J. R. M., Buydens, L. M. C., Kateman, G. *Chemomet. Intelligent Lab. Syst.*, 1994, 23: 267–291.

56 Smits, J. R. M., Melssen, W. J., Buydens, L. M. C., Kateman, G. *Chemomet. Intelligent Lab. Syst.*, 1994, 22: 165–189.

57 Clark, J.W., Ed. *Scientific Applications of Neural Nets*. Springer, Berlin, 1998.

58 Lisboa, P. G. J., Ed. *Neural Networks: Current Applications*. Chapman & Hall, London, 1992.

59 Murray, A. F., Ed. *Applications of Neural Networks*. Kluwer, Dordrecht, 1994.

60 Annunziato, M., Lucchetti, M., Pizzuti, S. in *Proceedings of the European Symposium on Intelligent Technologies, Hybrid Systems and their Implementation on Smart Adaptive Systems*. EUNITE, Aachen, 2002.

61 Glielmo, L., Milano, M., Santini, S. *IEEE-ASME Trans. Mechatron.* 2000, 5: 132–141.

62 Cundari, T. R., Deng, J., Zhao, Y. *Ind. Eng. Chem. Res.*, 2001, 40: 5475–5480.

63 Hou, Z.Y., et al. *Appl. Catal. A: General*, 1997, 161: 183–190.

64 Omata, K., Umegaki, T., Watanabe, Y., Yamada, M. *J. Jpn. Petroleum Institute*, 2002, 45: 192–195.

65 Nandi, S., Mukherjee, P., Tambe, S. S., Kumar, R., Kulkarni, B. D. *Ind. Eng. Chem. Res.*, 2002, 41: 2159–2169.

66 FLETCHER, R. *Practical Methods of Optimization, Vol. 1, Unconstrained Optimization, Vol. 2, Constrained Optimization,* Wiley, New York, 1980.

67 GILL, P. E., MURRAY, W., WRIGHT, M. H. *Practical Optimization,* Academic Press, London, 1981.

68 SCHITTKOWSKI, K. *Ann. Operations Res.,* 1985, 5: 485–500.

69 SUNDARAM, A., VENKATASUBRAMANIAN, V. *J. Chem. Information Comput. Sci.,* 1998, 38: 1177–1191.

7
Relative Quantification of Catalytic Activity in Combinatorial Libraries by Emissivity-Corrected Infrared Thermography

Guido Kirsten and Wilhelm F. Maier

7.1
Introduction

The application of combinatorial methods in materials and catalysis research is increasing rapidly [1]. Of fundamental importance in catalysis research is the availability of a reliable high-throughput technique for the rapid screening of catalytic activity or selectivity. Most general techniques, such as spatially resolved MS [2] or GC [3], are of sequential nature and thus the screening time increases proportionally with increasing sample number. In contrast, infrared thermography (IRT), which measures heat production on the surface of catalyst beds, is a truly parallel method and its screening time is independent of the sample number on a library [4]. It has been applied successfully to gas-phase [5] as well as liquid-phase [6] reactions as a qualitative indicator of reactivity differences. While originally its temperature resolution is poor [1] due to interferences of emissivity and reflectivity, with the use of emissivity correction (EC) [5] a reliable temperature resolution of 0.01 K is possible. Several potential problems of false positives or false negatives associated with changes in emissivity and reflectivity or activation–deactivation phenomena have been discussed [7]. ECIRT, therefore, seems to have the potential of a fast screening method. Such faster methods are essential for state-of-the-art catalyst developments [8].

Fig. 7.1 shows a typical image of emissivity-corrected infrared thermography (ECIRT), obtained in a search for active catalysts for the oxidation of propene with air at 100 °C. The image shows several active catalysts and many inactive materials. It also shows that there is a range of activity, i.e. among the active catalysts some produce more heat than others. This suggests that there is quantitative information hidden in the image. However, the origin of an infrared signal is complex (emissivity, reflectivity temperature) and exact quantification is practically impossible [9]. How to harvest this quantitative information for screening of new catalysts or materials is the subject of this chapter, which will also discuss the associated problems and potential pitfalls.

High-Throughput Screening in Chemical Catalysis
Edited by A. Hagemeyer, P. Strasser, A. F. Volpe, Jr.

ISBN: 3-527-30814-8

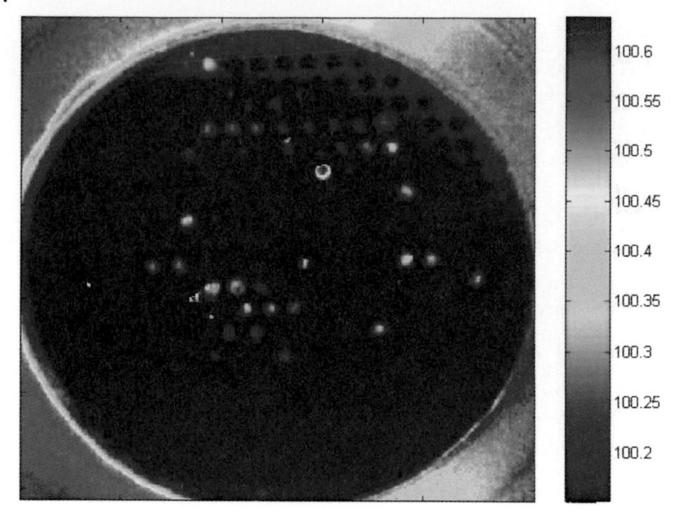

Fig. 7.1 IR thermography image of the heat of reaction of propene oxidation at 100 °C.

7.2
Theoretical Considerations

If a catalyst surface is exposed to a reaction gas mixture, heat is produced in exothermic reactions and consumed in endothermic reactions. The amount of heat produced will be proportional to the linear combination of catalytic activity and the enthalpy of all occurring reactions:

$$\delta Q \approx a_1 \Delta H_1 + \ldots + a_n \Delta H_n$$

Since the contributions of different reactions cannot be differentiated, only total heats are recorded. This is problematic in reactions with parallel and sequential side reactions, since the heat increase may not be associated with higher catalytic activity but with an increase in side reactions. A typical problem is selective oxidation, where undesired combustion as side reaction results in the most intensive heat signal. In such cases ECIRT can only be used to exclude inactive materials, but allows no judgement of selectivity, rendering signal quantification obsolete. Signal quantification is useful, if side reactions or sequential reactions can be excluded, such as combustion or CO-oxidation. If side or sequential reactions can be suppressed by selected reaction conditions, ECIRT may still be a useful screening tool. For example, we have used low reaction temperatures to suppress combustion in selective oxidations reactions by simultaneous monitoring of total CO_2 in the product gas with the help of CO_2 sensors.

If side and sequential reactions have been excluded, the heat should be proportional to the enthalpy of the desired reaction:

$$\delta Q \approx a_1 \Delta H_1$$

As the reaction enthalpy is constant at given reaction conditions for all catalysts on a library plate, differences in the heat produced must reflect differences of catalytic activity, especially if material composition and texture of the catalysts on the library is comparable. The produced heat cannot be measured directly; it has to be calculated from observable emission changes. These surface emission changes are recordable by ECIRT and can be converted into temperature data. These calculated temperatures will not reflect the bulk temperature of the catalyst. Since no heat is produced on the library substrate and on inactive materials most of the heat produced on the surfaces of active catalysts will be dissipated by transportation by conductance into the bulk and into the gas phase. Since heat capacities and conductances are not constant with different catalysts the scenario is highly complex. The catalyst bed in our set-up is not flow through, but only overflown by the reactant gases. The total heat produced therefore, is small compared with the heat capacity of the library as well as of the catalyst, which points to a rapid dissipation of most of the heat produced. Heat dissipation will be more effective with thin films than powdered beds, thus introducing another potential source of error. Furthermore, both heat capacity and heat conductance change with material composition, so that the steady-state surface temperature of a given reaction may vary not only with catalytic activity, but also with composition of catalysts and library. Since the mass of the library is much larger than that of the catalysts and the gas phase flow is slow, it is probably safe to assume that heat dissipation is very effective and the bulk catalyst temperature corresponds to that of the library, which means that small differences in heat capacity and conductance of the thin catalyst films can be neglected.

Because the spatial area with higher temperature on the catalyst surface of one of the samples of the library is very small the detection of catalytic activities through temperature measurement cannot be carried out by direct temperature measurements but only by non-contact methods such as pyrometry or IRT. The IR video camera used here measures the emission at every point of the library in parallel. The detector consists of a 256×256 pixel array of Pt-silicide-IR-sensors. Each pixel delivers a voltage-signal that depends on the infrared radiation and the sensitivity of that pixel (fixed pattern noise).

To calibrate the pixel sensitivities black body radiation is usually measured at different temperatures. Since a black body has an emissivity of 1 at every position, variations in detector pixel sensitivities are eliminated by a calibration function. As this IRT-method should be used here to quantify very small heat signals on combinatorial libraries with diverse materials, differences in emissivities have to be considered. Most materials are grey bodies with individual emissivities less than 1. Therefore, the calibration was not performed with a black body but with the library, as described before, a procedure that corrects not only for pixel sensitivity but also for emissivity differences across the library plate [5]. For additional temperature calibration, the IR-emission of the library is recorded at several temperatures in a narrow temperature window around the planned reaction temperature. By this procedure, emissivity changes, temperature dependence and individual sensitivities of the detector pixels can be calibrated in one step. After this calibration temperature changes can be determined from measured emission changes, whereby larger emission changes of materials with high emissivities are correctly appointed to the same temperature difference as smaller emission changes of materials with lower emissivities. After that calibration the library is brought to reaction temperature and the IR-image obtained can be assigned to a temperature range. The reaction is usually started at the reaction temperature by changing the feed gas. Since heat capacity and heat transfer to the gas phase also affect the IR-radiation entering the camera lens, it is best to calibrate the camera with a flow rate that provides a similar heat transport as is used later in the actual reaction. By application of reaction gas feed to this setup it is now possible to determine the catalytic activities from the temperature changes calculated from the emission measurement. A good procedure here is to use the same feed gas composition and flow as was used in the reaction, except that the reagent of lowest concentration is only added when the reaction actually starts.

7.3 Possible Pitfalls

With IRT temperatures are calculated from the measured emission and the calibrated emissivity data – every effect that leads to a change in emission or emissivity is a potential source for artefacts. Most prominent are adsorption and desorption processes, or surface oxidation or reduction. Binding between adsorbed molecules and the surface acts as a new radiation source. If the radiation is in the sensitivity range of the detector, increased emission may be misinterpreted as a higher temperature. To recognize such artefacts an online measurement of the product at the reactor outlet is necessary. If no product is detected, the hot spot originates from adsorption. Another potential pitfall is coking or catalyst oxidation or reduction. If the reaction gas reacts irreversibly with the material, the emissivity of the material changes, which is misinterpreted as a temperature change. Such errors can be identified by switching off the reaction gas mixture. Whereas hot spots from active catalysts disappear, these artefacts will remain. Artefacts may not only be introduced through the type of catalyst sample but also through temperature gradients in the library plate and in the heat the transport through the flowing medium. Both phenomena have been the subject of concern and are discussed below under reactor set-up.

7.4 Reactor and Setup

To achieve a homogeneous temperature distribution across the library is a challenging problem. Homogeneity is essential due to the temperature dependence of chemical reactions (Arrhenius law). With increasing reactor diameters, temperature gradients increase. Fig. 7.2 (see page 180) shows the temperature distribution on a slate plate (diameter 10 cm) measured manually on the surface of a library plate by a thermocouple. Here, in our original design, a spiral of heating wire was integrated as heating device in the reactor bottom.

At a reactor temperature of 300 °C the temperature gradient between reactor centre and edge was about 40 °C and thus not acceptable. Different heat sources had been studied. The best solution proved to be a commercial heating plate used in Ceran top stoves. For libraries of 10 cm diameter a heating plate of 1,2 kW and outer diameter of 17 cm has been used. This heat source heats an air pad, which in turn heats the reactor bottom mounted above the heater (see Fig. 7.9 below). With this setup the temperature gradient has been reduced to <4 °C at 300 °C (Fig. 7.3), which was found acceptable for a primary screening technique.

Another potential source of error is the flowing gas under steady state reaction conditions, which transports heat to or from the reactor. Here, for simplicity, a "flow-over" concept has been realized, in which the gas mixture does not flow through the catalyst bed but only over it. Thus, most of the gas–surface interaction takes place on the upper surface of each catalyst probe. Since the total heat produced will be calculated from the temperature changes of the geometrical area in the IR-image associated with each catalyst probe, an identical diameter for each of the catalyst cavities is essential to obtain comparable heat readings. Since the flowing gas carries heat in and out of the reactor it is also essential to have a slow and homogeneous flow distribution across the library. An even flow was approached by having not one but a series of gas inlet holes on one half of the reactor and a series of outlet holes on the other half of the reactor. As indicated above, the flow rate has to be chosen in such a way that the heat transport by the flow gas is identical during calibration and measurement. Another problem is the development of concentration gradients on libraries with too many active catalysts. In such a case catalytic activity decreases visibly across the library from the gas inlet side to the gas outlet side. In such a case, the total conversion (measured at the outlet of the high-throughput reactor) has to be reduced significantly by well-known parameters (change of concentrations, temperature, flow rate, smaller catalyst amount). It is therefore advisable to keep the total conversion across the library as low as possible, preferably below 15%.

Another concern is the camera setup geometry. The IR-camera can only record that part of IR-radiation which enters the lens of the camera. The greater the distance to the catalyst surface, the smaller is the angular section and the smaller is the relative amount of IR-radiation entering the camera compared with the total emission from the catalyst surface. This will significantly reduce the absolute IR-sensitivity and affect the observable temperature resolution. Since black body radiation is temperature-dependent, this problem is most prominent at low reaction temperatures, while above 400 °C a large distance is advantageous to avoid saturation of the detector by the high flux of IR-photons.

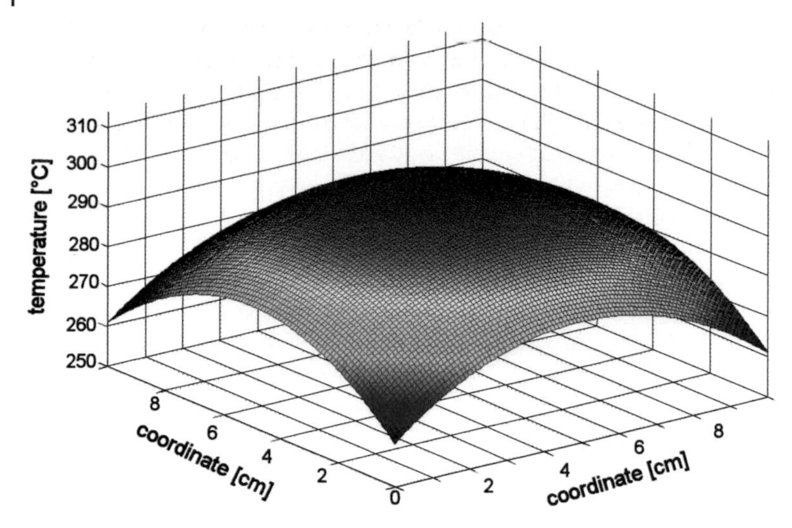

Fig. 7.2 Temperature distribution of a reactor heated with a heat wire.

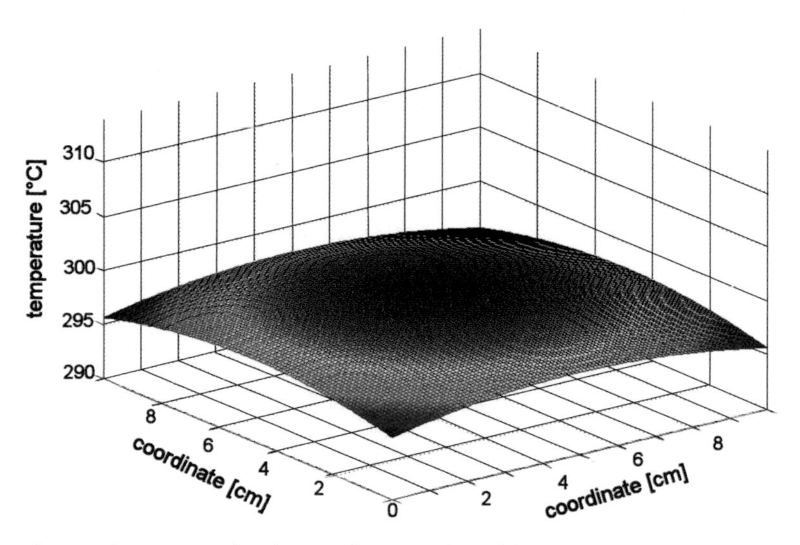

Fig. 7.3 Temperature distribution of a reactor heated by an air pad.

7.5 Calculation of Catalytic Activities

The task was to obtain catalytic activities directly from the IR-image of the active spots on a library under reaction conditions (Fig. 7.1) by numerical integration. Because of problems associated with heat measurements by IRT, as outlined above, it was not attempted to provide absolute heat data. For primary screening purposes the heat measurements should only assist in a ranking of relative activities on an individual library. Therefore, it seemed sufficient to normalize all heat assignments by the hottest spot on the library (100% peak). By such an approach all catalysts on the library are assigned %-values relative to the hottest spot. Comparison between libraries is simplified, if an identical reference material has been placed on both libraries.

All catalysts were analysed simultaneously under identical reaction conditions, which should help to cancel most of the above-mentioned potential error sources.

Reaction temperatures in our reactors can vary from room temperature to about 500 °C, thus covering a huge range of infrared-emission intensity.

As a test-reaction for developing of the integration algorithm the oxidation of CO with air was chosen (free of side- or sequential reactions). A library with several hundred potential catalysts was prepared, placed in the reactor and measured for heats of reaction of CO oxidation by calibrated and corrected IRT (ECIRT) at 50 to 200 °C. As has already been observed in this and other reactions, the stability and homogeneity of the temperature background of the library plate decreases with increasing temperature resolution (background is not constant across the library and undergoes fluctuations). This was not the effect of a temperature gradient and was attributed to fluctuations during calibration or during the measurements. Fig. 7.4 shows a schematic projection of a hot spot on a constant background, whose integration would result in a number associated with total heat change detected. Taking into account background fluctuations or inhomogeneous background the integrating algorithm must be capable of automatic background recognition in the direct neighborhood of the spot to be integrated. A second problem is often a lack of symmetry in the temperature spots. Therefore integration cannot be analytical by taking advantage of the hottest central pixel of a hot spot. For a reliable quantification only the numerical inte-

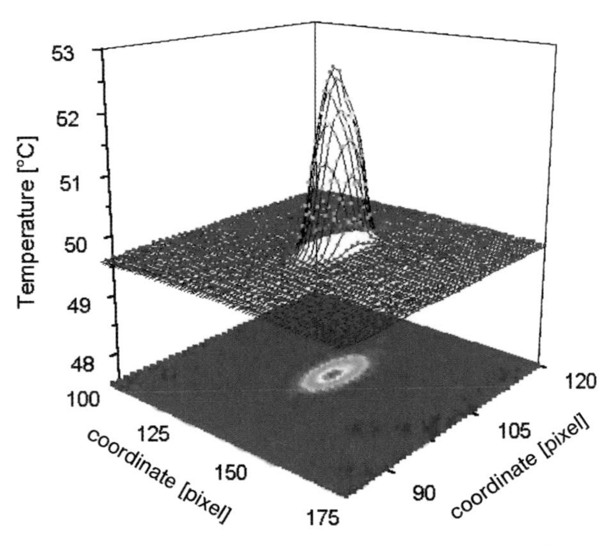

Fig. 7.4 Schematic projection of a hot spot on a constant background.

gration of the peak volume delivers sufficient results. An algorithm has been developed that starts at an arbitrary point of the hot spot. From this point all gradients to the neighboring points are calculated. The point is shifted in the direction of the most positive gradient until all gradients are zero or negative (peak centre). From this new starting point the peak volume is calculated by going outside from the centre stepwise. If the volume increase by another step drops below an arbitrary chosen limit the algorithm terminates, because the background is reached. By this algorithm relative quantification can be carried out readily. Either the hottest spot or the spot of a reference material is set to 100%. All other data are then calculated relative to this reference.

By comparing the data with the data obtained from identical catalysts, that were prepared and measured conventionally, good agreement is often found, with an error of around 5–10%. Such an error may be high for conventional measurements, but it is tolerable for the evaluation of screening data. The good agreement also confirms that quantitative evaluation of catalytic activity from ECIRT is a viable and good method for primary screening of catalysts.

7.6
Applications

To explore the limits and the potential of this algorithm several studies have been carried out. In a first experiment materials with a different content (0–33 mol%) of RuO_2 as active component in TiO_2 were synthesized by a sol-gel procedure [10]. The catalytic activity of RuO_2 for CO-oxidation has been discovered by Kisch et al. [11]. The materials were filled in the library well positions and ECIRT images were recorded under reaction conditions. Fig. 7.5 documents the activity increase with increasing temperature. The temperature rise caused by the oxidation of CO was measured at a reactor temperature of 50 °C (Fig. 7.8). In the range of 15–35% RuO_2 content the heats of the catalysts were integrated and plotted versus the Ru-content (Fig. 7.5). An almost perfect linear dependency of heat production from Ru-content not only documents a clear first order of the reaction in Ru, it also documents a rather sensitive and reliable heat determination by ECIRT, which correlates very well with catalytic activity.

In the range of 0–15% active component no reaction was detected. This is readily explained by the reaction kinetics of steady-state reaction conditions, where the heat transport by the gas flow is apparently larger than the heat produced by the catalysts (no ignition). Only with higher Ru-content the heat production is higher than the heat removal, leading to ignition. This hypothesis is confirmed qualitatively by the two ECIRT images shown in Fig. 7.7.

At the relatively low temperature of 50 °C the two catalysts with the highest Ru-loading were highly active, three more catalysts showed low activity, while the remaining five catalysts were inactive. At 200 °C five catalysts were highly active and two more catalysts showed a lower activity, while only the three catalysts with lowest Ru-content did not produce enough heat to ignite the reaction.

In Fig. 7.6 the plots of integrated heat against Ru-content of the same libraries at 50 °C and at 150 °C show ignition–extinction curves typical of steady-state exothermic reactions.

The required Ru-content for reaction ignition shifts to lower values with increasing temperature, in agreement with conventional knowledge. The obvious advantage of this method is that the measurements are carried out in a single experiment per temperature and all catalysts are examined under the same reaction conditions.

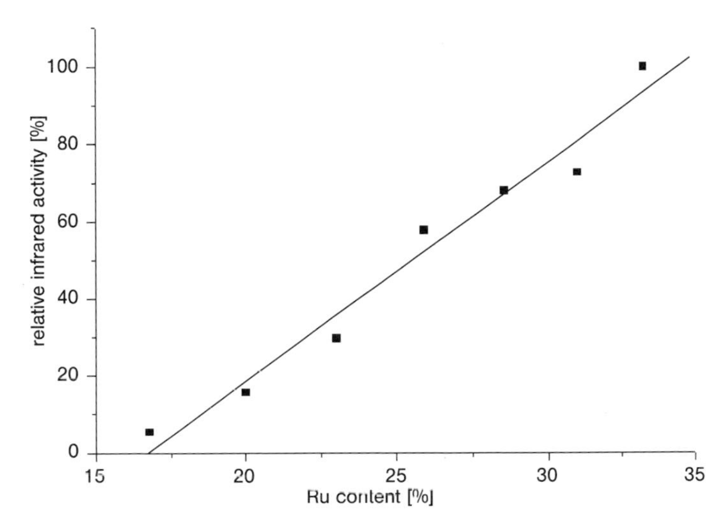

Fig. 7.5 Normalized heat formation of the same catalysts plotted against Ru-content.

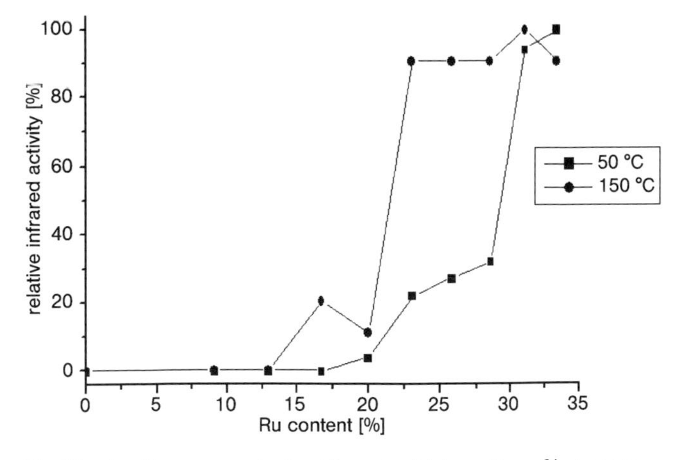

Fig. 7.6 Ignition curves obtained from heat integration of hot spots from the library shown in Fig. 7.7 (at 50 and 150 °C).

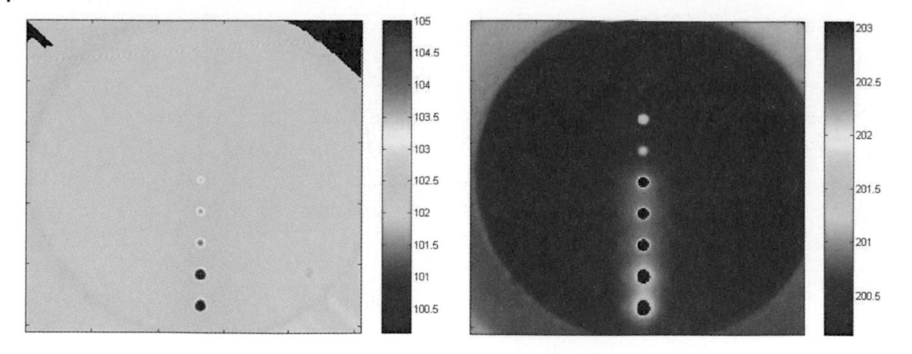

Fig. 7.7 ECIRT images of the same $(RuO_2)_xTiO_2$ catalyst library under CO oxidation conditions at 50 °C (left) and 200 °C (right).

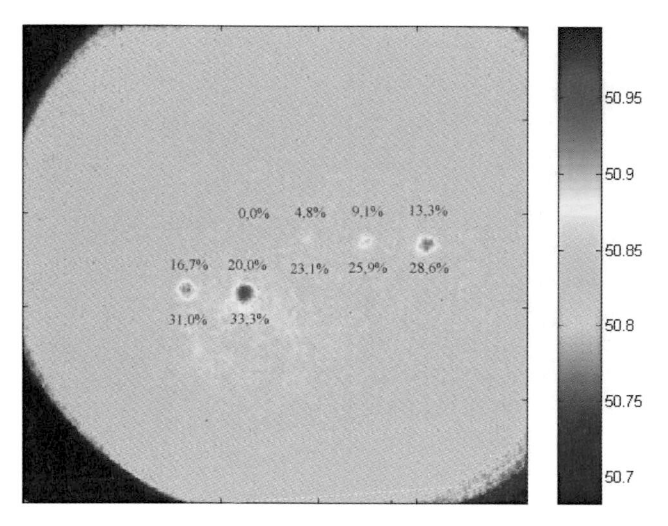

Fig. 7.8 ECIRT image of the heat of reaction of $(RuO_2)_xTiO_2$-catalyzed CO oxidation at 50 °C.

As stated in other papers, catalysts optimisation can be accelerated by application of ECIRT as a quantitative screening tool [12].

7.7 Experimental

Measurements were carried out with an infrared video camera (PtSi-crystal detector with 256×256 pixel) from AIM (Aegäis). This detector is cooled by a Stirling cooler to 75 K; it is sensitive to IR radiation in the wavelength range of 3–5 μm. The image was focused by Ge-lenses with a focus of 28 mm.

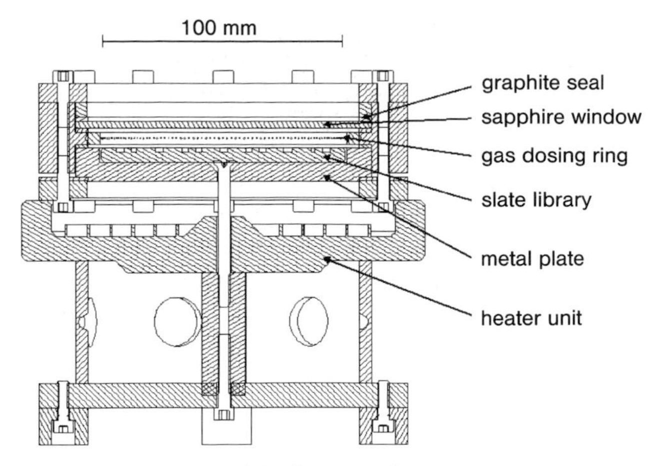

Fig. 7.9 Cross section of the thermography reactor.

Catalysts libraries were made from slate (thickness 6 mm, diameter 99 mm, 64, 207 or 763 wells, 2 or 3.5 mm in diameter, 2 mm deep). The catalysts were either positioned as powders or synthesized and calcined in the wells. The reactor (cross section shown in Fig. 7.9) was made of brass; it housed the library. The reactor bottom was positioned on top of a heating plate (commercial heating plate from a Ceran top stove). Above the library a gas inlet and outlet ring was positioned, which provided a homogeneous distribution of the gas flow over the area of the library by a large number of small drill holes for gas inlet and gas outlet. On top of the reactor, above the library, a large sapphire window was positioned for IR-imaging of the library surface. Heating was controlled by a steady-state controller from Juno

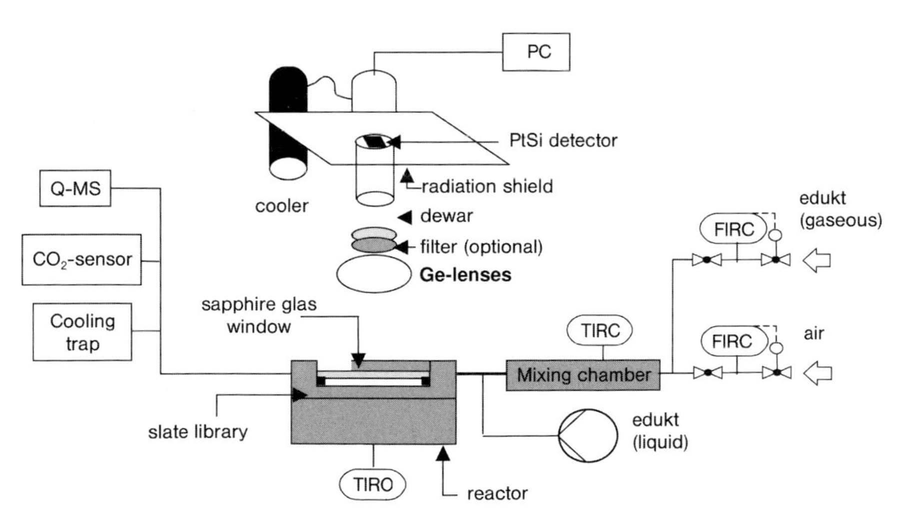

Fig. 7.10 Experimental setup for thermography measurements.

model dtron04 and a thermocouple type Pt100 positioned at the bottom of the library substrate. The gas flow was controlled by mass flow controllers (Fig. 7.10).

7.8 Conclusions

By integration of the volume of the temperature peaks over the associated library area it has been shown that, under carefully chosen reaction conditions, temperature images obtained by a commercial IR-camera can be used to estimate relative heats of reactions of materials on libraries, which correlate with relative catalytic activity. This direct relationship is very valuable for a direct ranking of catalytic activity of potential catalysts on libraries in a primary screening approach. The relative activities obtained are useful for a selection of hits on large libraries, but are by no means reliable data. It is essential that the catalytic activity of individual materials is confirmed by conventional measurements. Potential error sources of this method have been outlined and should be kept in mind by the user. The advantage of the method is that it is a parallel method, which simultaneously determines the catalytic activity of all catalysts on a library with high time resolution, which in addition to relative activity determination allows one to follow changes in catalytic activity with time simultaneously for all materials on the library. For practical applications and internal standards, it is advisable to always have empty wells in addition to those filled with a well-characterized reference catalyst present on each library. The temperature sensitivity of the method presented allows its application in many other areas of property examinations on libraries, such as heats of adsorption or absorption or emissivity changes upon surface reactions as an indicator of surface interaction with gas molecules.

7.9 Acknowledgments

We thank H.W. Schmidt and R. Richter for the construction of the reactors, A. Holzwarth for initial discussions, and the Fonds der Chemischen Industrie for support.

7.10 References

1 Senkan, S., Combinatorial heterogeneous catalysis – a new path in an old field. *Angew. Chem., Int. Ed.*, 40 (2001) 312.

2 Urschey, J., et al., *Solid State Sci.*, 5 (2003) 909.

3 Hoffmann, C., H.-W. Schmidt, F. Schüth, A multipurpose parallelized 49-channel reactor for the screening of catalysts: methane oxidation as the example reaction. *J. Catal.*, 198 (2001) 348.

4 Moates, F.C., et al., Infrared thermographic screening of combinatorial libraries of heterogeneous catalysts. *Ind. Eng. Chem. Res.*, 35 (1996) 4801.

5 Holzwarth, A., H.-W. Schmidt, W.F. Maier, Detection of catalytic activity in combinatorial libraries of heterogeneous catalysts by ir thermography. *Angew. Chem., Int. Ed.*, 37(19) (1998) 2644.

6 Reetz, M.T., et al., A method for high-throughput screening of enantioselective catalysts. *Angew. Chem., Int. Ed.*, 38(12) (1999) 1758.

7 Holzwarth, A., W.F. Maier, Catalytic phenomena in combinatorial libraries of heterogeneous catalysts. *Platinum Metals Rev.*, 44(1) (2000) 16.

8 Cawse, J.N., Experimental strategies for combinatorial and high-throughput materials. *Acc. Chem. Res.*, 34(3) (2001) 213.

9 Karstädt, D., et al., Sehen im infrarot – grundlagen und anwendungen der thermographie. *Physik Z.*, 29(1) (1998) 6.

10 Kirsten, G., PhD thesis, University of the Saarland, Saarbruecken 2003.

11 Zang, L., H. Kisch, Katalytische luftoxidation von kohlenmonoxid bei raumtemperatur durch rutheniumdioxid-hydrat. *Angew. Chem.*, 112(21) (2000) 4075–4076.

12 Kirsten, G., W.F. Maier, Strategies for the discovery of new catalysts with combinatorial chemistry. *Appl. Surf. Sci.* 223 (2004) 87.

8
Gas Sensor Technology for High-Throughput Screening in Catalysis

Yusuke Yamada and Tetsuhiko Kobayashi

8.1
Introduction

High-throughput screening (HTS) of heterogeneous catalysts is a fundamental technology for combinatorial catalysis [1–3]. Much effort has been devoted to developing HTS technologies. Moates [4] and Maier [5] (see also Chapter 7) have developed IR thermography, which detects the heat of catalytic reaction independently. The method screens many catalysts rapidly because the analysis evaluates two-dimensionally arrayed catalysts, though it provides no information on catalytic products. Two-dimensionally arrayed catalysts have also been evaluated by fluorescence spectroscopy, developed by Yeung [6, 7], and IR imaging spectroscopy, developed by Lauterbach [8, 9]. REMPI can also be used to evaluate two-dimensionally arrayed catalysts [10, 11]. In this method, a UV or visible laser selectively excites targeted molecules in multiple effluents. Of the one-by-one methods, mass spectrometry has often been utilized due to its wide dynamic range and rapid response [12–16]. A commercially available high-performance micro-gas chromatograph provides detailed information on the products [17]. The quality of quantification and separation of products is almost same as that obtained by conventional gas chromatograph. A scanning electron microscopy has also been used to evaluate catalysts that produce structured carbon materials [18]. Currently, we can select a suitable HTS method for each reaction system.

Gas sensor technology also supports the HTS of heterogeneous catalysts. The advantages of gas sensors are their small-size, rapid response, easy parallelization, and they are less expensive. Various kinds of gas sensors have been developed for the product analysis on catalytic reactions. Semiconductor-type gas sensors are useful in detecting odor compounds. Potentiometric and combustible gas sensors have been used for CO detection [19]. IR gas-sensors have been employed to detect CO, CO_2, ethylene, alcohol, aldehydes etc. [20]. Quartz crystal microbalance sensors have been used for the quantification of cresol and benzoquinone [21]. Dye compounds, which change color in the presence of a specific chemical, have been employed to detect hydrogen evolution, MeOH formation and decomposition [22], aniline [23], and NOx [24]. How the color change is detected, i.e., change in visible light absorbance, fluorescence intensity, reflectance change etc., depends on the system em-

High-Throughput Screening in Chemical Catalysis
Edited by A. Hagemeyer, P. Strasser, A. F. Volpe, Jr.

ISBN: 3-527-30814-8

ployed. Here we review recent developments of gas sensor technologies for the high-throughput screening of heterogeneous catalysts classified by catalytic reaction.

8.2
Evaluation of CO Oxidation Catalysis

As CO is highly toxic, and causes respiratory problems, its elimination is quite important. Industrially, CO elimination is required for CO_2 laser application, and the production of pure hydrogen for fuel cells. Catalysis testing of the reaction has been performed, independently, with an IR gas-sensor by Hoffmann et al. and with a combustible CO gas sensor by Yamada et al. [25, 26]. In both groups, CO_2 produced by CO oxidation was quantified by an IR gas-sensor.

8.2.1
With a CO Combustible Gas Sensor

Dilute CO elimination from air has been rapidly evaluated using CO gas sensors [25, 26]. The oxidation of 1 vol% CO in air was carried out under atmospheric pressure from 30 to 150 °C with a ramp rate of 2 °C min^{-1}. The CO concentration of an effluent from the reactor was continuously determined by the CO gas sensor. Fig. 8.1 shows the output signal and CO conversion determined by the CO gas sensor, and the CO conversion determined by FID-GC, for CO oxidation over Rh/TiO_2. This indicates that the CO concentration decreases with increasing temperature, i.e., the CO conversion rate increases at higher temperature. The closed circles in Fig. 8.1 indicate the CO conversion determined by gas chromatography. The conversion curve calculated from the output signals of the CO gas sensor is

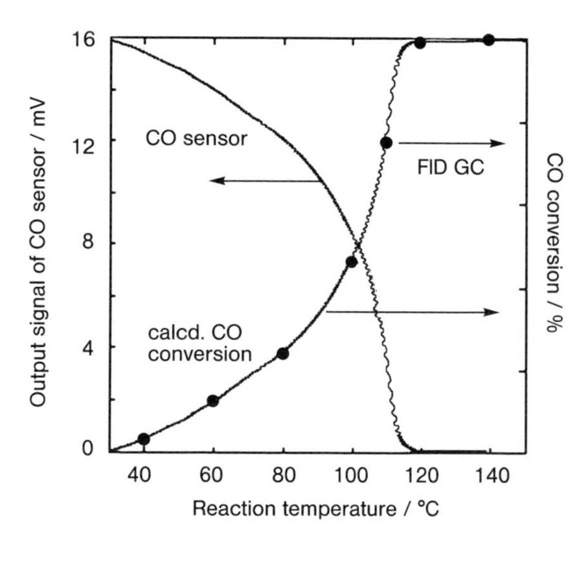

Fig. 8.1 CO concentration in the effluent from Rh/TiO_2 determined by CO-combustible gas sensor. CO conversion calculated from output of the sensor was compared with that obtained by FID-GC (filled circles) (reproduced by permission of Elsevier from [19]).

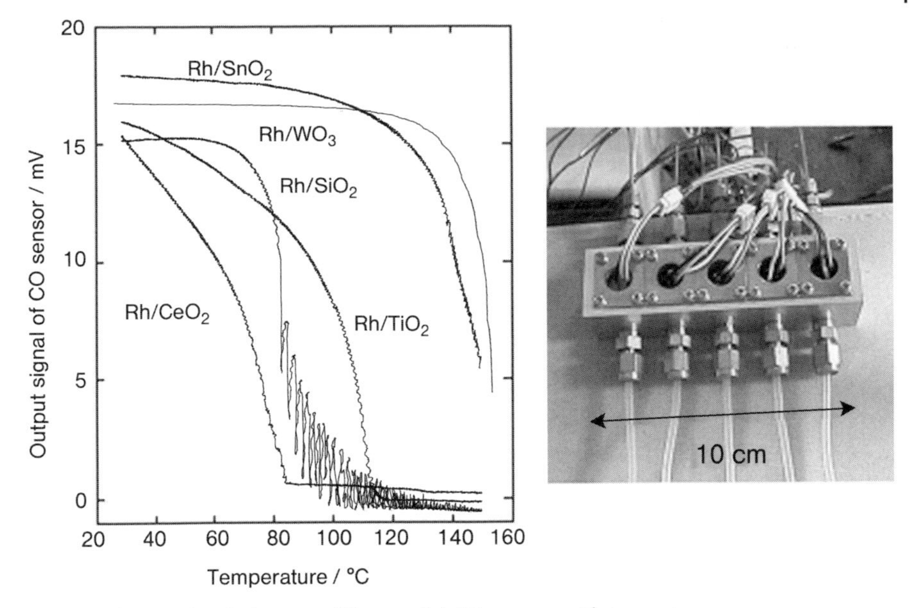

Fig. 8.2 Output signal changes of five parallel CO sensors with increasing catalytic reaction temperature. The catalysts whose effluents were analyzed by CO gas sensors were Rh/SnO_2, Rh/WO_3, Rh/SiO_2, Rh/TiO_2, Rh/CeO_2 (reproduced by permission of Elsevier from [19]).

tracing all the filled circles. This clearly shows that the evaluation with the CO sensor is as same as that with the gas chromatograph.

CO oxidation catalysis of a series of Rh/MOx has been evaluated simultaneously using five CO gas sensors. The evaluation time for each sample was 1/5 of that for a one-by-one method. Reaction tubes were placed together in an oven to control the reaction temperature. Fig. 8.2 shows the output signals from the five CO sensors in the temperature range 30–150 °C. The Rh/CeO_2 catalyst brought about a decrease in the CO concentration at the lowest temperature. The order of reactivity, $Rh/CeO_2 > Rh/SiO_2 > Rh/TiO_2 > Rh/SnO_2 > Rh/WO_3$, is clearly demonstrated in Fig. 8.2. Notably, the gas sensor precisely monitors the oscillation of CO oxidation on Rh/SiO_2. This continuous gas analysis is useful in studying dynamic phenomena. Although the lack of gas selectivity is the drawback of a catalytic combustion-type sensor, gas selectivity is not required for the evaluation during simple combustion. When compared with a conventional single line flow reactor equipped with an automatic gas chromatograph, a higher throughput of more than 20 times was obtained for the five parallel reactors with five sensors. Fig. 8.3 shows the screening results of 72 catalysts designed by the combination of precious metals and metal oxides for CO oxidation. CO conversion at 80 °C is indicated for each catalyst. The combination of a support metal oxide and an additive sometimes induces high catalytic ability; however, this seems to be difficult to understand intuitively.

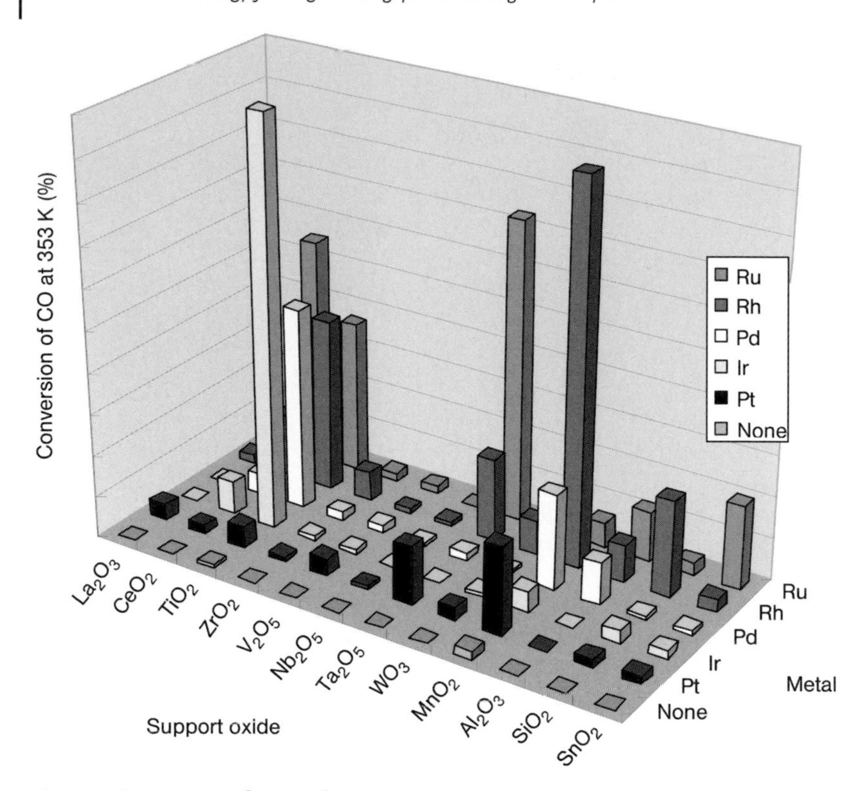

Fig. 8.3 Comparison of CO oxidation activity at 353 K among 12 metal oxides with/without one of five precious metal ions.

8.2.2
With an IR Gas Sensor

Hoffmann et al. reported the use of IR-type CO and CO_2 gas sensors for the evaluation of CO oxidation catalysts [25]. They prepared eight gold-supported catalysts on cobalt oxide with different concentrations of gold (0.4–26 wt%). They checked the effect of catalyst washing after calcination on CO oxidation catalysis. Catalysis was evaluated at 25, 70, 125 and 180 °C with a 16-channel monolith reactor. Washing after calcination was found to activate the Au/Co_3O_4 catalysts for each Au concentration. They also examined the effect of TiO_2 structure on the CO oxidation catalysis of Au/TiO_2, and found that the anatase structure (BET surface area: $180\ m^2\ g^{-1}$) seems to be the best among TiO_2 tested. This result matches the data previously reported [27].

8.3
Evaluation of Selective Oxidation Catalysis

Selective conversion of alkanes or aromatics into corresponding oxygenated compounds is a challenging target for catalysis researchers. Direct conversion of natural gas into oxygenated compounds such as corresponding alcohol or aldehydes remains commercially unavailable. For instance, the design of the catalyst for direct conversion of methane into methanol is still an open problem, although the process is thermodynamically favorable. Combinatorial catalysis should address the problems of this field. The desired products of alcohol and aldehydes have specific odors, while lower alkanes, CO and CO_2 have no smell. Gas sensors developed for odor discrimination have been utilized as HTS for propane-selective oxidation [19]. The desired products of oxygenated compounds were selectively quantified with semiconductor-type gas sensors. Both CO and CO_2 as undesired products were detected by a potentiometric CO gas sensor with an active carbon filter and an ND-IR CO_2 sensor. The sensor system screens product distributions within 1 min. For HTS of ethane oxidation catalysts, Ueda et al. have reported the use of photoacoustic spectroscopy (PAS), which is a kind of IR-using gas sensor [20]. The PAS sensor quantified CO, CO_2, ethylene, acetaldehyde, ethanol and acetic acid within 1 min. Also, the selective oxidation of aromatics is important to reduce waste compounds from the chemical industry. For example, toluene oxidation to cresol in the gaseous phase is desired to reduce harmful waste produced during solution processes. To detect oxygenated aromatics, Potyrailo and May have reported the potential of acoustic-wave sensors for oxygenated aromatics [21]. They quantified cresol or benzoquinone in various solvents by acoustic-wave sensors in 2 min.

8.3.1
Propane Oxidation

Products obtained by propane-selective oxidation have been analyzed by gas sensor systems [19, 26]. Usually, several or multiple kinds of compounds are produced during the selective oxidation of propane. The formation of CO, CO_2, aldehydes such as acrolein, and ketone were observed over iron-silica catalysts [28, 29]. During the initial stage of catalyst investigation, the conversion of propane and the selectivity toward useful oxygenate products as chemical resources are of interest. Semiconductor-type gas sensors selective toward the oxygenate were employed to estimate the yield of oxygenate products, with a combination of the potentiometric CO sensor and the ND-IR CO_2 sensor [30].

Fig. 8.4 (a) shows the response of the oxygenate sensor-1 (SnO_2 sensitized with TiO_2) towards an alcohol (1-propanol), aldehyde (propanal), ketone (acetone), carbon monoxide (CO) and propane. The sensor is sensitive to the alcohol, aldehyde and ketone but not to CO and propane. Conversely, oxygenate sensor 2 (SnO_2 sensitized with 13 wt% SiO_2/Al_2O_3) is less sensitive to the alcohol than aldehyde (Fig. 8.4 b). Alcohol formation can thus be estimated from a comparison of the output signals of oxygenate sensors 1 and 2.

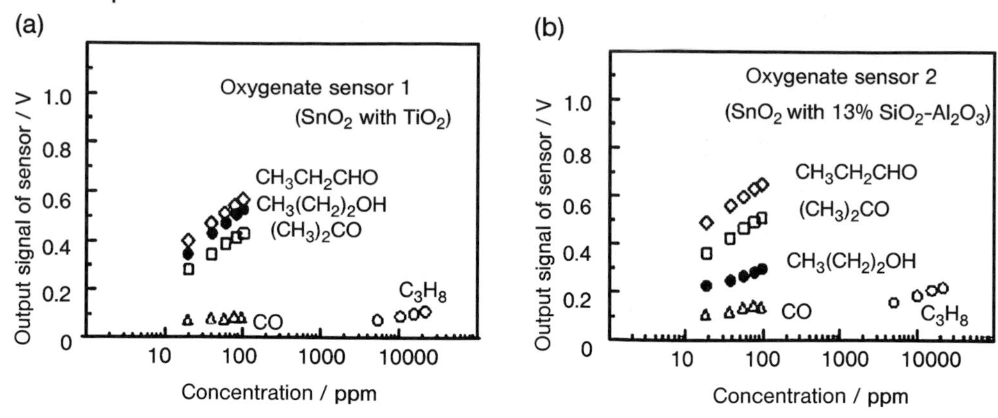

Fig. 8.4 Response of oxygenate sensors toward C_3-oxygenated organic compounds, CO and C_3H_8. Oxygenate sensors 1 (a) and 2 (b) (reproduced by permission of Elsevier from [19]).

The concentration of propane is nearly 100× higher than those of CO and other oxygenates during the reaction as the conversion of propane is kept at <15% for the selective oxidation. The sensors used were selective enough to confirm oxygenate formation in the reaction gas. The effects of produced water, hydrogen and olefin on the output signals of the oxygenate sensors 1 and 2 are negligible under the reaction conditions employed. The potentiometric CO sensor requires an active carbon filter as an absorber for propane and oxygenates, which disturbs the precise determination of CO concentration. The selectivity of the NDIR CO_2 sensor is high enough for this purpose.

Fig. 8.5 shows the change in output signals from oxygenate sensors 1 and 2, the potentiometric CO sensor and the ND-IR CO_2 sensor after the introduction of the reaction gas into the gas sensor system. A 90%-response requires 20, 90 and

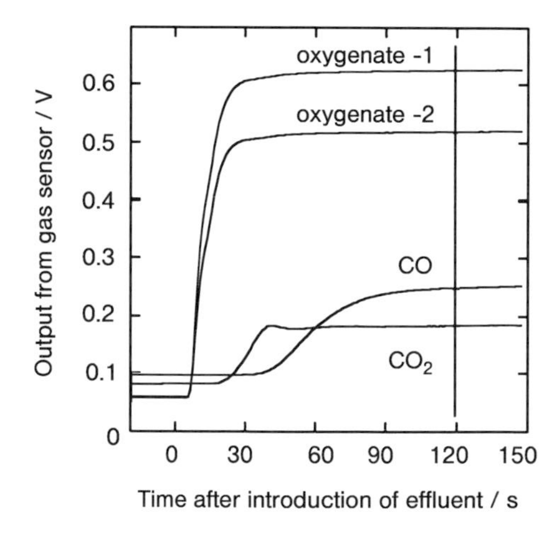

Fig. 8.5 Time courses of output signals of semiconductor-type oxygenate sensors, potentiometric CO sensor and ND-IR CO_2 sensors. The delay of CO gas sensor was due to the resistance of active carbon filter which removes oxygenate compounds (reproduced by permission of Elsevier from [19]).

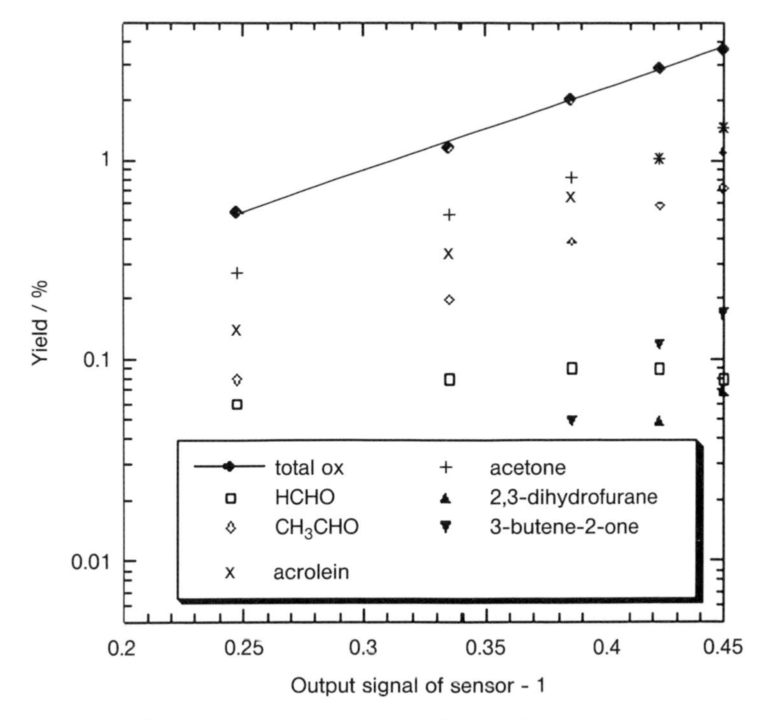

Fig. 8.6 Calibration plot of oxygenate sensor-1 for total oxygenate compounds produced by catalytic propane oxidation; the yield of each oxygenate was determined with a conventional FID gas chromatograph (reproduced by permission of Elsevier from [19]).

40 s for the oxygenate, CO and CO_2 sensors, respectively. The slower response of the CO sensor is due to the presence of the active carbon filter. After 120 s, the output signal of each gas sensor reached 99% for the oxygenate sensors, 97% for the CO sensor and 100% for the CO_2 sensor. The response time of this gas sensor system is less than 2 min, even for a transient change in the products. This is much faster than the 40–50 min needed for precise gas analysis by conventional FID-GC.

The effluent of the propane oxidation over Cs:Fe:SiO_2 (1:0.05:100) [28, 29, 31] at 400 to 500 °C was analyzed by the gas sensor system and also by the FID-GC equipped with a methanator. Fig. 8.6 shows the correlation between the concentration evaluated from the FID-GC and the output signals of the gas sensors. The output signals increased in proportion to the sum of the oxygenate yield on a logarithmic scale. The obtained fitting shown in Fig. 8.6 was to calculate the sum of oxygenate yields on propane oxidation over alkali:Fe:SiO_2 (=1:0.05:100).

Fig. 8.7 shows the product distributions determined by (a) a gas sensor system and (b) gas chromatography for propane oxidation over alkali:Fe:SiO_2 (=1:0.05:100). Since little alcohol was produced, there was no large difference between the signals from oxygenate sensors 1 and 2. When we compared the oxyge-

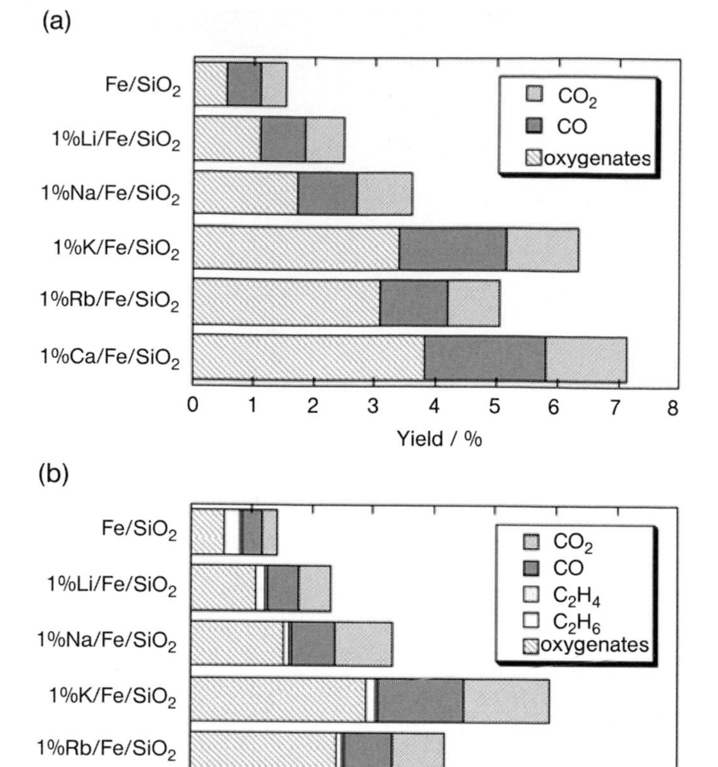

Fig. 8.7 Comparisons of screening result by (a) gas sensor system with (b) conventional FID-GC for the propane oxidation effluents over Fe/SiO_2 and alkali-modified Fe/SiO_2 (reproduced by permission of Elsevier from [19]).

nate yield estimated from the gas sensor with that from FID-GC, the largest difference was found in $Rb/Fe/SiO_2$, with a 2.4% yield by the sensor and 3.0% yield by FID-GC. Although the difference is not small, the evaluation results obtained with gas sensor system are as same as those with FID-GC, in the order of good catalysis. Both results indicated that the addition of potassium or cesium induces a high propane conversion and a high oxygenate yield. The gas sensor system has the advantage of rapid evaluation of the selective oxidation catalysts.

8.3.2
Ethane Oxidation

Ueda et al. reported the quantification of CO, CO_2, ethylene, acetaldehyde, ethanol and acetic acid produced by selective ethane oxidation with photoacoustic spectroscopy (PAS) [20]. PAS is a kind of gas sensor using IR light. The detection mechanism is as follows: a molecule excited with IR light increases in volume and then reduces for a while if the light is turned off. Intermittent irradiation induces a continuous volume change of the molecule, which can be detected as a pressure change by a microphone in terms of sound volume change. The PAS sensor used by Ueda has five filters, corresponding to the quantification of CO, CO_2, ethylene, acetaldehyde, separately, and the yield sum of ethanol and acetic acid.

A catalyst library of 112 catalysts was screened. Seven metal oxides, Al_2O_3, SiO_2, TiO_2, ZnO, Ga_2O_3, ZrO_2 and La_2O_3 were selected as supports and 16 metal or metal oxides, Mn, Fe, Co, Ni, Ru, Rh, Pd, Ir, Pt, Cu, Zn, Ag, Au, Ga, In and Sn, were selected as additive to support oxides. An additive metal or metal oxide was

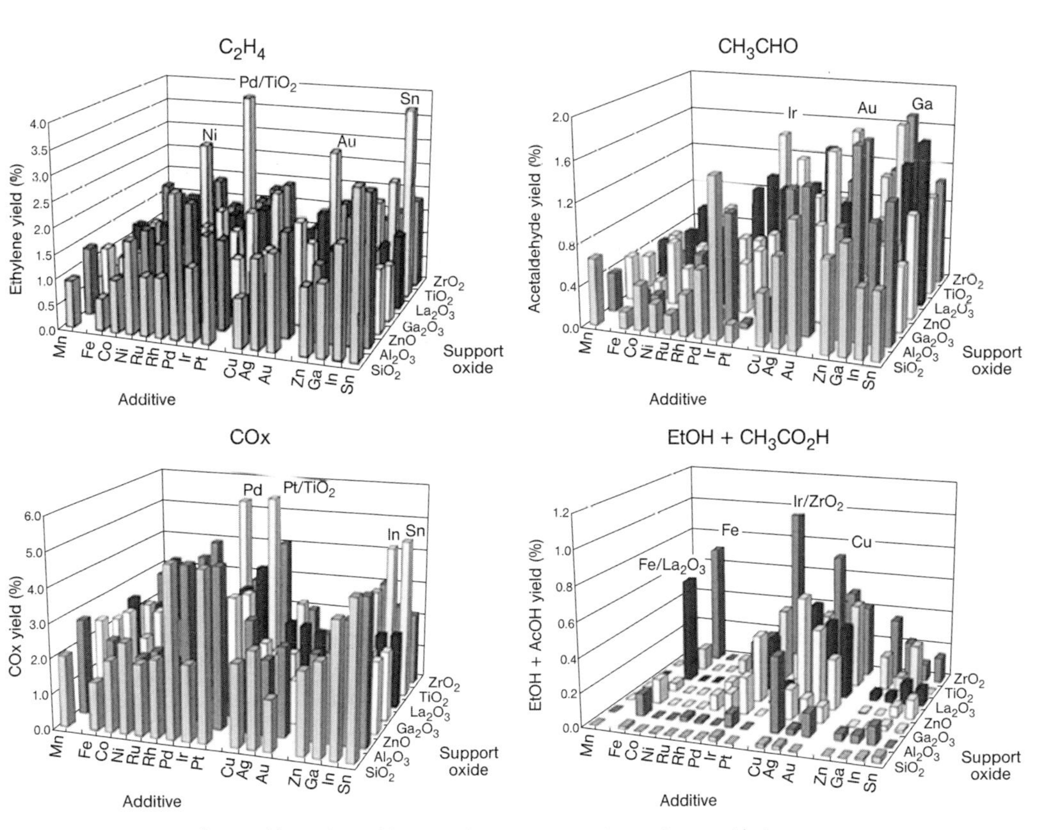

Fig. 8.8 Product yields evaluated by PAS detector on catalytic ethane oxidation over NO_2-treated catalysts. NO_2 gas was flowed onto each catalyst before reaction to produce active site (reproduced by permission of Elsevier from [20]).

loaded on a support by impregnation of corresponding precursor. An obtained catalyst was exposed to 2% NO_2 at 300 °C for 30 min. TPD (temperature-programmed desorption) measurement of Fe/TiO_2 indicates that the NO_2-bound Fe species are stable up to 600 °C. Adsorbed NO_2 on the catalyst surface appears to provide an active site. A gas mixture of 30% C_2H_6 and 15% O_2, and 55% He, was passed through a catalyst (200 mg) at 50 mL min^{-1}. Fig. 8.8 shows the yields of each product of the reaction at 450 °C. Formations of COx, ethylene and acetaldehyde were detected for all tested catalysts. Conversely, the formation of acetic acid and ethanol was observed with only a limited number of catalysts. The yield sum of ethanol and acetic acid is highly dependent on additives, and is less influenced by support metal oxide.

8.3.3 Aromatic Oxygenates

Sensors using quartz crystal are very sensitive and can detect samples of the order of µg. Usually, an organic thin film is pasted on quartz surface since the crystal surface hardly absorbs any chemical species. The organic thin film provides the potential to detect various kinds of volatilities with high selectivity and sensitivity. The principle of the gas sensor is based on Eq. (1) [32]. The quartz oscillator has a specific resonance frequency with an oscillating circuit. Its frequency is decreased by the absorption of volatilities on the quartz surface due to the increase in mass. The frequency shift caused by exposure to a volatile depends on the amount adsorbed. With a 9 MHz quartz oscillator, the frequency is decreased by 400 Hz upon adsorption of 1 µg of a compound. A resonance oscillator with a higher resonance frequency can detect smaller amounts.

$$\Delta F = k\frac{F^2 \times \Delta W}{A} \tag{1}$$

where

F = resonance frequency
ΔF = the decrease in frequency
ΔW = the increase in mass upon adsorption
A = surface area of the adsorbent

For an organic adsorbent with long-term stability and high selectivity in volatilities, quartz gas sensors would be the most promising means of analysis for catalyst evaluation because of their high sensitivity and small size.

Potyrailo and May used the acoustic-wave sensor to quantify cresol and benzoquinone [21]. They tested sensors for the quantification of cresol and benzoquinone in mixtures with multivariate regression analysis tools. Only 2 mL of solution was used for analysis, which included cresol and/or benzoquinone on the order of µg. After the sensor had been calibrated with a standard library of pure cresol or benzoquinone solution, 19 model libraries, including mixtures of cresol and benzoquinone, were analyzed (Fig. 8.9). A linear correlation occurs between actual

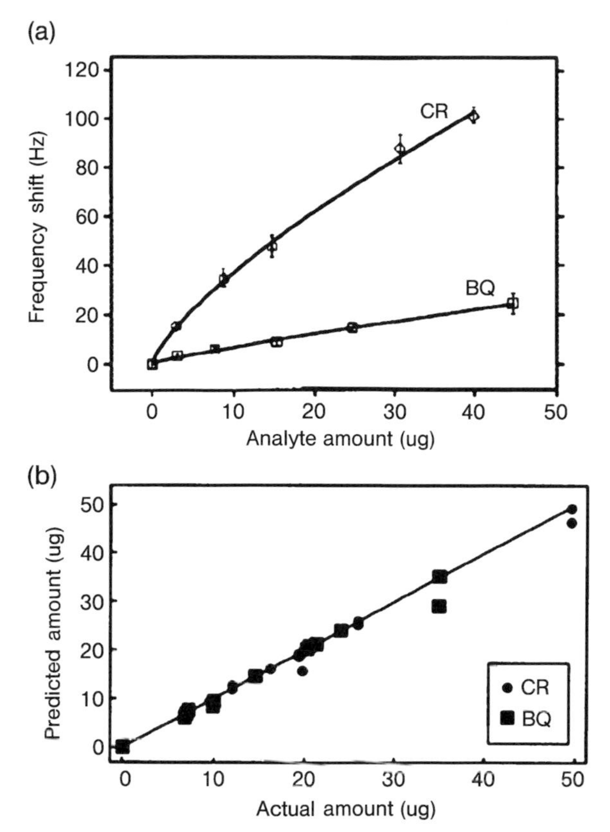

Fig. 8.9 (a) Calibration plot for determination of cresol (CR) and benzoquinone (BQ) with TSM sensor and (b) multivariate validation plot between actual and predicted concentrations of cresol and benzoquinone in complex mixtures (reproduced by permission of AIP from [21]).

and predicted amounts. The highly selective and sensitive detection of the mixture indicates that the sensor is promising for the screening of real catalyst libraries.

8.4 Evaluation of Oxidative Dehydrogenation Catalysis

Oxidative dehydrogenation (ODH) is an important process for converting ethane or propane into more valuable ethylene or propylene. Ethylene has a specific IR absorption band around 950 cm^{-1}, which has been utilized by two research groups, using IR-based gas sensors, in the HTS of ethane ODH. Cong et al. at Symyx have used photothermal deflection [12, 13] and Johann et al. at the Max Planck Institute used a PAS sensor [24]. Johann et al. reported that position-sensitive quantification was possible by using a fast-response microphone. They succeeded in the parallel screening of eight catalysts. Semiconductor-type gas sensors were also tested for ethylene ODH because ethylene has specific odors, while educts, CO and CO_2, which are by-products, have no smell. A semiconductor-type gas sensor developed for odor discrimination has been used to quantify successfully ethylene in the presence of ethane [19].

8.4.1 With a Semiconductor-type Gas Sensor

Ethylene quantification has been performed with a semiconductor gas sensor in the presence of a large excess ethane [19]. Fig. 8.10 (a) shows the correlation between the output signals from the olefin-selective sensor and the concentration of C_2H_6, C_2H_4 and CO in N_2. An output signal of 1–2 V was obtained in the presence of 1000 to 5000 ppm C_2H_4. An output signal of 0.5–1.2 V was obtained for 1–5 vol% C_2H_6. The presence of CO did not affect the output signal. Although the sensor is more sensitive towards C_2H_4 than C_2H_6, the effect of C_2H_6 on the output signal is not negligible at lower conversions of C_2H_6. Fig. 8.10 (b) shows the relationship between the output signal of the sensor and the concentration of C_2H_4 in C_2H_6. The total flow rate of C_2H_4 and C_2H_6 was maintained at 42 mL min^{-1}. The gases were diluted with 958 mL min^{-1} of dry air before introduction into the sensor system. The output signal increased in proportion to the concentration of C_2H_4 with a linear correlation. This indicates that the gas sensor is useful for both the determination of C_2H_4 concentration and the evaluation of ODH catalysis. The concentration of C_2H_4 in a reaction effluent was evaluated by the correlation curve.

Ethane ODH catalysis by mixed oxides of nickel and iron has been evaluated by both the gas sensor and an FID gas chromatograph. Fig. 8.11 (a) shows the C_2H_4

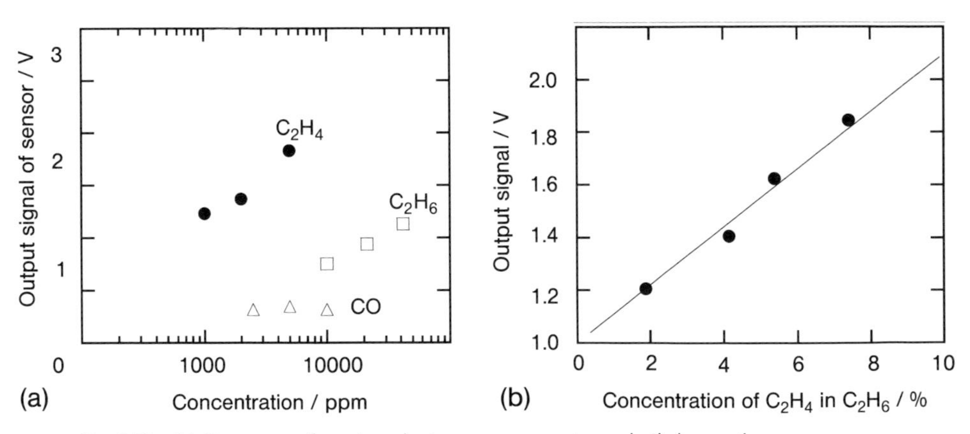

Fig. 8.10 (a) Response of semiconductor gas sensors toward ethylene, ethane and CO and (b) calibration plot for determination of ethyl ene in ethane (reproduced by permission of Elsevier from [19]).

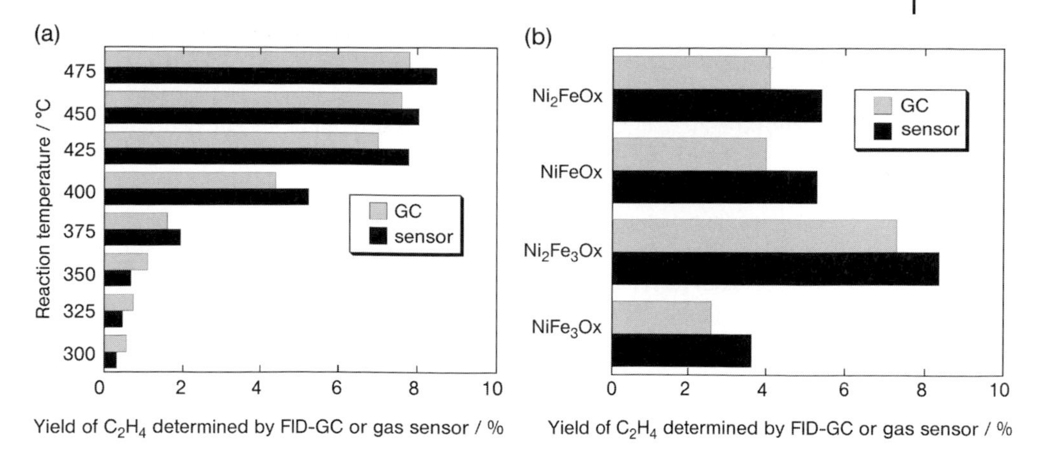

Fig. 8.11 Comparisons of ethylene concentration in effluent by semiconductor gas sensor with that by conventional gas chromatograph. (a) Effluents obtained by changing reaction temperature from 300 to 475 °C and (b) effluents from Ni-Fe-mixed oxide with different composition at 450 °C (reproduced by permission of Elsevier from [19]).

yield determined by the gas sensor compared with that by FID-GC on ODH over NiFeOx in the temperature range 300–500 °C. The output signals of the gas sensor were reproducible despite the higher production of water vapor at high conversion. Although the gas sensor overestimated the C_2H_4 yields, the difference is less than 0.8 and the tendency, i.e., high yield of C_2H_4 at high temperature, was correctly found. The gas sensor was useful in investigating the temperature-dependent catalysis of NiFeOx. Fig. 8.11 (b) shows the evaluation of the mixed oxide of nickel and iron with different Ni/Fe ratios. The reaction was performed at 400 °C. The ratio of Ni/Fe was changed, ranging from 1/3 to 2/1: $NiFe_3Ox$, $NiFe_2Ox$, NiFeOx and Ni_2FeOx. The order of better catalysts estimated from the output signal of the gas sensor is the same as that estimated from gas chromatography. These results indicate that the gas sensor would be useful in optimizing the ODH catalysis of the mixed oxides of nickel and iron in terms of their composition and reaction temperature.

The sensor used here, SnO_2 covered with 28 wt% SiO_2/Al_2O_3, is sensitive towards hydrogen and aldehydes as well as ethylene. Since neither hydrogen nor aldehyde is produced under the present reaction conditions, reliable yields of ethylene were obtained from the sensor output. When aldehydes are produced in the reaction, an adsorbent such as an active carbon filter is required in front of the sensor to remove the aldehydes from the analyzed gas. A hydrogen sensor would also be necessary to compensate the output signal from the olefin-selective sensor. In the next section, a combination of gas sensors is employed to evaluate the reaction selectivity during propane oxidation.

8.4.2
With a PAS Detector

Fig. 8.12 shows the setup of a laser beam and microphone for ethylene quantification with a PAS detector [24]. In the array, the eight reaction tubes are arranged linearly. A pulsed laser is passed through each effluent from the reactors to excite ethylene molecules. The pulsed laser used emitted at 943–950 cm^{-1} (where ethylene has a strong absorption) – a 10 or 100 Hz modulated 25 W laser with a pulse length of 35 or 25 μs. A microphone with a fast response time and decay was used. The ethylene concentration of each effluent was determined by the volume and response time. The signal from the most distant tube is weak so that the signals were accumulated for 2.5 s. Data presented in the reference are shown in Fig. 8.13. Ethylene concentrations were determined for the effluent from the mixed-oxide catalyst consisting of La, Ba, Pb, Th, Mn, Ni and Cu.

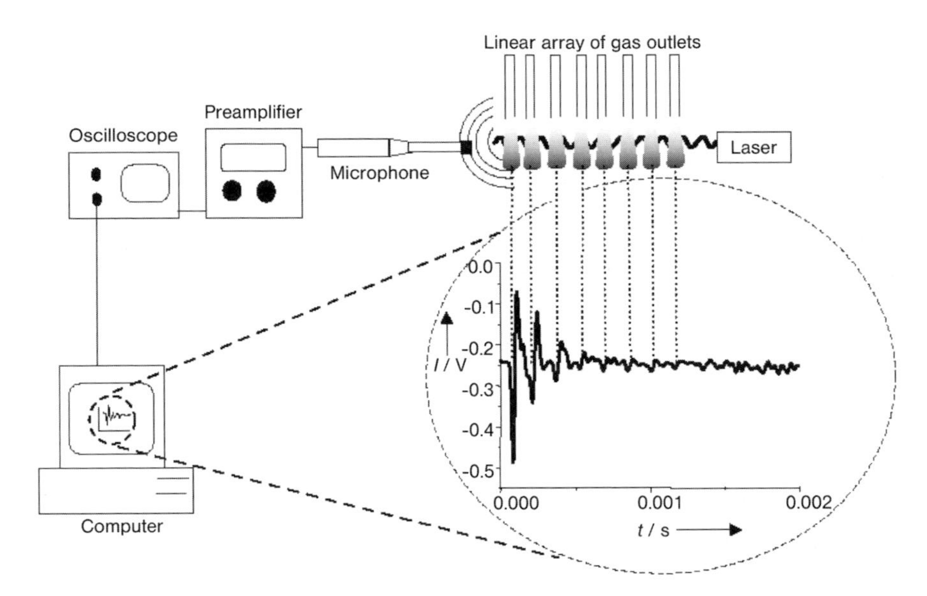

Fig. 8.12 Setup for photoacoustic free-field measurement. The diagram shows a real signal of voltage versus time when 2% ethylene flowed through all channels (reproduced by permission of Wiley-VCH, Weinheim from [24]).

8.4.3
With a PTD Detector

Another type of IR gas-sensor has been reported by Cong et al. for the HTS of ethane ODH [12, 13]. They used a photothermal deflection (PTD) detector with a mass spectrometer because an electron-impact mass spectrometer cannot detect C_2H_4 concentrations at the ppm level in the presence of high concentrations of

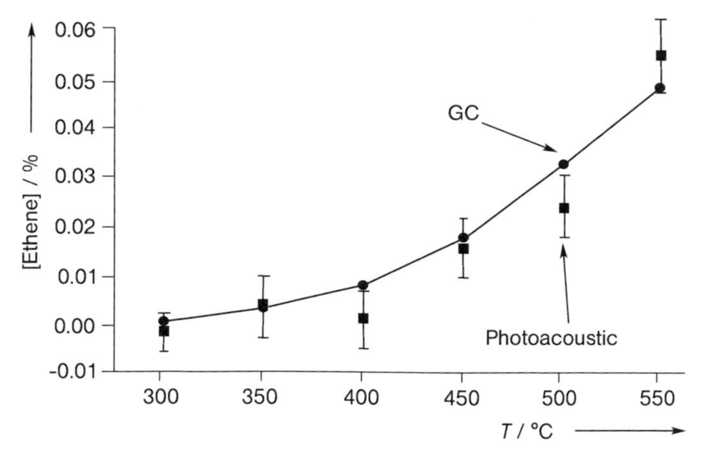

Fig. 8.13 Ethene concentration in the oxidative dehydrogenation of ethane. The concentration estimated by PAS was compared with that by GC (reproduced by permission of Wiley-VCH, Weinheim from [24]).

C_2H_6. The sensitivity of the PTD detector toward C_2H_4 is below 0.1 ppm. The detector can estimate ethylene yields of each reaction within 90 s.

The library, on a quartz substrate, consisted of almost 50 catalysts (ca. 200 μg). Each catalyst was heated by CO_2-laser irradiation to between 300 and 400 °C before reaction. Then, a gas mixture of N_2, C_2H_6 and O_2 (molar ratio 5:4:1) was blown to each catalysts. A library consisting of Mo-V-Nb with 10% compositional increments per matrix element was prepared and screened. The binary catalysts of Mo-V-O show low conversion and ethylene selectivity. The presence of Nb increases both activity and selectivity dramatically. The superior catalysts found by the screening were scaled up and showed similar catalysis.

8.5 Evaluation of Hydrogen Evolution Catalysis

Hydrogen is the promising fuel of the 21st century since the waste produced by burning it is only water, a non-harmful gas with no greenhouse effect. Also, the proton exchange membrane fuel cell (PEMFC) is regarded as a promising power source. At present pure hydrogen without CO must be supplied for PEMFC because CO strongly poisons Pt on the anode catalyst. Hydrogen produced by reforming hydrocarbons always includes a certain amount of CO, which is removed by successive water-gas shift reactions at high and low temperatures and selective oxidation. If we can produce hydrogen from water, the hydrogen can be used directly for the fuel cell. Unfortunately, the electrical decomposition of water is the only realistic process for obtaining hydrogen from water at present. The catalytic decomposition of water by a less energy-consuming process is expected.

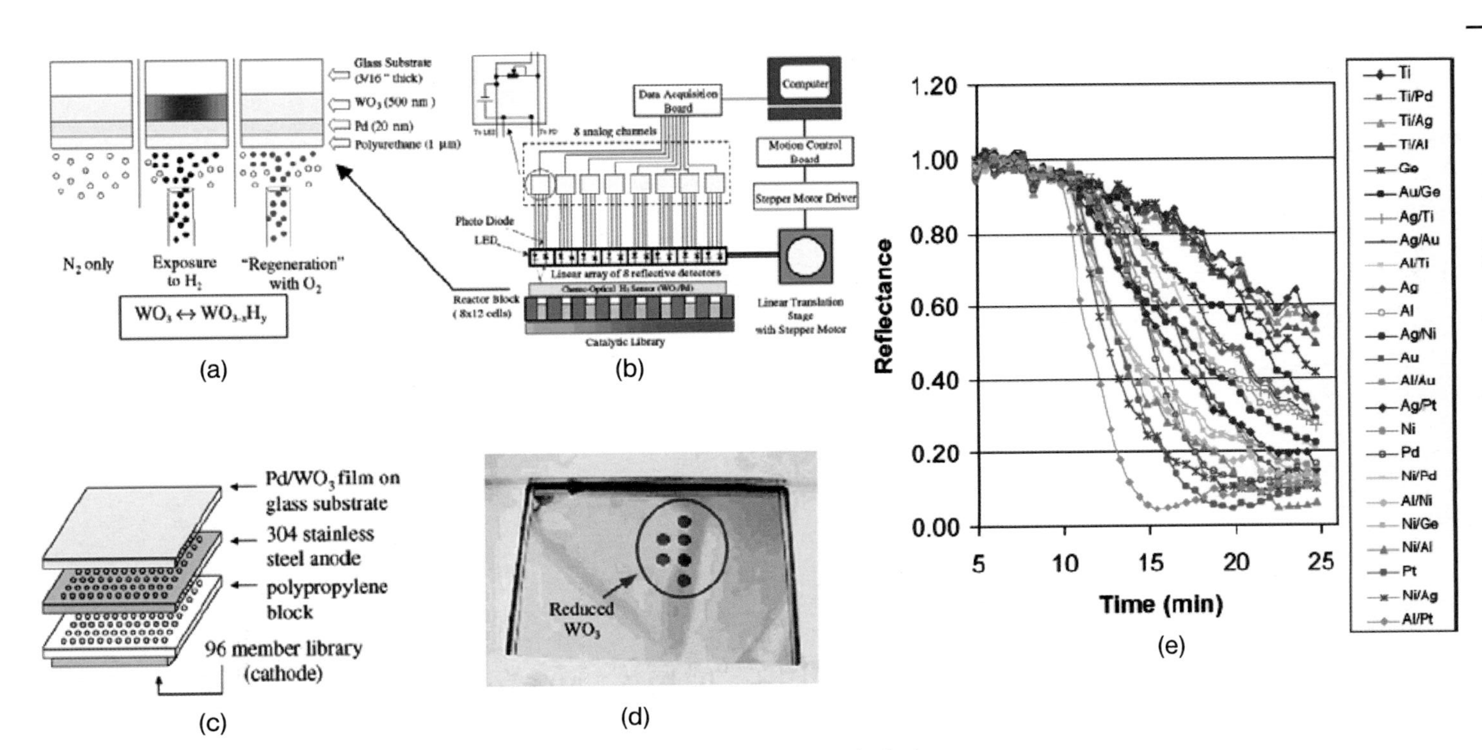

Fig. 8.14 (a) Conceptual drawings of H_2 detection system with color change. (b) HTS rig for hydrogen production composed of reactor parts and detector parts. (c) Details of reactor and sensor parts. (d) Color change after typical reaction. (e) Raw reflectance data taken every 30 s for 96 catalysts (24×4). The sudden decrease of reflectance observed in several catalysts is correlated to the high H_2 production rate (reproduced by permission of American Chemical Society from [33]).

HTS of hydrogen production catalysts by water splitting has been performed with a hydrogen sensor by Jaramillo et al. [33]. The sensor used was a deposited WO_3 film coated with Pd. When hydrogen reaches the Pd surface, it splits atomically. The atomic hydrogen then penetrates into WO_3 to produce $HyWO_3$-x by reduction. The color change of WO_3 by the reduction can be detected as the reflectance change is more sensitive than the absorbance change. The sensor can quantify hydrogen concentrations of 0.1 to 1% with a 935 nm laser.

The tested catalyst library of 24 samples was prepared by electron beam deposition of the corresponding metal sources on a substrate. The catalysts consisted of pure metals of Ti, Pt, Ni, Au, Pd, Al, Ag, Ge or mixed alloy of the metals in the ratio of 2:8. The plate for the high-throughput assembly consisted of a polyurethane block, a 304 stainless steel block, and a Pd/WO_3 film on a glass substrate (Fig. 8.14 a). The deposited metals or their alloys were cathode catalysts; a polypropylene block was mediator and 304 stainless steel worked as anode catalyst. Each well was filled with an aqueous solution of 0.2 M sodium acetate. Electrolysis in the sodium acetate solution was conducted for 20 min with the library at –2.8 V with respect to a common 304 stainless steal anode. The reflectance change was recorded every 30 s for 30 min. A high reflectance change rate corresponds to a high production rate of hydrogen. Among single-element catalysts, Pt, Pd and Ni showed high catalytic performance, as expected. Al-Pt alloy showed much higher hydrogen production than single-element catalysts (Fig. 8.14 b).

8.6 Evaluation of DeNOx Catalysis

Reductive elimination and storage of dilute NOx has been an important technology to achieve cleaner exhaust gas from engines. At present, Pt, Pd and Rh are often used for the so-called three-way catalyst. The demand for a high-performance DeNOx catalyst with low cost and non-harmful materials is growing as the environmental issues expand worldwide.

Busch et al. have reported dye utilization for HTS of DeNOx catalysts [34]. They impregnated a dye called ABTS [2,2′-azinobis(3-ethylbenzthiazolin-6-sulfonic acid)] on filter paper, which changes from colorless to green or blue-green upon exposure to NO or NO_2 (Fig. 8.15 a). The filter paper was placed at the open ends of SUS pipes leading to a 49-channel parallel path flow reactor. Gases of three different compositions were used for the catalyst test: (1) 1000 ppm NO, 1000 ppm propylene and 5% oxygen; (2) 1000 ppm NO and 1000 ppm propylene; and (3) 1000 ppm NO in nitrogen. The total flow rate of the reaction gas was 490 mL min^{-1}, which corresponds to ca. 10 mL min^{-1} flown onto each catalyst. The author did not clearly describe the exposure time for each catalyst test; however, the threshold concentration for NO detection was ca. 800 ppm at an exposure time of 300 s (Fig. 8.15 b). The absorbance of ABTS increased linearly in proportion to NO concentration above threshold. Longer exposure times, or impregnating more ABTS on the filter paper, would realize a more sensitive detection.

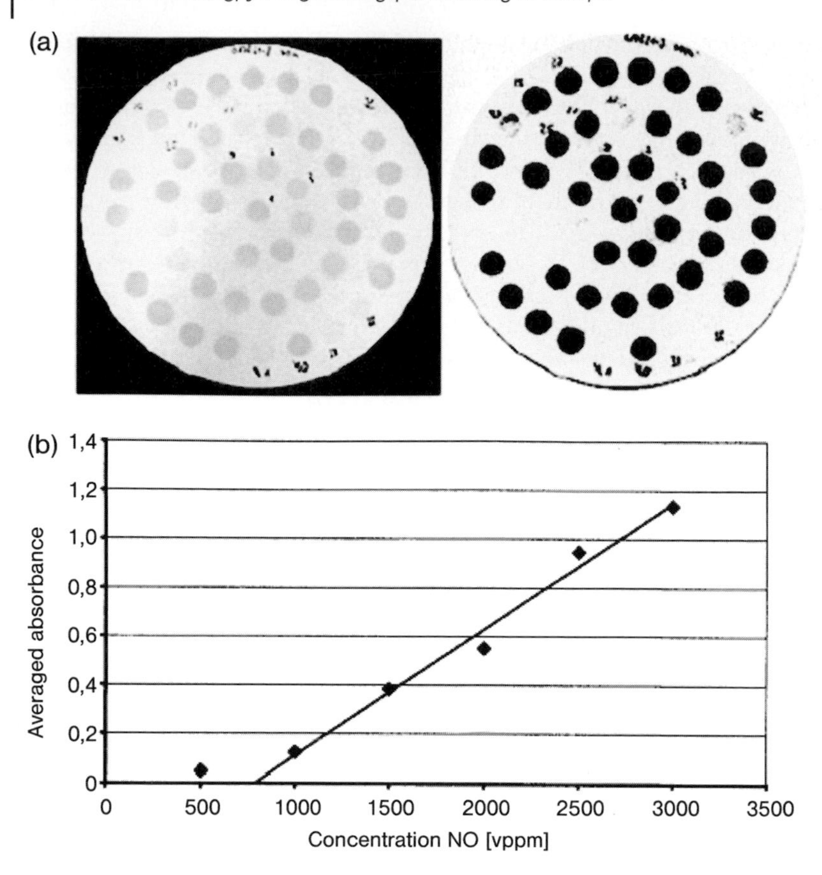

Fig. 8.15 (a) Color changes of organic dye exposed to gas including NOx. Each dot corresponds to a location where the reactor effluent passed through. The position corresponding to reference catalyst (Pt/Al_2O_3) remained white. (b) Average absorbance of ABTS at 412 nm depending on nitric oxide concentration (reproduced by permission of American Chemical Society from [34]).

The compositions of tested catalysts were determined by mixing 10 elements, Fe, Co, Cr, Cu, Ni, Mn, Ni, Zn, La and Pb, with the ratio at random. By using the high-throughput apparatus, four suitable catalyst compositions were found for the reducing condition and three for the oxidizing condition among 77 catalysts.

8.7 Evaluation of Methanol Production Catalysis

Omata et al. have investigated catalysts for methanol synthesis from CO_2-rich synthetic gas under pressurized conditions [22]. They tested 96 mixed-oxide catalysts composed of Cu, Zn, Al, Cr, Zr, Ga at random ratio. Each catalyst was placed in a

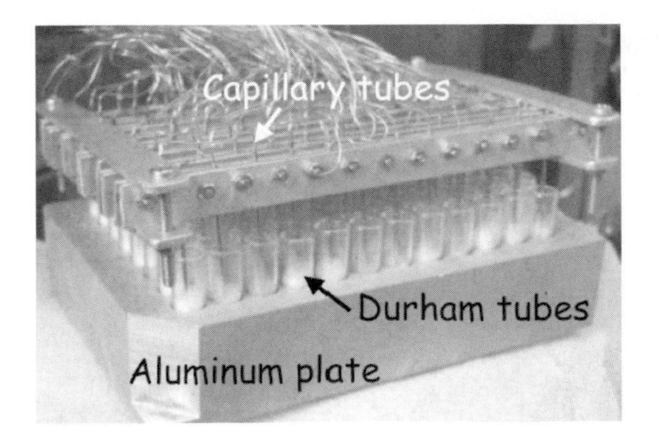

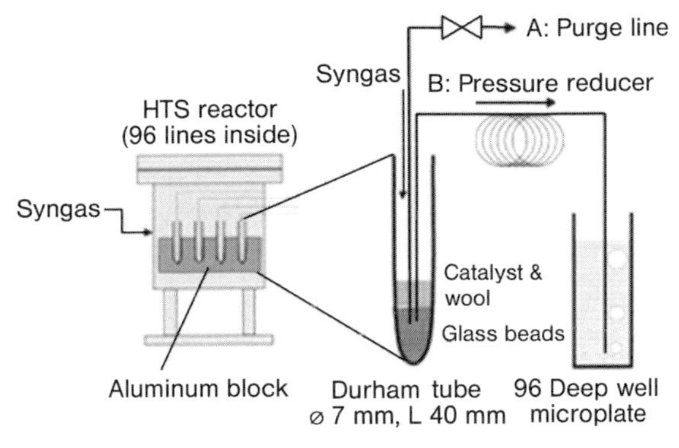

Fig. 8.16 (a) Parallel reactor with 96 tubes for methanol synthesis from Syngas, including high level CO_2 under pressurized conditions. The out gas was bubbled into water to absorb formed MeOH. The MeOH concentration in each solution was determined by the addition of inorganic dye to the solution and the absorbance was determined with a microplate reader (reproduced by permission of The Japan Petroleum Institute from 34]).

glass tube with glass beads (Fig. 8.16). The catalyst array was put in an autoclave under 1 MPa at 498 K. A gas mixture of H_2, CO, CO_2 and N_2 (43 : 22 : 30 : 5) filled the autoclave. The products and educts of each catalyst were then passed for 2 h through a certain amount of water, via a capillary, to absorb formed MeOH. To this solution an aqueous dye solution of potassium bichromate was added. The dye changes color by chromium reduction, from orange to green, when exposed to MeOH. Quantification of MeOH was performed by absorbance change detected by a 96-microplate reader at 595 and 655 nm. The screening results gave an optimized composition of $Cu_{0.44}Zn_{0.15}Al_{0.18}Cr_{0.18}B_{0.21}Zr_{0.28}Ga_{0.08}$ for methanol synthesis from synthetic gas under CO_2-rich conditions.

8.8 Concluding Remarks

As described above, gas sensors as HTS tools in catalysis have been applied in several reactions due to their small size, easy parallelization and high sensitivity. We expect the importance of gas sensors to increase for HTS in catalysis. Recently, Simon et al. reported the development of gas sensors with a combinatorial approach [35]. Developing tailor-made gas sensors for specific reactions, this technology will become increasingly important for catalysis researchers.

8.9 References

1 B. Jandeleit, H.W. Turner, T. Uno, J.A.M. vanBeek, W.H. Weinberg, *Cat. Technol.* **1998**, *2*, 101.

2 B. Jandeleit, D.J. Schaefer, T.S. Powers, H.W. Turner, W.H. Weinberg, *Angew. Chem., Int. Ed.* **1999**, *38*, 2494.

3 S. Senkan, *Angew. Chem., Int. Ed.* **2001**, *40*, 312.

4 F.C. Moates, M. Somani, J. Annamalai, J.T. Richardson, D. Luss, R.C. Wilson, *Ind. Eng. Chem. Res.* **1996**, *35*, 4801.

5 A. Holzwarth, P.W. Schmidt, W.E. Maier, *Angew. Chem., Int. Ed.* **1998**, *37*, 2644.

6 H. Su, E.S. Yeung, *J. Am. Chem. Soc.* **2000**, *122*, 7422.

7 H. Su, Y.J. Hou, R.S. Houk, G.L. Schrader, E.S. Yeung, *Anal. Chem.* **2001**, *73*, 4434.

8 C.M. Snively, J. Lauterbach, in: *Combinatorial Catalysis and Highthroughput Catalyst Design and Testing*, NATO Science Series, C560, **1999**, 311 pp.

9 C.M. Snively, G. Oskarsdottir, J. Lauterbach, *Catal. Today* **2001**, *67*, 357.

10 S.M. Senkan, *Nature* **1998**, *394*, 350.

11 S.M. Senkan, S. Ozturk, *Angew. Chem., Int. Ed.* **1999**, *38*, 791.

12 P. Cong, A. Dehestani, R. Doolen, D.M. Giaquinta, S. Guan, V. Marcov, D. Poojary, K. Self, H. Turner, W.H. Weinberg, *Proc. Natl. Acad. Sci. USA* **1999**, *96*, 11077.

13 P. Cong, R.D. Doolen, Q. Fan, D.M. Giaquinta, S. Guan, E.W. McFarland, D.M. Poojary, K. Self, H.W. Turner, W.H. Weinberg, *Angew. Chem., Int. Ed.* **1999**, *38*, 483.

14 Y.M. Liu, P.J. Cong, R.D. Doolen, H.W. Turner, W.H. Weinberg, *Catal. Today* **2000**, *61*, 87.

15 S. Senkan, K. Krantz, S. Ozturk, V. Zengin, I. Onal, *Angew. Chem., Int. Ed.* **1999**, *38*, 2794.

16 P. Claus, D. Honicke, T. Zech, *Catal. Today* **2001**, *67*, 319.

17 Y. Yamada, A. Ueda, Z. Zhao, T. Kobayashi, *Res. Chem. Int. Med.* **2002**, *28*, 397.

18 A.M. Cassell, S. Verma, L. Delzeit, M. Meyyappan, J. Han, *Langmuir* **2001**, *17*, 260.

19 Y. Yamada, A. Ueda, Z. Zhao, T. Maekawa, K. Suzuki, T. Takada, T. Kobayashi, *Catal. Today* **2001**, *67*, 379.

20 A. Ueda, Y. Yamada, T. Kobayashi, *Appl. Catal. A* **2001**, *209*, 391.

21 R.A. Potyrailo, R.J. May, *Rev. Sci. Instrum.* **2002**, *73*, 1277.

22 K. Omata, Y. Watanabe, T. Umegaki, M. Hashimoto, M. Yamada, *J. Jpn. Petrol. Inst.* **2003**, *46*, 328.

23 A. Hagemeyer, B. Jandeleit, Y.M. Liu, D.M. Poojary, H.W. Turner, A.F. Volpe, W.H. Weinberg, *Appl. Catal. A Gen.* **2001**, *221*, 23.

24 T. Johann, A. Brenner, M. Schwickardi, O. Busch, F. Marlow, S. Schunk, F. Schüth, *Angew. Chem., Int. Ed.* **2002**, *41*, 2966.

25 C. Hoffmann, A. Wolf, F. Schüth, *Angew. Chem., Int. Ed.* **1999**, *38*, 2800.

26 Y. Yamada, M. Ando, A. Ueda, T. Kobayashi, K. Suzuki, T. Maekawa, T. Takada, in: *Combinatorial Catalysis and Highthroughput Catalyst Design and Testing* (Ed.: E.G. Derouan), NATO Science Series, C560, **1999**, 283 pp.

27 M. Haruta, N. Yamada, T. Kobayashi, S. Iijima, *J. Catal.* **1989**, *115*, 301.

28 Y. Teng, T. Kobayashi, *Chem. Lett.* **1998**, 327.

29 Y. Teng, T. Kobayashi, *Catal. Lett.* **1998**, *55*, 33.

30 Y. Yamada, A. Ueda, M. Ando, T. Kobayashi, T. Maekawa, K. Suzuki, T. Takada, in: *Chemical Sensors IV* (Eds.: M. Butler, N. Yamazoe, P. Vanysek, M. Aizawa), The Electrochemical Society, Inc., Pennington, **1999**, 143 pp.

31 K. Nakagawa, Y. Teng, Z. Zhao, Y. Yamada, A. Ueda, T. Suzuki, T. Kobayashi, *Catal. Lett.* **1999**, *63*, 79.

32 G. Sauerbrey, *Z. Phys.* **1959**, *155*, 206.

33 T.F. Jaramillo, A. Ivanovskaya, E.W. McFarland, *J. Comb. Chem.* **2002**, *4*, 17.

34 O.M. Busch, C. Hoffmann, T.R.F. Johann, H.-W. Schmidt, W. Strehlau, F. Schüth, *J. Am. Chem. Soc.* **2002**, *124*, 13527.

35 U. Simon, D. Snaders, J. Jockel, C. Heppel, T. Brinz, *J. Comb. Chem.* **2002**, *4*, 511.

9
Parallel Approaches to the Synthesis and Testing of Catalysts for Liquid-phase Reactions

Paolo P. Pescarmona, Jan C. van der Waal, Leon G. A. van de Water, and Thomas Maschmeyer

9.1
Introduction

The fascinating idea behind the first applications of combinatorial chemistry methods was to mimic nature's evolutionary approach in the search of pharmaceutically active compounds. The concept was to combine the variation of many synthetic parameters to produce and evaluate vast numbers of different compounds simultaneously and/or more focussed sets in parallel [1]. Just the active species would 'survive' this screening and these would be studied further, refining the selection for active drug candidates. Starting at the end of the 1980s, combinatorial chemistry rapidly developed, both from the methodological and technical point of view, in such a way that these methods are now increasingly considered as a standard tool for the discovery of novel drugs. The success in the pharmaceutical area has stimulated application of these techniques to other fields of chemical research and, particularly, to catalysis [2–14].

When performing a synthetic combinatorial chemistry experiment, several basically different strategies may be followed to create a library of compounds. The most commonly used are: *mix&split* (or *split and pool*) synthesis [1] *masking* strategies [15, 16] and *parallel* synthesis. In this chapter, the attention is focussed on the application of parallel synthesis to catalysis in the liquid phase.

In the parallel synthesis approach the various reactions take place in separate vessels; typically, robotic equipment is used to pick and mix the reactants in different miniature vessels or wells, so that an array of distinct products is obtained. The library of products has to be screened to determine the active compound(s). In contrast to other combinatorial synthetic methods, the identification of the active species is straightforward, since separate wells are used. However, finding a suitable way to test all the isolated products commensurate with the high-throughput synthesis speeds presents a principal challenge. If the screening process of the library is performed one well at a time it can be very expensive and time consuming and would become a bottleneck, nullifying some of the speed advantages of combinatorial synthesis. To avoid this, integrated systems including fast and affordable analytical methods, usually referred to as high-throughput experimentation (HTE) or high-speed experimentation (HSE) techniques [17] are being devel-

High-Throughput Screening in Chemical Catalysis
Edited by A. Hagemeyer, P. Strasser, A. F. Volpe, Jr.

ISBN: 3-527-30814-8

oped by many groups. Current HTE technology is based on the use of miniaturised, automated and parallel versions of tools conventionally employed to screen the compounds under study, such as chromatographic, spectroscopic or thermographic techniques. In some cases, completely new methods have been developed specifically [2, 4, 11, 13, 18] (for example, mass spectrometry for enantioselectivity assays using isotopically labelled substrates, with a throughput of about 1000 samples per day) [19].

The development of new catalysts is a challenging task: in many cases the correlation between their features (structural, electronic) and their performance (activity, selectivity, lifetime) is not easily established, especially for supported heterogeneous catalysts. Therefore, an iterative process of 'design', synthesis and testing is usually followed to improve catalyst performance. HTE techniques can accelerate this process considerably, allowing for the simultaneous evaluation of a large number of candidates. Many of the operations involved in the development of a catalyst lend themselves to the application of HTE techniques. The synthesis of both homogeneous and heterogeneous catalysts, their screening for activity and selectivity in test reactions, and also the determination of the optimal process parameters for a specific reaction can in most cases be conducted with much greater speed and efficiency employing the miniaturised, automated combinatorial/HTE procedures as compared with conventional methods.

In HTE techniques, a multi-dimensional and scientific approach – based on literature data, computational modeling, personal chemical knowledge and intuition – is essential to determine which parameter space has to be investigated to gain the desired information from the experiment. These techniques may be seen as a new and useful tool for chemical research in the same way as, for example, spectroscopic techniques or quantum mechanical calculations. The applicability of HTE techniques is often determined by the technical limitations of the automated workstations employed. These limitations may range from the preparation of the samples (for example, inefficient mixing of reaction mixtures in the small HTE vessels) to their analysis (for example, equipment for the high-throughput screening of samples is not yet available for every spectroscopic technique).

In general, HTE screenings generate leads together with a correlation between their properties and their structural and synthetic characteristics. Once leads and correlations of sufficient quality have been produced, a more detailed study at a conventional laboratory scale follows. Repeating the experiment in a conventional manner is an important check regarding the reproducibility of the HTE results on a larger-volume scale. Results should never be published without such a check, since the ability to independently validate a particular result should not be dependent on the HTE equipment used [20].

In this chapter, one homogeneous and one heterogeneous catalysis example involving HTE parallel methods are presented.

9.2
Exploration of the Synthesis of Silsesquioxane Precursors for Epoxidation Titanium Catalysts by Means of High-Throughput Experimentation

Titanium centres dispersed on, and supported by, various forms of silica are catalysts with remarkable properties for partial oxidations [21]. Although the best – on the basis of activity per gram of titanium – heterogeneous catalysts for epoxidations of alkenes with peroxides all have tetra-coordinated Ti-centres [22–24] it is still unclear whether the active catalytic species requires the specific configuration of four siloxy groups, or if being partially hydrolyzed to, say, one hydroxy and three siloxy ligands is equally good. Chemical modeling studies by various groups [25–28] have shown that Ti-catalysts with fewer than four siloxy groups should also be active as catalysts. These considerations indicate that there is still some residual uncertainty regarding the active site, despite the many experimental and computational studies carried out to resolve it [29–33]. Hence, different types of Ti-centres might exhibit the desired characteristics.

For these heterogeneous catalysts, which are difficult to characterise, hybrid inorganic–organic compounds have proved to be useful models. Especially, silsesquioxanes [34] $(RSiO_{1.5})_a(H_2O)_{0.5b}$, where R is a hydrogen atom or an organic group and a and b are integers with $a+b=2n$ and $b \leq a+2$, have been a focus of attention as model compounds for silica. Of particular interest for applications in catalysis are the incompletely condensed silsesquioxanes ($b \neq 0$), which contain Si–OH groups and, therefore, can coordinate to catalytically active centres [34–36]. The titanium complexes of silsesquioxane *a7b3* are reportedly very active homogeneous catalysts for the epoxidation of alkenes (Fig. 9.1) [25–28]. There is considerable interest in these catalysts, but commercial introduction is severely hindered by the long and expensive preparation method of the silsesquioxane precursor *a7b3* [37, 38]. Therefore, the identification of a new, faster and efficient way to

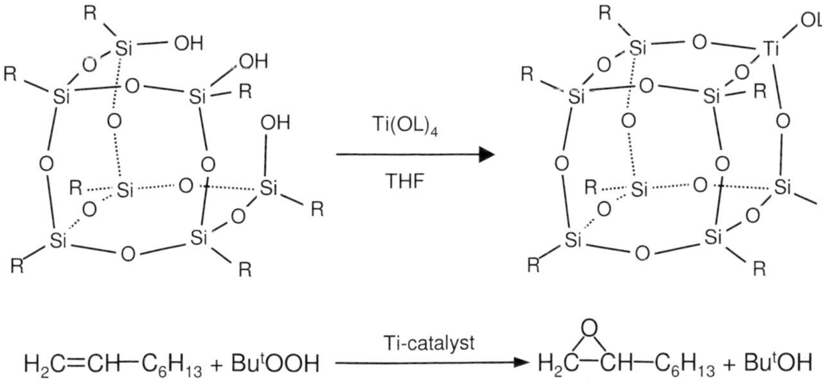

Fig. 9.1 Complexation of titanium to the incompletely condensed silsesquioxane $R_7Si_7O_{12}H_3$ (R=cyclohexyl, cyclopentyl), yielding a catalyst active in the epoxidation of alkenes with TBHP.

synthesise silsesquioxane precursors for titanium catalysts active in the epoxidation of alkenes was set as the first goal of this research.

9.2.1 High-Throughput Experimentation Approach

On the basis of existing knowledge, some general conclusions can be drawn:

- Silsesquioxanes are usually synthesised by the hydrolytic condensation of organosilanes $RSiX_3$ (Scheme 9.1): this is a multiple-step reaction for which an overall mechanism is not available.
- Many parameters influence the hydrolytic condensation and determine which silsesquioxane species are formed and in what amounts.
- Hydrolytic condensation of organosilanes usually produces a mixture of completely and incompletely condensed silsesquioxanes.
- Incompletely condensed silsesquioxanes different from *a7b3* may also act as precursors for titanium complexes that are catalytically active in the epoxidation of alkenes.

$$RSiX_3 + 3H_2O \rightarrow RSi(OH)_3 + 3HX \tag{1}$$

$$aRSi(OH)_3 \rightleftarrows (RSiO_{1.5})_a(H_2O)_{0.5b} + (1.5a - 0.5b)H_2O \tag{2}$$

Scheme 9.1 Synthesis of silsesquioxanes via hydrolytic condensation of organosilanes.

From these considerations, the synthesis of silsesquioxanes was optimised, by means of HTE, as a function of the activity of the catalysts obtained after titanium coordination to the silsesquioxane structures. Therefore, this approach aimed at producing any incompletely condensed silsesquioxane that would result in active catalysts after titanium coordination rather than a specific structure (like silsesquioxane *a7b3*). The epoxidation of 1-octene with tert-butyl hydroperoxide (TBHP) as the oxidant was chosen as test reaction for the activity of the catalysts [26].

Experiments were performed on an automated parallel synthesis workstation coupled with a personal computer supplied with software enabling the workstation to be programmed. Samples were prepared in a rack containing a 6×4 array of glass tubes. Catalytic activities were measured by means of gas chromatography. The reported activities are the averages of the results from different experiments.

9.2.2 Effect of the Nature of the R Group and of the Solvent

The first step of this screening was to decide which parameter space had to be screened. It was hypothesised that the solvent and the R group have the most relevant role among the parameters influencing the hydrolytic condensation of organosilanes [39]. Therefore, the parameter space was defined by the combination of four solvents and eight R groups. The set of R groups consisted of eight trichlorosilanes $RSiCl_3$ (R=cyclohexyl, cyclopentyl, phenyl, methyl, ethyl, tert-butyl, n-octyl and allyl). Since the reaction includes water among the reagents, only water-miscible solvents were selected: acetone, acetonitrile, methanol and tetrahydrofuran (THF). Solvents were chosen on a varying scale of polarity to check the influence of this property on the synthesis of silsesquioxanes. The parameter space was screened for the activity of the catalysts obtained by coordination of titanium isopropoxide, $Ti(OPr^i)_4$, to the silsesquioxane precursors. Since the goal was to identify a more efficient way to synthesise silsesquioxane structures, the reaction time for the hydrolytic condensation was set at 18 h, a much shorter time than that commonly required to synthesise silsesquioxane *a7b3* [37, 38]. Other parameters were chosen as: reaction temperature=50 °C, X=Cl, solvent:water=4:1, trichlorosilane concentration=0.136 M.

The epoxidation activity of the titanium catalysts, as a function of the different solvents and R groups varied in the synthesis of the silsesquioxanes precursors, is reported in Fig. 9.2. Values are normalised to the activity of the complex obtained by reacting $Ti(OPr^i)_4$ with the pure cyclopentyl silsesquioxane *a7b3* in THF. The results show some general trends:

- Catalysts derived from silsesquioxane structures synthesised in acetonitrile as solvent show the highest catalytic activity, followed, in decreasing order of activity, by those from acetone, methanol and from tetrahydrofuran (Fig. 9.2). This trend applies to all R groups except ethyl and tert-butyl for which silsesquioxanes synthesised in THF generate more active catalysts than those synthesised in acetone and methanol. Interestingly, acetonitrile is more effective than acetone, the solvent commonly reported for the synthesis of incompletely condensed silsesquioxane *a7b3* [37, 38]. The best results with acetonitrile can be explained on the basis of its high polarity (of the solvents used, acetonitrile has the highest dipole moment and dielectric constant). In the presence of a polar molecule the activation barrier for the condensation reactions towards the formation of silsesquioxanes is proposed to be reduced and, therefore, the synthesis is speeded up [40, 41]. Moreover, a highly polar solvent might stabilise incompletely condensed silsesquioxanes by interaction with their silanol groups, therefore favoring the synthesis of these silsesquioxanes over that of the less polar, completely condensed species.

- With respect to the silanes employed, the highest epoxidation activities are observed in the order cyclopentyl > cyclohexyl > phenyl > tert-butyl > ethyl > methyl > allyl ~ n-octyl ~ 0, for all solvents apart from THF, for which tert-butyl- and ethyltrichlorosilanes yield more active catalysts than phenyltrichlorosilanes (Fig. 9.2). This trend is in good agreement with the literature, where cyclopentyl- and cyclohexyltrichlorosilanes are reported to form incompletely condensed silsesquioxanes in high yields [37, 38]. In principle, the nature of the organic group might influence which silsesquioxane structures are produced and in which ratios [34], as well as the electronic and spatial properties of the actual catalyst. Because organic groups are linked to the titanium centre through a Si–O unit, electronic effects should have a negligible influence on the catalytic properties and, since the organic groups point away from the titanium centre, steric effects are probably also not relevant. The principal effect of the organic group is, therefore, considered to be its role in determining which silsesquioxane structures are formed during the hydrolytic condensation and in what amounts. The trend identified seems to be related to the size of the organic substituent: bulky groups probably hinder the formation of completely condensed silsesquioxanes, which are not able to bind to titanium centres and, therefore, can not generate active catalysts.

From a methodological point of view, the fact that different points in the parameter space screened show very different activities confirms the hypothesis that the chosen parameters strongly influence the hydrolytic condensation: this means that a representative parameter space was studied. An important question that arose from this experiment was whether the parameter space was narrowed too much using prior knowledge as the starting point. In this respect, the coincidence of finding only cyclopentyl and cyclohexyl as good candidates, while they are also the only ones reported in the literature for the synthesis of silsesquioxane *a7b3* [37, 38], might be due to bias for the synthesis conditions particularly suited for these silanes. This remark together with the trend indicating a beneficial effect of

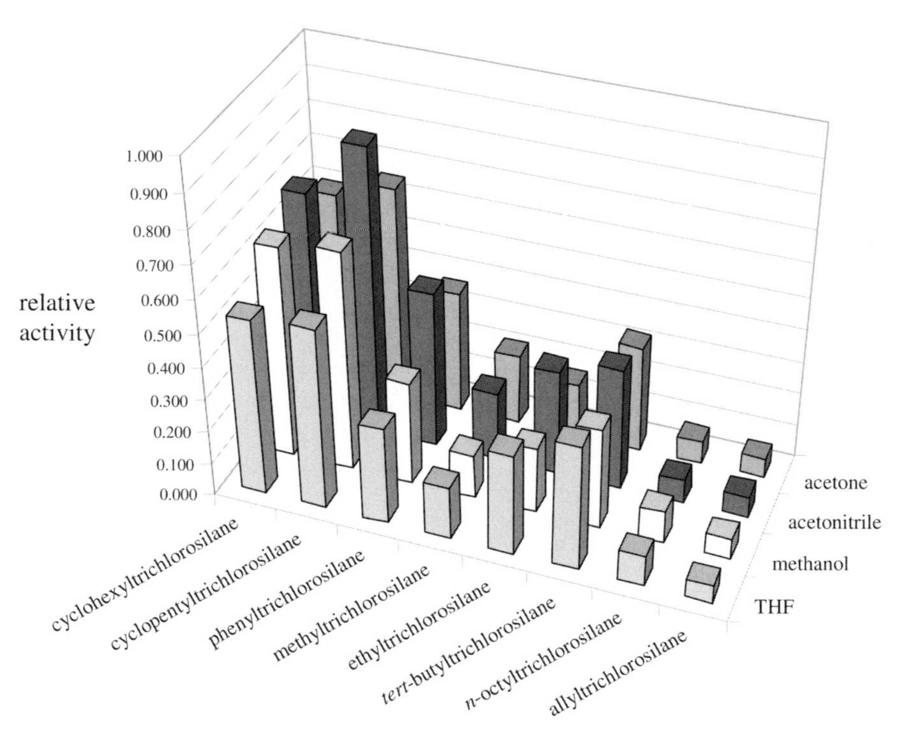

Fig. 9.2 Catalytic activity in the epoxidation of 1-octene with TBHP of the Ti-silsesquioxanes in the screened parameter space.

polarity suggest that it could be interesting in further research to broaden the screened parameter space to other polar solvents (Section 9.2.4).

The Ti-catalyst derived from the silsesquioxanes synthesised by the hydrolytic condensation of cyclopentyltrichlorosilane in acetonitrile presents the highest catalytic activity (Fig. 9.2). This activity is 87% of that of the Ti catalyst obtained using pure silsesquioxane *a7b3* as precursor. The relevance of this result lies in the fact that the synthesis of these silsesquioxane precursors does not require any purification process and is much less time-consuming than the synthesis of silsesquioxane *a7b3*.

9.2.3 Effect of Trichlorosilane Concentration

Besides leading to the identification of a new route to synthesise a promising Ti-catalyst and to an additional knowledge about the synthesis of silsesquioxane structures, the HTE exploration described in Section 9.2.2 provided a starting point for further investigation. One of the directions that can be taken is that of studying other parameters influencing the catalytic activity of titanium silsesquioxanes. The initial choice of investigating the effect of the nature of the solvent and of the R group on the synthesis of silsesquioxane precursors for Ti-catalysts was based on the assumption that these two parameters were the most relevant in determining which silsesquioxane structures were formed and in what amounts. The validity of this hypothesis can be verified (or falsified) [42, 43] by studying the effect of the other parameters affecting the synthesis of silsesquioxanes [34]. Here, the effect of varying the initial concentration of the trichlorosilane employed to synthesise the silsesquioxane precursors is presented. For this HTE screening, the parameter set was reduced from 12 (8 R-groups + 4 solvents) to 4 (3 R-groups + 1 solvent) elements: cyclopentyl-, cyclohexyl- and phenyltrichlorosilane with acetonitrile as solvent, i.e., the combinations that led to the highest catalytic activities (Section 9.2.2). This reduced parameter space was then used as a basis to create a larger parameter space by varying the initial concentration of the trichlorosilanes. To investigate the effect of the R groups and of the solvents, the silsesquioxanes products were reacted with $Ti(OPr^i)_4$ and subsequently screened for catalytic activity in the epoxidation of 1-octene with TBHP. The reported activities are normalised to the activity of $(c\text{-}C_5H_9)_7Si_7O_{12}TiOPr^i$ (Fig. 9.3).

For most of the trichlorosilane concentration tested, the order of activity is still cyclopentyl > cyclohexyl > phenyl. At low concentration the order between cyclohexyl and phenyl reverses but cyclopentyl still results in the best activity. This confirms the starting hypothesis that the R group is a more relevant factor than the concentration of the silane in influencing the hydrolytic condensation. The maximum activity corresponds to a different silane concentration for each of the three R groups. The maximum for cyclopentyl as R group – corresponding to a concentration of silane of 0.181 M – displays an activity that is 95% of that of pure Ti-(cyclopentyl silsesquioxane *a7b3*): this is a relevant improvement compared with the optimum obtained previously (87%) [39]. These results show that the experi-

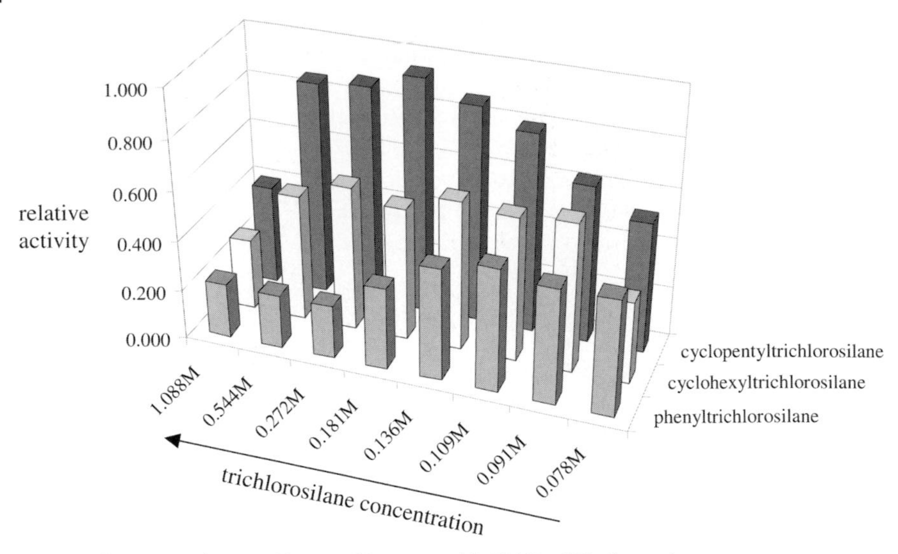

Fig. 9.3 Activity in the epoxidation of 1-octene with TBHP of Ti-silsesquioxane catalysts as a function of the trichlorosilane concentration in the initial silsesquioxane synthesis mixture.

ment worked as a fine-tuning of the synthesis of silsesquioxane precursors. Methodologically, it is important that, in the wide concentration range applied, no activity low enough to reject any of the three R groups – if such trichlorosilane concentration would have been chosen in the initial experiment – was obtained. This significantly strengthens the assumption that the conditions of the first experiment were sufficiently discriminative towards the R groups screened.

9.2.4 Effect of Highly Polar Solvents

One of the observed trends in the HTE screening reported in Section 9.2.2 indicated a favorable effect induced by high-polarity solvents [39, 44]. This stimulated the investigation of the synthesis of silsesquioxanes in very polar solvents [45].

A new parameter space for the synthesis of silsesquioxane precursors was defined by six different trichlorosilanes (R=cyclohexyl, cyclopentyl, phenyl, methyl, ethyl and tert-butyl) and three highly polar solvents [dimethyl sulfoxide (DMSO), water and formamide]. This parameter space was screened as a function of the activity in the epoxidation of 1-octene with tert-butyl hydroperoxide (TBHP) [26] displayed by the catalysts obtained after coordination of $Ti(OBu)_4$ to the silsesquioxane structures. Fig. 9.4 shows the relative activities of the titanium silsesquioxanes together with those of the titanium silsesquioxanes obtained from silsesquioxanes synthesised in acetonitrile. The values are normalised to the activity of the complex obtained by reacting $Ti(OBu)_4$ with the pure cyclopentyl silsesquioxane *a7b3* $[(c\text{-}C_5H_9)_7Si_7O_{12}TiOC_4H_9]$.

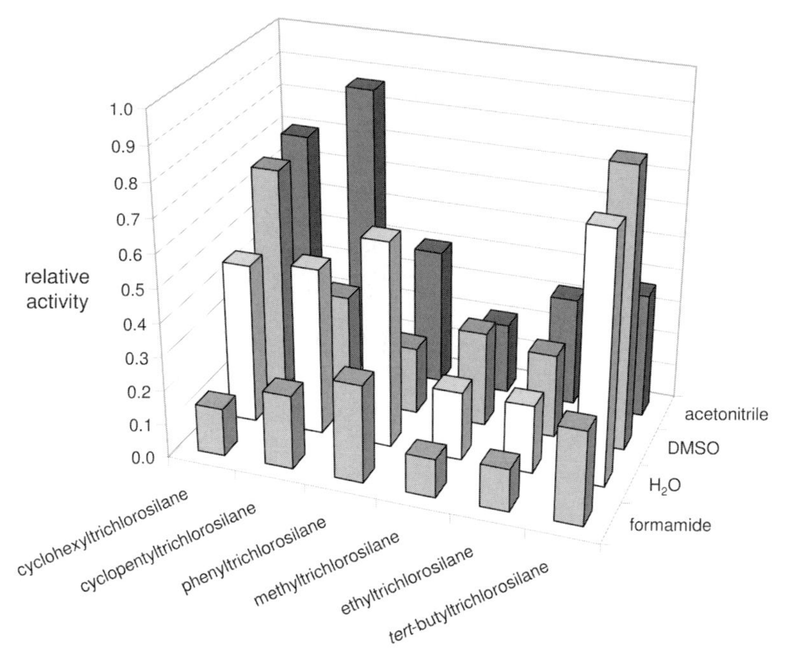

Fig. 9.4 Screening of the epoxidation activity of the titanium silsesquioxanes as a function of the solvent and of the trichlorosilanes used in the synthesis of the silsesquioxane precursors.

The highest catalytic activities were found for the titanium complexes obtained from tert-butyl silsesquioxanes synthesised in DMSO and water, with respectively 84% and 74% of the activity of the reference catalyst $(c\text{-}C_5H_9)_7Si_7O_{12}TiOC_4H_9$. [With the experimental conditions employed, $(c\text{-}C_5H_9)_7Si_7O_{12}TiOC_4H_9$ gives complete and selective conversion of TBHP towards the epoxide: therefore, the relative activities of the reported catalysts correspond to their TBHP conversions towards 1,2-epoxyoctane.] These two catalysts exhibited almost the same activity as the previous best HTE catalyst (87%) obtained from cyclopentyl silsesquioxanes synthesised in acetonitrile [39, 44, 46] and are the first reported examples of tert-butyl silsesquioxanes as precursors for very active titanium catalysts. Relevant catalytic activities were also obtained with cyclohexyl silsesquioxanes synthesised in DMSO (67%) and with phenyl silsesquioxanes synthesised in H_2O (61%).

Besides the identification of these leads, the HTE screening provided additional information about the system under study. No regular trend can be identified: for each solvent the order of activity as a function of the R groups is different. This can be explained on the basis of the different nature of the solvents employed. Water acts both as solvent and reagent, and the fact that silsesquioxanes with only a very low level of condensation are soluble in water [47] causes the fast precipitation of condensed structures from solution. Formamide has a high boiling point (220 °C), requiring high temperatures to remove the solvent from the silsesquioxane products. At these high temperatures, the silsesquioxanes tend to condense further to produce completely condensed structures, which are not suitable for coordinating metal centres [34]. This can explain the weak activity observed for the catalysts obtained from silsesquioxanes synthesised in formamide.

Finally, all the methyl and ethyl silsesquioxanes are poor precursor for titanium-based epoxidation catalysts, in agreement with the initial experiments [39, 44].

9.2.5
Up-scaling and Characterisation of the HTE Leads

As shown in the previous sections, HTE techniques proved to be a powerful tool to identify leads and gain knowledge about the system under study, but they did not provide information about the actual silsesquioxane structures that were synthesised [46]. To gain such information, selected HTE leads need to be prepared on a conventional laboratory scale and characterised by appropriate analytical techniques (such as spectroscopy and chromatography) [45, 46, 48, 49]. Here, the results of the up-scaling of the synthesis of cyclopentyl silsesquioxanes in acetonitrile (Sections 9.2.2 and 9.2.3) and of the synthesis of tert-butyl silsesquioxanes in water (Section 9.2.4) are reported.

9.2.5.1 Cyclopentyl Silsesquioxanes Synthesised in Acetonitrile

First, the hydrolytic condensation of cyclopentyltrichlorosilane in acetonitrile (Section 9.2.2) was repeated on a 125 ml scale (50 times up-scaling). Next, the silsesquioxane products were reacted with a titanium alkoxide, yielding a catalyst with the same epoxidation activity of the HTE lead and, therefore, confirming the applicability of HTE techniques to the synthesis of silsesquioxanes. The silsesquioxane products obtained prior to reaction with the titanium centre could be divided into two fractions: one as a precipitate (A) and the other as solute in the reaction mixture (B). Fraction B was dried under reduced pressure and redissolved in tetrahydrofuran.

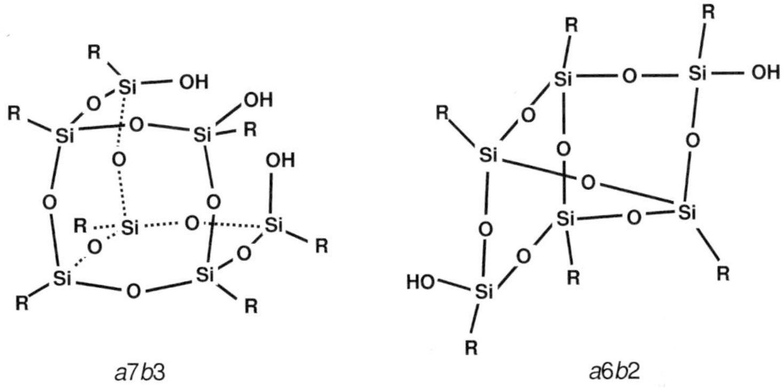

a7b3 *a6b2*

Fig. 9.5 Silsesquioxanes *a7b3* and *a6b2* (R=cyclopentyl).

Both fractions were characterised by NMR spectroscopy and mass spectrometry. Fraction A mainly consists of silsesquioxane *a7b3* (Fig. 9.5) [38], while fraction B is a mixture of different silsesquioxanes, mostly incompletely condensed species, with the main species assigned to silsesquioxane structure *a6b2* (Fig. 9.5). Finally, both fractions were reacted with a titanium alkoxide and tested for catalytic activity in the epoxidation of 1-octene as a function of the reaction time and the results compared with those of HTE lead (all three catalysts are homogeneous) (Fig. 9.6).

The activity of the titanium catalyst derived from fraction A (TOF=0.97 $mol_{epo} \cdot mol_{Ti}^{-1} \cdot min^{-1}$) is comparable to that of the Ti catalyst obtained with pure silsesquioxane *a7b3*, which after 4 h of reaction gives complete conversion of TBHP towards the epoxide. The titanium catalyst derived from fraction B shows a much lower, but still significant, activity (TOF=0.16 $mol_{epo} \cdot mol_{Ti}^{-1} \cdot min^{-1}$): this confirms the validity of the initial assumption that incompletely condensed silsesquioxanes other than *a7b3* can be precursors for active titanium catalysts. For fraction A, the conversion reaches a maximum after 30 min, while for fraction B the conversion slowly increases throughout the 4 h and does not reach a plateau. The behavior of the HTE lead (TOF=0.43 $mol_{epo} \cdot mol_{Ti}^{-1} \cdot min^{-1}$) is an average of that of the two fractions.

The white precipitate (fraction A) obtained after 18 h of reaction at 50 °C contained silsesquioxane *a7b3* [$(c\text{-}C_5H_9)_7Si_7O_9(OH)_3$] in ~50% yield. [Relative to silsesquioxane *a7b3* (the yield is calculated by dividing the number of grams of product by the number of grams of silsesquioxane *a7b3* that would be obtained if the reaction quantitatively gave silsesquioxane *a7b3*).] Small amounts of impurities, predominantly the cubic completely condensed silsesquioxane *a8b0*, $(c\text{-}C_5H_9)_8Si_8O_{12}$, were present in the precipitate. The synthesis was further optimised according to the fine tuning of the reaction conditions described in Section 2.3 and on the basis of the observation of Feher et al. that the yield of the synthesis of cyclopentyl silsesquioxane *a7b3* in acetone can be significantly improved by

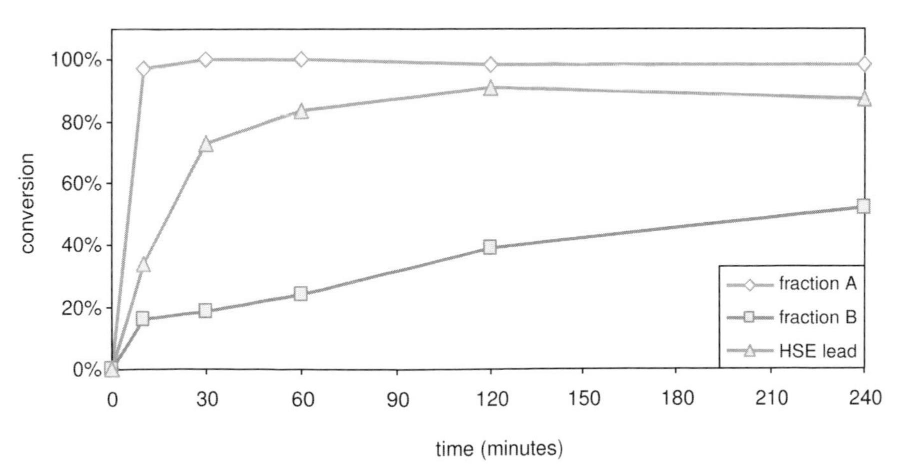

Fig. 9.6 Activity in the epoxidation of 1-octene with TBHP as a function of the reaction time.

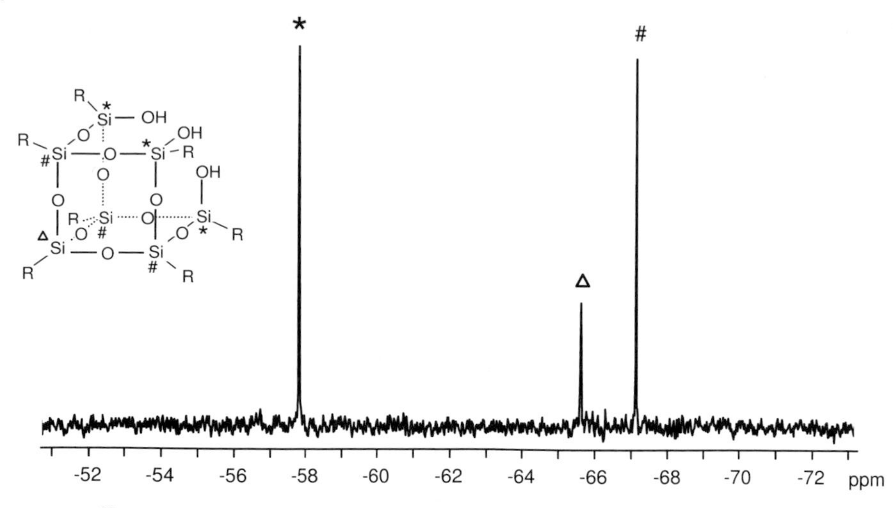

Fig. 9.7 ^{29}Si NMR spectrum of cyclopentyl silsesquioxane *a7b3* (THF as solvent, δ –57.83, –65.71, and –67.24 (3:1:3)].

refluxing the reaction mixture [38]. Therefore, the reaction temperature was increased from 50 °C to reflux conditions. This led to an increase of the yield to 81% after 18 h of reaction, with the high selectivity towards silsesquioxane *a7b3* being preserved. Silsesquioxane *a7b3* was then purified by pyridine extraction [37, 38] and the pure compound (as determined by ^{29}Si NMR, see Fig. 9.7) could be isolated in an overall yield of 64% – a significant improvement on the previously reported yield of 29% in 3 days [38]. The increased yield is mainly ascribed to the effect of acetonitrile as very polar and reactive solvent. This method provides a fast, high-yield way to prepare silsesquioxane *a7b3*, therefore making this compound more readily available for its many applications.

To shed light on the mechanism of formation of silsesquioxane *a7b3*, to identify the species formed during the process, and to try to explain the high selectivity towards structure *a7b3* of the optimised synthetic method described above (64% yield in 18 h), the synthesis of cyclopentyl silsesquioxane *a7b3* was monitored by electrospray ionisation mass spectrometry (ESI MS) [50–52] and in situ attenuated total reflection Fourier-transform infrared (ATR FTIR) spectroscopy [53, 54]. Spectroscopic data from the latter were analysed using chemometric methods to identify the pure component spectra and relative concentration profiles.

9.2.5.1.1 **Monitoring by Mass Spectrometry**

The mechanistic study by means of mass spectrometry was performed by analysing samples of the reaction mixture at regular intervals throughout the experiment. The MS spectra recorded between $t=0$ and 1440 min show peaks corresponding to cyclopentyl silsesquioxane structures with $1 \leq a \leq 13$.

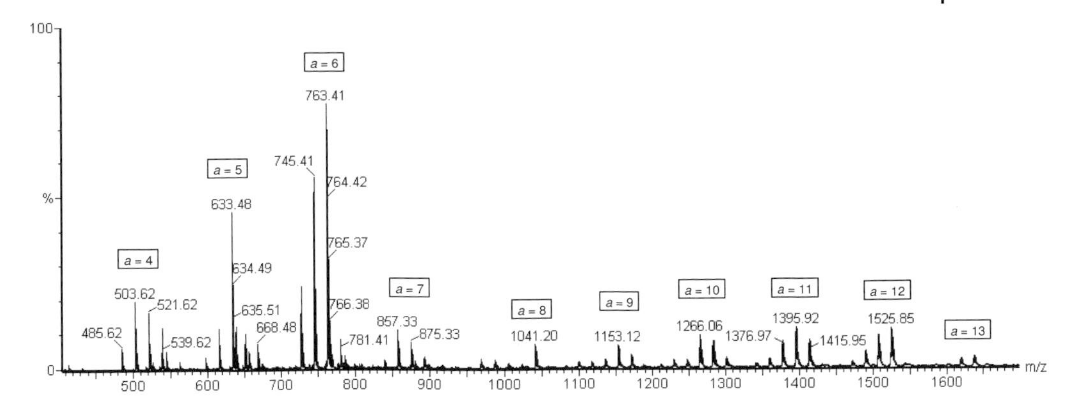

a	*b*	*m/z*	*a*	*b*	*m/z*	*a*	*b*	*m/z*	*a*	*b*	*m/z*	*a*	*b*	*m/z*
1	3	148.06	4	0	484.16	6	0	726.24	7	1	856.29	8	0	968.32
			4	2	502.17	6	2	744.25	7	3	874.30	8	2	986.33
2	0	242.08	4	4	520.19	6	4	762.27	7	5	892.32	8	4	1004.35
2	2	260.10	4	6	538.21	6	6	780.29	7	7	910.33	8	6	1022.37
2	4	278.11				6	8	798.30	7	9	928.35	8	8	1040.38
			5	3	632.22							8	10	1058.40
3	3	390.14	5	5	650.24									
3	5	408.16	5	7	668.25									

Fig. 9.8 ESI MS plot after 90 min reaction and list of possible cyclopentyl silsesquioxanes with $1 \leq a \leq 8$. Analytical parameters for MS measurements were set as: flow rate (of the syringe pump) = 40 μl min^{-1}, RF lens = 0.31 V, capillary = 3.20 kV, cone = 30 V, extractor = 4 V, source block temperature = 80 °C, desolvation temperature = 300 °C, nebuliser gas flow = 85 l h^{-1}, desolvation gas flow = 450 l h^{-1}.

Peaks due to $a=7$ species ($m/z=839.3$, 857.3, 875.3, 893.3) are present in all the recorded spectra (see Fig. 9.8 for an example). The relative abundance of these peaks compared to that of the peaks of other silsesquioxane species is rather low at any reaction time, indicating a low concentration in solution. Knowing that the precipitate produced by the synthesis is mainly silsesquioxane *a7b3*, it can be inferred that this compound has a low solubility in the reaction mixture and tends to precipitate as it is formed. Its lower solubility with respect to other cyclopentyl silsesquioxanes is probably due to the tendency of silsesquioxane *a7b3* to form dimers [37, 55], which are insoluble in polar solvents such as acetonitrile. This is in agreement with the observed formation of a white precipitate after 1 h of reaction.

The concentration of the $a=7$ species in solution can, therefore, be considered approximately constant during the course of the reaction. This assumption allows one to normalise the intensity of the peaks of the other silsesquioxanes relative to the $a=7$ peaks, making it possible to compare the spectra measured at the various reaction times. (A normalisation of the MS data is necessary because the mass spectrometer produces plots in which the intensities of the peaks are not absolute but normalised to the most intense peak, the value of which is set at 100%.) The relative concentrations of the principal silsesquioxane species with increasing *a* are plotted against reaction time in Fig. 9.9. This plot shows how various silsesquioxane species formed and disappears over time. At $t=0$, silsesquioxane species with $1 \leq a \leq 6$ were present in considerable amounts. No cyclopentyltrichlorosilane was detected, indicating that the hydrolysis step (1) (Scheme 9.1) was complete before the first sample was injected in the mass spectrometer [55]. The presence of various species at $t=0$ shows that the process of condensation of silsesquioxane *a*1*b*3, $(c\text{-}C_5H_9)Si(OH)_3$ (step (2) in Scheme 9.1), leading to the formation of larger silsesquioxane structures, is also very fast. At $t=0$, species with $a=4$ were the most abundant. Silsesquioxanes with $1 \leq a \leq 4$ can be seen as the building blocks for larger silsesquioxane structures (see Fig. 9.10 for a proposed synthesis scheme). They are likely to be reactive species that tend to condense further, as confirmed by their rapidly diminishing concentrations as the reaction proceeded. The smaller the structures, the faster they disappeared to form larger ones. At $t=30$ min, silsesquioxane *a*1*b*3 had already disappeared from the reaction mixture. Silsesquioxanes with $a=2$ disappeared after 1 h and those with $a=3$ reached a very low concentration after 1 h of reaction, and could no longer be detected after 6 h.

The concentration of silsesquioxanes with $a=4$ decreased less rapidly. These species were still present in small amounts after 10 h and were only fully consumed close to the end of the reaction, after 24 h.

For $a=1$, the only possible structure is the trisilanol *a*1*b*3 [34]. For $a=2$ and 3, more structures are possible but *a*2*b*4 and *a*3*b*5 are the only two without a large geometrical strain. For $a=4$, two structures are the most likely: *a*4*b*4 and *a*4*b*6, the first being a ring and the second a linear structure. The ring structure is part of the structure of the final product *a*7*b*3, and so it can be assumed that *a*4*b*4 [56] is the precursor species. These structures are schematised in Fig. 9.10, together with their most likely paths of reaction.

For most of the reaction, silsesquioxanes with $a=5$ and $a=6$ were the two major species in solution. Their concentrations increased at the beginning of the reaction to reach a maximum after 1 h and then gradually decreased. This happened more rapidly for species with $a=5$, which were only present in very small amounts at the end of the reaction, than for silsesquioxanes with $a=6$, which were still present in relevant amounts after 24 h. The two most likely structures with $a=5$ and $a=6$ are silsesquioxanes *a*5*b*5 and *a*6*b*4, which are represented in Fig. 9.10, together with their possible pathways of formation.

For silsesquioxanes with $a=8$, two groups of peaks with different relative concentration profiles exist. The different behavior as a function of reaction time suggests that the two groups of peaks belong to two different species with $a=8$. The first group presented a main peak at $m/z=987.15$, which corresponds to an *a*8*b*2 silsesquioxane. This species has a low concentration during the entire reaction, with only a slightly positive slope towards the end (Fig. 9.9, top plot). The second group of peaks presented a main peak at $m/z=1041.20$; its intensity reached a maximum at $t=30$ min and then decreased rapidly to disappear after 3 h of reaction (Fig. 9.9, bottom plot). The intensity profile for these species is very similar to those of species with $9 \leq a \leq 13$, indicating a maximum species concentration at $t=30$ min for $a=9$, 10 and at $t=60$ min for $11 \leq a \leq 13$. Given the short lifetime of the species with $8 \leq a \leq 13$,

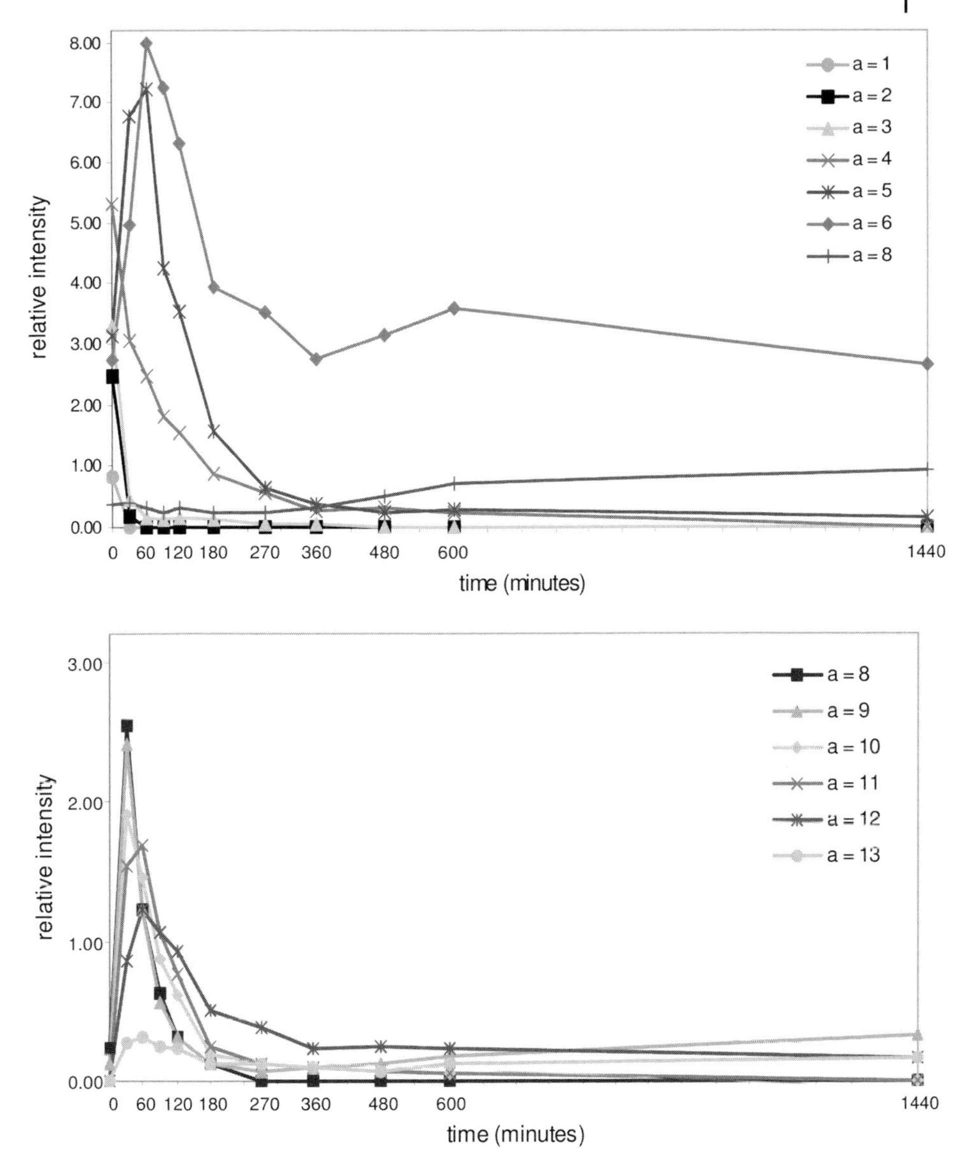

Fig. 9.9 Relative intensity of the ESI MS peaks of the various silsesquioxane species present in solution as a function of the reaction time. All intensities are normalised to the intensity of the peaks of $a=7$ species.

and that they were present in the early stages of the reaction, it can be suggested that they are instable aggregates of two smaller silsesquioxanes. (A similar effect can be seen measuring the mass spectrum of pure cyclopentyl silsesquioxane *a7b3* under similar experimental conditions but at two different cone voltages: with cone= 30 V, two main peaks are present, at $m/z=$ 875.33 and at $m/z=$1749.89; with

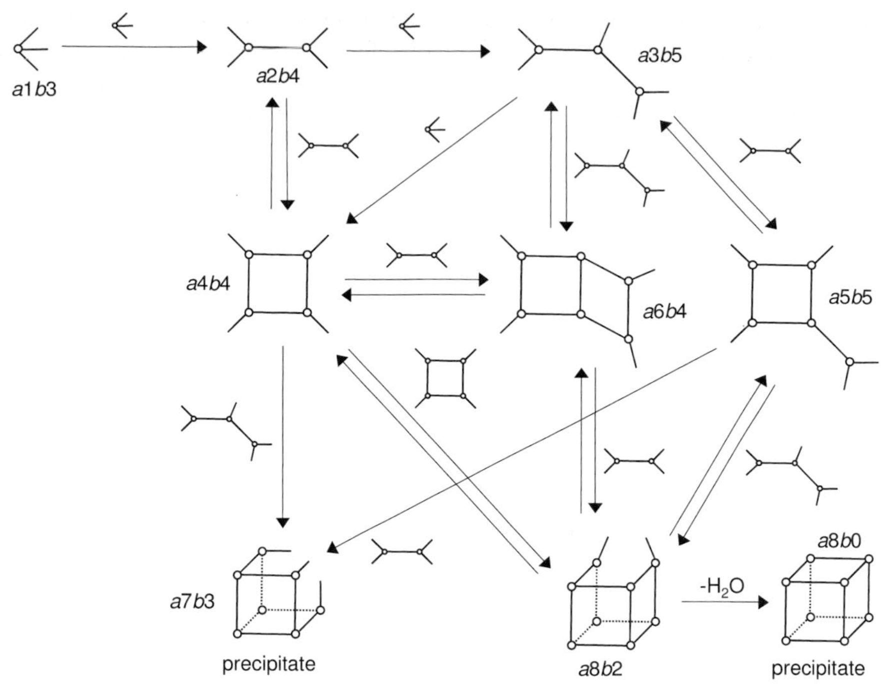

Fig. 9.10 Proposed mechanism for the synthesis of cyclopentyl silsesquioxane *a7b3*. The silsesquioxane structures are represented schematically. Each circle symbolises a siloxane unit [(c-C_5H_9)SiO_3]: silicon atoms are represented by the circles and the oxygen atoms by the lines; non-bridging lines represent –OH groups; cyclopentyl groups not shown.

cone = 65 V, the peak at $m/z = 1749.89$, due to the *a7b3* dimer, disappears.) For example, species with $a = 8$ are probably dimers of $a = 4$ species, those with $a = 9$ are constituted by $a = 4$ and $a = 5$ species, and so on.

On the basis of all the information collected from this MS study, it is possible to propose a mechanism for the formation of silsesquioxane *a7b3* (Fig. 9.10). Silsesquioxanes with $1 \leq a \leq 4$ are formed in the early stages of the synthesis and react very quickly to form more condensed species. It is assumed that the very reactive *a1b3* will not be formed again by hydrolysis reactions of more condensed species: this means that the reactions in which *a1b3* takes part are effectively irreversible and that the compound will only be available for the reactions in the early stages of the synthesis. Silsesquioxanes with $2 \leq a \leq 4$ are then going to react with each other to form the more condensed silsesquioxanes with $4 \leq a \leq 8$. These reactions are more easily reversible, meaning that the structures that are formed by condensation can be hydrolyzed back to the original compounds or to other less condensed species.

As mentioned above, silsesquioxane *a7b3* is less soluble than other silsesquioxanes present in the reaction mixture and precipitates as a white solid. This will influence the equilibrium composition and consequently drive the reactions that involve its formation towards the product; therefore, these reactions are considered irreversible. Silsesquioxanes with $a=6$ exhibit a maximum concentration in solution after 1 h of reaction; they are then rapidly consumed up until 3 h, after which the reaction proceeds more slowly. Since no $a=6$ silsesquioxane is present in the precipitate, it is inferred that the compound is slowly hydrolyzed to smaller species that then recombine to give the product *a7b3* [38, 57]. The absence of an $a=6$ silsesquioxane product is, therefore, not due to the instability of the compound but rather to its higher solubility in the reaction mixture. The completely condensed silsesquioxane *a8b0*, obtained by condensation of *a8b2*, is only present in traces in the precipitate. This means that, although silsesquioxane *a8b0* might be expected to be the most thermodynamically favored structure, its formation is unfavorable compared with that of silsesquioxane *a7b3*, suggesting that the reaction is kinetically controlled.

9.2.5.1.2 **Monitoring by Infrared Spectroscopy**

Attenuated total reflection (ATR) FTIR spectroscopy allows in situ monitoring of liquid-phase reactions by collecting the infrared spectra for solution species directly in contact with the infrared probe [58]. In the general case of a reaction in which various compounds are present, each ATR FTIR spectrum will consist of the overlapping spectra of all the pure components present in solution at a specific reaction time. During the reaction, the concentration of reagents and products will change, thus influencing the spectrum profile. This may allow appropriate chemometric methods to be used to deconvolute the pure component spectra and relative concentration profiles as a function of time. If the IR spectra of some components are known in advance (for example, the solvent, reagents and selected products) it is possible to use them as references to improve the fidelity of the deconvolution.

In the ATR FTIR study of the synthesis of cyclopentyl silsesquioxane *a7b3*, in situ ATR FTIR spectra of the reaction mixture were collected every 2 min during the reaction. The spectra obtained were plotted as a function of reaction time (Fig. 9.11). Pure component spectra and relative concentration profiles were subsequently recovered using a multivariate curve resolution (MCR) [59] technique based on a modified target factor analysis algorithm [60].

Principal component analysis (PCA) [61] was first used to determine the number of independently varying chemical species present and to provide initial estimates of the spectral shapes resulting from these species and of their concentration profiles. Reference ATR FTIR spectra for several components (the solvent acetonitrile and water, the reagent cyclopentyltrichlorosilane and the product *a7b3*) were measured to assist in the deconvolution of the data. Frequency windows were selected that allowed the best discrimination between the reference compounds (725–775 cm^{-1} for acetonitrile, 850–900 cm^{-1} for water and the silsesquioxane). Finally, the MCR technique was applied to the data in the selected frequency windows to find the component spectra and relative concentration profiles that best fit the observed spectra.

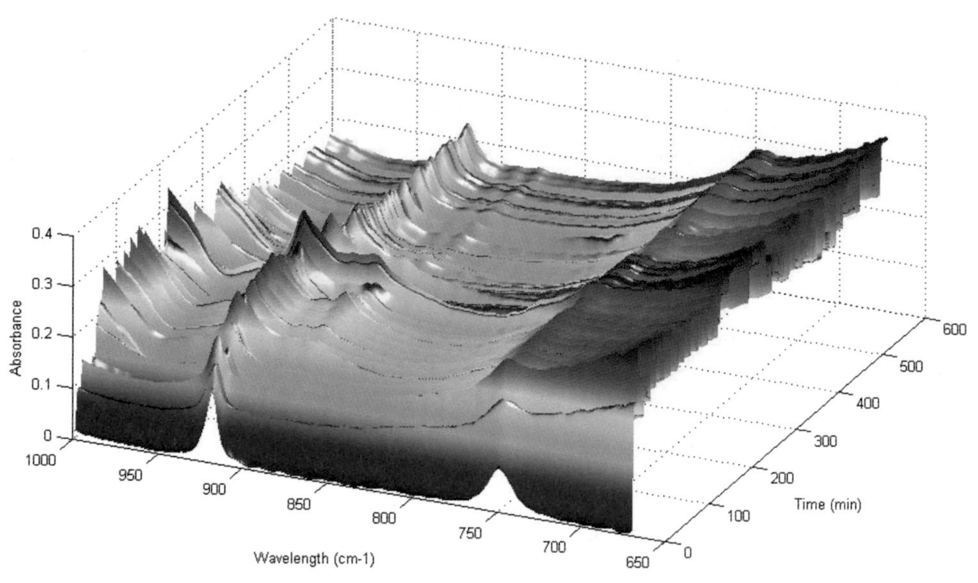

Fig. 9.11 In situ ATR FTIR spectra of the silsesquioxane reaction mixture as a function of the reaction time for a window of 670–1000 cm^{-1}. Analytical parameters were: IR analysis (670–4000 cm^{-1}) was performed for 10 h with the acquisition of a spectrum every 2 min, 16 scans per spectrum and a resolution of 4 cm^{-1}.

From Fig. 9.12 it can be seen that the pure component spectra obtained by the MCR technique closely match the reference spectra for acetonitrile (with a correlation coefficient $R=0.987$), water ($R=0.993$) and silsesquioxane ($R=0.953$). A poor match ($R=0.214$) was obtained for the cyclopentyltrichlorosilane reference spectra, indicating that cyclopentyltrichlorosilane cannot be detected at any stage during the reaction. This supports the MS results that also suggest that trichlorosilane is immediately hydrolyzed once water is added to the solution.

Relative concentration profiles for the identified species are given in Fig. 9.13. Acetonitrile and water concentrations were reasonably constant during the reaction, reflecting the fact that both liquids were present in large quantities, and that any change in concentration is effectively negligible.

The relative concentration profile of the silsesquioxane component, which has a slightly less accurate fit to the silsesquioxane *a7b3* reference spectrum ($R=0.953$), reaches a maximum after 2 h and becomes approximately constant after 4 h. From the MS study (Section 9.2.5.1.1) it is inferred that silsesquioxane *a7b3* has a very low and approximately constant concentration in solution throughout the entire reaction, since the compound is only sparingly soluble, and precipitates at higher concentrations. This suggests that the silsesquioxane monitored by ATR FTIR is not represented by structure *a7b3*. Other incompletely condensed cyclopentyl silsesquioxanes are expected to have infrared spectra very similar to that of *a7b3* in the studied spectral regions [50, 55]. (This similarity was confirmed by

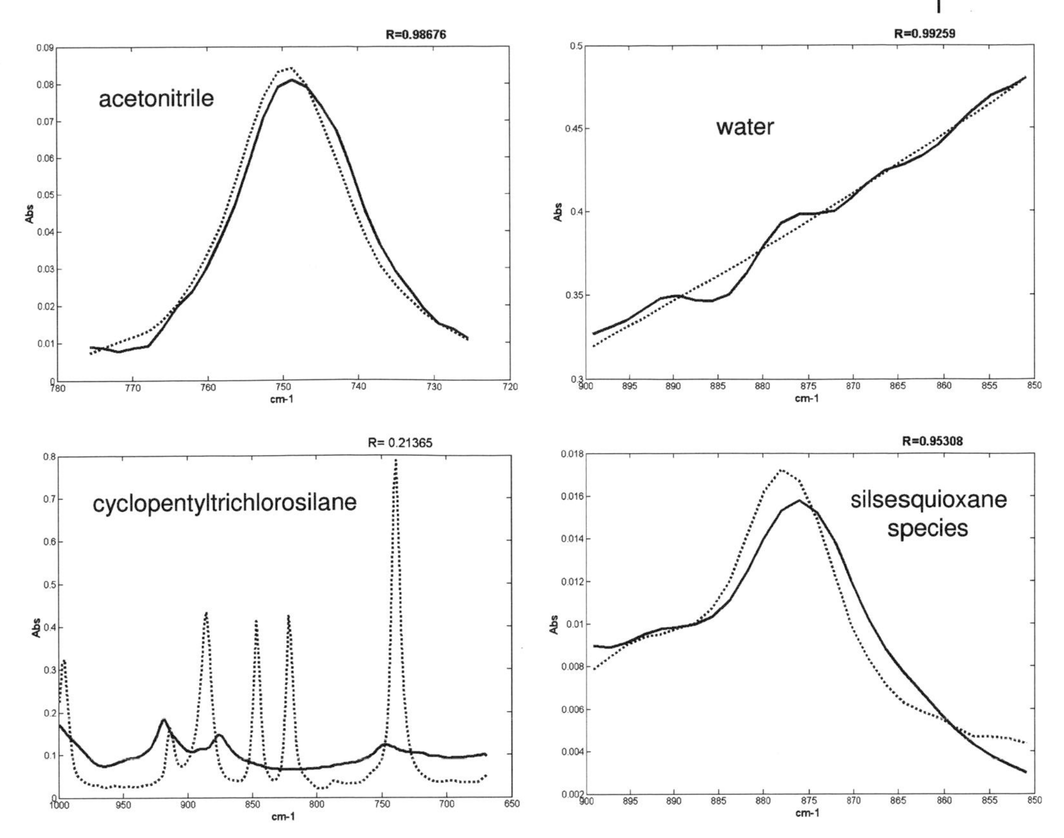

Fig. 9.12 Comparison of the measured reference spectra (dotted line) with the pure component spectra (solid line) obtained by chemometric deconvolution of the complex ATR FTIR spectra reported in Fig. 9.11.

measuring and comparing the transmission IR spectra (KBr pellet) of cyclopentyl silsesquioxane *a7b3* with that of the mixture of silsesquioxanes (mainly containing $a=5$ and $a=6$ species) obtained by removing the solvent from the reaction mixture after 18 h of reaction at 50 °C (instead of reflux conditions). The two spectra present corresponding sets of peaks (within 6 cm^{-1} in the region 670–1000 cm^{-1}).

Therefore, it is proposed that the species identified by MCR is an $a=5$ or an $a=6$ structure, which presents an MS concentration profile similar to that obtained by ATR FTIR, or a mixture of more silsesquioxane structures. The hypothesis of an $a=6$ structure is supported by the fact that the ATR FTIR concentration profile for the silsesquioxane species as a function of time is rather similar to the MS concentration profile obtained for the $a=6$ silsesquioxane (cf. Figs 9.9 and 9.13). The slightly different position of the maximum in the two plots is considered to be due to the longer time required in the ATR FTIR experiment to reach the reflux temperature. (This explanation was confirmed by repeating part of the MS study of the synthesis of silsesquioxane *a7b3* at 50 °C (instead of reflux tem-

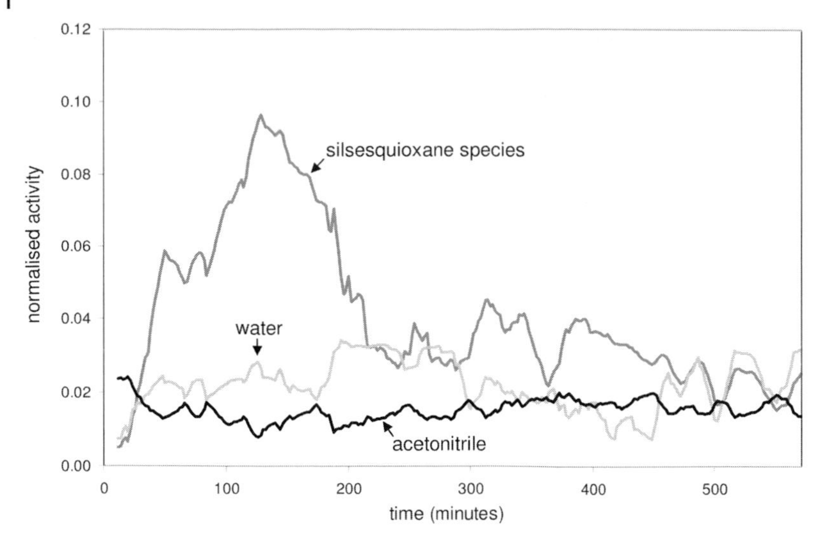

Fig. 9.13 Relative concentration profiles of the three species identified by chemometric deconvolution of the complex ATR FTIR spectra reported in Fig. 9.11.

perature). Under these conditions the hydrolytic condensation is slower and, though the same silsesquioxane species were present in solution, species with $1 \leq a \leq 4$ disappeared more slowly and those with $a=5$ and $a=6$ reached the maximum of the concentration profile after a longer reaction time.)

The combined MS and ATR FTIR results point to the fact that the concentration of the silsesquioxane species in solution becomes almost constant after 4 h of reaction. This might imply that a reaction time shorter than 18 h could be sufficient to obtain silsesquioxane *a7b3* in high yield. Therefore, the synthesis was repeated with the same conditions but with a reaction time of 6 h: silsesquioxane *a7b3* was isolated with a yield of 54%. After collection of the precipitate, the reaction mixture was allowed to react for a further 12 h, yielding more precipitate, and a total yield of 64% was achieved. Although the yield after 6 h is lower than that obtained after 18 h, the experiment confirmed that most of the yield of the desired *a7b3* structure is generated within the first hours of reaction. This finding would certainly be of practical importance for a potential up-scaling of silsesquioxane *a7b3* production.

9.2.5.2 tert-Butyl Silsesquioxanes Synthesised in H_2O

The hydrolytic condensation of tert-butyltrichlorosilane Bu^tSiCl_3 in water was performed in a 25-fold up-scaling of the HTE synthesis. Bu^tSiCl_3 is a solid at room temperature: since the HTE workstation employed can only handle liquids, the trichlorosilane was dissolved in a minimum of acetonitrile. The same procedure was used for the up-scaled synthesis. After 18 h of reaction at 50 °C, the reaction mixture contained a white precipitate, which was isolated by filtration (fraction A).

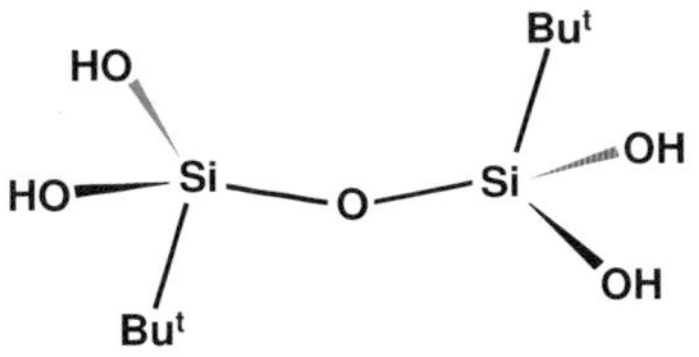

a2b4

Fig. 9.14 tert-Butyl silsesquioxane *a2b4*, $Bu^t_2Si_2O(OH)_4$.

Upon drying the filtrate, a white solid was obtained (fraction B). The two fractions were exsiccated in an oven at 100 °C. Both fractions contained the same single silsesquioxane structure, which was assigned to silsesquioxane $Bu^t_2Si_2O(OH)_4$ (*a2b4*, Fig. 9.14) on the basis of NMR data and of single-crystal X-ray diffraction analysis, which provided the same cell parameters of the tert-butyl silsesquioxane *a2b4* as reported in ref. [62]. The synthetic procedure reported here allows complete and selective conversion of Bu^tSiCl_3 into silsesquioxane $Bu^t_2Si_2O(OH)_4$. Further investigation showed that only 7 h of hydrolytic condensation at 50 °C are necessary to obtain silsesquioxane *a2b4* as the only product. This method is more straightforward and results in a higher isolated yield (90%) than the literature method (65%) in which the compound is synthesised by addition of Bu^tSiCl_3 in diethyl ether to a mixture of KOH, water and methanol [62]. The selectivity of the synthesis is ascribed to the bulkiness of the tert-butyl group, which can hinder further reaction of $Bu^t_2Si_2O(OH)_4$, and to the use of water as the solvent, which disfavors condensation.

Besides the intrinsic value of the identification of a new, selective and high yield method to synthesise silsesquioxane $Bu^t_2Si_2O(OH)_4$, this experiment proved that tert-butyl silsesquioxane *a2b4* is a suitable precursor for titanium catalysts. Thus, silsesquioxane structures different from the known precursor silsesquioxane *a7b3* [$R_7Si_7O_9(OH)_3$] [25, 26, 28] can effectively coordinate titanium centres to yield almost equally active epoxidation catalysts.

To investigate how the titanium centre coordinates to tert-butyl silsesquioxane *a2b4* and which is the optimum number of titanium centres that the structure can accommodate, two catalysts with different titanium-to-silsesquioxane ratios were prepared, characterised and tested for epoxidation activity: catalysts **I** and **II**, prepared by reacting $Ti(OBu)_4$ with tert-butyl silsesquioxane *a2b4* in 1:1 and 2:1 molar ratios, respectively. Characterisation by ^{13}C and ^{29}Si NMR and gel permeation chromatography showed that these catalysts consist of various titanium centres and silsesquioxanes units that are linked in different arrangements [45]. The two catalysts displayed higher activity (per mole of titanium) than the HTE lead (74% TBHP conversion towards 1,2-epoxyoctane after 4 h of reaction): catalyst **I** gave 80% conversion after 3 h and reached a plateau at 93% conversion (TOF = 0.44 $mol_{epo} \cdot mol_{Ti}^{-1} \cdot min^{-1}$); catalyst **II** gave 90% conversion after 3 h and reached a plateau at 97% conversion (TOF = 0.54 $mol_{epo} \cdot mol_{Ti}^{-1} \cdot min^{-1}$) (Fig. 9.15). Both catalysts exhibited 97% selectivity towards 1,2-epoxyoctane. The difference in activity between these

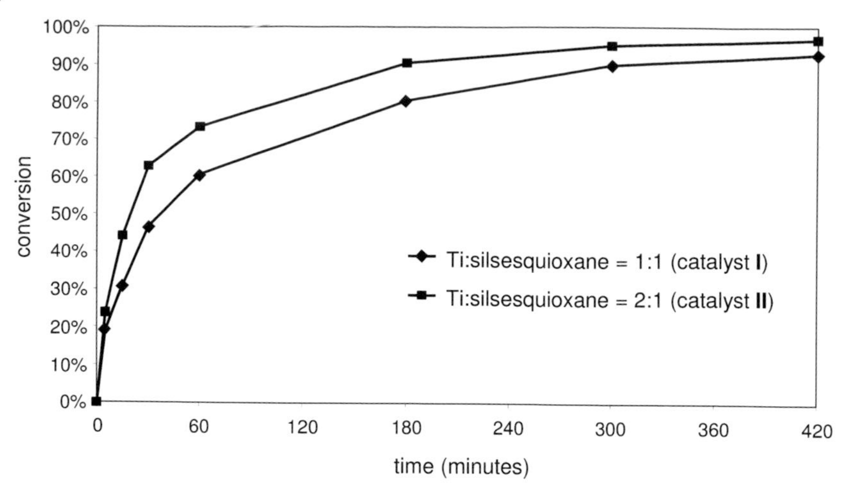

Fig. 9.15 Activity in the epoxidation of 1-octene with TBHP, for Ti-*a2b4* complexes with 1 : 1 and 2 : 1 titanium-to-silsesquioxane ratios. The concentration of titanium and the titanium-to-TBHP ratio were the same in the two catalytic tests.

two catalysts probably originates from the different nature of the titanium centres (different accessibility) [45].

Finally, the concept was broadened by supporting these titanium silsesquioxanes on silica: remarkably, the heterogeneous silica-supported catalysts displayed an epoxidation activity per mole of titanium (94% TBHP conversion after 3 h) similar to that of the homogeneous titanium-silsesquioxane *a2b4* complexes, although with a lower selectivity towards 1,2-epoxyoctane (92%). These heterogeneous catalysts did not leach active species and proved to be recyclable [45].

9.3
Screening of the Catalytic Activity of Germanium-containing Zeolites

Zeolites [63] are extensively used as shape-selective solid acid catalysts in many industrial processes [64]. Their acidic properties stem from the presence of trivalent elements, such as Al, in the zeolite framework. The strength of these acid sites is one of the main features that determine the catalytic properties of a zeolite catalyst. Substitution of the Al atoms by other trivalent elements, such as Ga, Fe, and B, alters the strength of these acid sites, and hence also the catalytic properties of a zeolite. The possible effect of the partial substitution of the tetravalent Si atoms (which, in principle, do not create acid sites in zeolites) by Ge atoms (also tetravalent) on the catalytic properties of zeolite ZSM-5 [65] is presented here. The idea is that the different electronic and geometric properties of Ge, compared with Si, may influence the acid sites related to the Al atoms, and thereby the catalytic properties of ZSM-5.

9.3.1
High-Throughput Experimentation Approach

HTE techniques played an important role in this study, both at the synthesis stage and in the screening the catalytic properties of Ge-ZSM-5. The most suitable synthetic procedure towards Ge-ZSM-5 could be identified by the variation of the relative amounts and of the nature of the starting materials within one set of experiments [66]. This approach allowed a substantial reduction of both the required amount of starting materials and time. The optimised synthetic procedure for Ge-ZSM-5 was then repeated on a larger scale to prepare a series of Ge-ZSM-5 samples with similar Al contents [(Ge + Si)/Al $\approx$ 50] and with varying Ge contents (Ge/(Ge+Si)=0 to 0.10). The catalytic activity of these zeolite samples was then investigated in several test reactions using HTE [67]. Gas-phase experiments were performed in a heated reactor block containing 16 miniaturised gas flow reactors, which required only small amounts (25 mg) of catalyst. The reactor block was coupled to an on-line gas chromatograph, allowing the analysis of each of the gas flows at regular intervals and at variable, regulated temperatures. In this way, the catalytic performance of up to 16 zeolite samples could be screened simultaneously overnight at a range of reaction temperatures and at prolonged times on stream.

9.3.2
Catalyst Screening Using HTE Techniques

The catalytic properties of Ge-ZSM-5 have been compared with ZSM-5 for various acid-catalyzed test reactions. Here, results are reported about the dehydration reaction of 2-butanol in the gas phase, and about the Friedel-Crafts acylation of anisole in the liquid phase. For the dehydration of 2-butanol, the gas feed into the heated reactor block consists of a 2-butanol–N_2 mixture. HTE techniques allowed the screening of many materials with different composition for this reaction [67]. Also, 2-butanol dehydration at different gas flow rates and experiments with regenerated catalyst samples could be performed with relatively small amounts of catalyst in a short time. Fig. 9.16 shows the conversion values of 2-butanol over a range of catalysts (with varying Ge contents) from the HTE screening as a function of temperature and time on stream. Ge and Al contents of each sample are given in parentheses: Ge(0.04)ZSM-5(49) indicates a Ge/(Ge+Si) ratio of 0.04 and a (Ge+Si)/Al ratio of [49]. The selectivity towards butenes (1-butene, *cis*-2-butene and *trans*-2-butene) is 100% for all catalysts and at all investigated temperatures. The distribution of butene products is very similar for the different zeolites and is not a function of the reaction temperature, the 1-butene:*cis*-2-butene:*trans*-2-butene ratio being around 11:55:34.

As can be concluded from Fig. 9.16, the conversion values over all catalysts increase upon heating from 80 to 130 °C, following a common Arrhenius equation. Above 130 °C the catalysts start to behave differently: the conversion over Ge(0.09)ZSM-5(36) increases further (from 92 to 100%) upon increasing the tem-

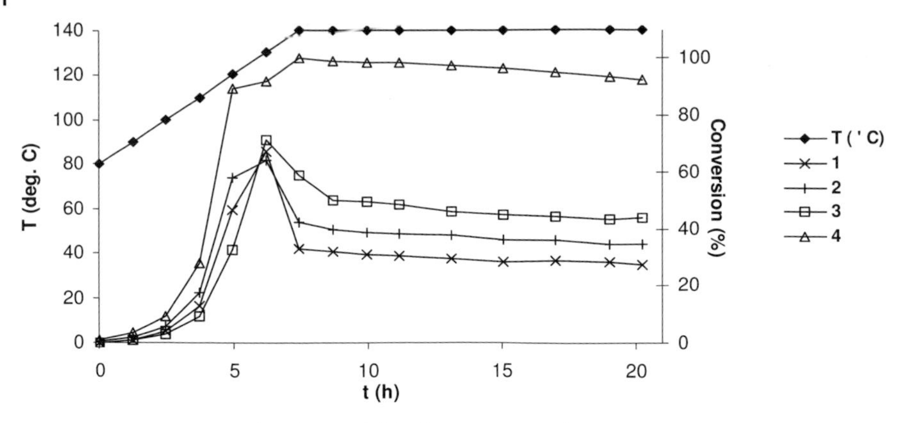

Fig. 9.16 Conversions in the 2-butanol dehydration reaction by (1) ZSM-5(47), (2) Ge(0.04)ZSM-5(49), (3) Ge(0.06)ZSM-5(58) and (4) Ge(0.09)ZSM-5(36) at prolonged reaction time (at 140 °C). WHSV = 1.5 h^{-1}.

perature from 130 to 140 °C, while the conversion over the catalysts with lower Ge contents drops at 140 °C.

As different conversion levels only arise at high reaction temperature, no significant difference in acidic strength of the catalytically active sites of these samples can be anticipated. The catalytic activity of high Ge-content materials decreases to a much smaller extent than that of the zeolites with less Ge, indicating that deactivation through coke formation [68] has only a small effect on Ge(0.09)ZSM-5(36). Analysis of the porous properties of this series of zeolites revealed that an increasing Ge content in ZSM-5 induces increasing levels of meso- and macroporosity in the structure [66]. This difference is due to the different crystallisation behavior of Ge-ZSM-5 compared with ZSM-5. The additional larger pores in Ge-ZSM-5 prevent quick deactivation of the catalyst, as some of the active sites are still accessible for reaction after blocking of most of the micropores by coke residues.

The set of catalysts selected for the dehydration of 2-butanol was also tested for the Friedel-Crafts acylation of anisole [69, 70]. The catalytic test was performed in the liquid phase due to the high boiling points of the reactants and products of this reaction. Anisole was reacted with acetic anhydride at 120 °C in the absence of solvent. In principle, acylation can occur on both the ortho and para positions of anisole. The main product (>99%) over all catalysts in this study was *para*-methoxyacetophenone, indicating that the reaction predominantly takes place inside the zeolite micropores. The same trend in catalytic activity as in the 2-butanol dehydration reaction is observed: the conversion of anisole into *para*-methoxyacetophenone increases upon increasing Ge content of the catalyst (Fig. 9.17) [67]. The main cause of deactivation for this reaction is accumulation of the reaction products inside the micropores of the zeolite. The different behavior of Ge-ZSM-5, compared with ZSM-5, may therefore be due to improved diffusional properties of the former, as the presence of additional meso- and macropores allows for

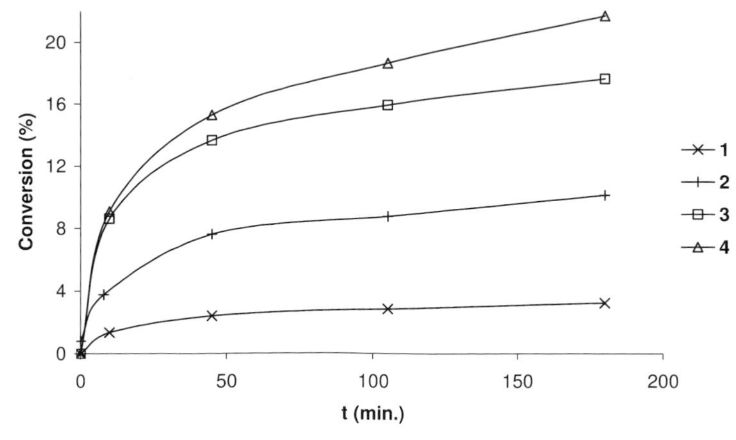

Fig. 9.17 Anisole conversions in the acylation reaction at 120 °C by (1) ZSM-5(47), (2) Ge(0.04)ZSM-5(49), (3) Ge(0.06)ZSM-5(58) and (4) Ge(0.09)ZSM-5(36).

more efficient diffusion of reactant molecules into, and product molecules out of the micropores.

The Ge-ZSM-5 catalysts have also been screened in other test reactions [67]. In general, materials with high germanium content appeared to suffer to a much smaller extent from deactivation than samples with lower germanium levels for most of the acid-catalyzed reaction studied. In this way it was shown that the difference in catalytic performance originates from the presence of Ge in these samples and that the improved catalytic stability is similar in several acid-catalyzed test reactions.

9.4 Conclusions

The application of HTE techniques for studying liquid-phase reactions has been demonstrated for several catalytic reactions. Good results are obtained when the design of the catalyst library, typically a small, full-factorial library of 24 catalysts, is coupled to an extensive literature survey that pinpoints the major variables of the reaction system under study. Combining less important parameters into the reduced and optimised catalyst library can be used to further fine-tune the catalyst.

The exploration of silsesquioxane-based titanium catalysts has resulted in valuable insight into the fundamentals of silsesquioxane formation. Furthermore, an improved synthesis method for the important precursor $R_7Si_7O_{12}H_3$, with R = cyclopentyl, was identified. Finally, new catalytic systems based on $R_2Si_2O_7H_4$, with R = tert-butyl, were discovered.

Ge-containing ZSM-5 zeolites displayed enhanced stability in several acid-catalyzed reactions. Examination of the catalyst library showed an increased mesoporosity and higher external surface with increasing Ge content; therefore, it was concluded that the enhanced catalytic properties are due to slower pore blocking.

These examples reveal the wide applicability of HTE techniques in studying catalyst systems for liquid-phase reactions.

9.5 References

1 N.K. Terret, M. Gardner, D.W. Gordon, R.J. Kobylecki, J. Steele *Tetrahedron*, 1995, *30*, 8135.
2 P.P. Pescarmona, J.C. van der Waal, I.E. Maxwell, T. Maschmeyer, *Catal. Lett.*, 1999, *63*, 1.
3 J.M. Newsam, F. Schüth, *Biotechnol. Bioeng. (Comb. Chem.)*, 1998/1999, *61*, 203.
4 B. Jandeleit, D.J. Schaefer, T.S. Powers, H.W. Turner, W.H. Weinberg, *Angew. Chem., Int. Ed.*, 1999, *38*, 2494.
5 R. Schlögl, *Angew. Chem., Int. Ed.*, *1998*, *37*, 2333.
6 T. Bein, *Angew. Chem., Int. Ed.*, 1999, *38*, 323.
7 W.F. Maier, *Angew. Chem., Int. Ed.*, 1999, *38*, 1216
8 J.M. Thomas, *Angew. Chem., Int. Ed.*, 1999, *38*, 3589.
9 K.D. Shimizu, M.L. Snapper, A.H. Hoveyda, *Chem. Eur. J.*, 1998, *4*, 1885.
10 R.H. Crabtree, *Chem. Commun.*, 1999, 1611.
11 M.T. Reetz, *Angew. Chem., Int. Ed.*, 2001, *40*, 284.
12 A. Holzwarth, P. Denton, H. Zanthoff, C. Mirodatos, *Catal. Today*, 2001, *67*, 309.
13 S. Senkan, *Angew. Chem., Int. Ed.*, 2001, *40*, 312.
14 J.M. Newsam, T. Bein, J. Klein, W.F. Maier, W. Stichert, *Micropor. Mesopor. Mater.*, 2001, *48*, 355.
15 T.X. Sun, *Biotechnol. Bioeng. (Comb. Chem.)*, 1998/1999, *61*, 193.
16 X.-D. Xiang, *Biotechnol. Bioeng. (Comb. Chem.)*, 1998/1999, *61*, 227.
17 M.F. Asaro, R.B. Wilson, *Chem. Ind.*, 1998, *19*, 777.
18 M.T. Reetz, *Angew. Chem., Int. Ed.*, 2002, *41*, 1335.
19 M.T. Reetz, M.H. Becker, H.-W. Klein, D. Stöckigt, *Angew. Chem., Int. Ed.*, 1999, *38*, 1758
20 M. Baerns, C. Mirodatos, *NATO Science Series, Ser. II, 2002, Vol. 69*, 469.
21 B. Notari, *Adv. Catal.*, 1996, *41*, 253.
22 G. Bellussi, M.S. Rigutto, *Stud. Surf. Sci. Catal.*, 1994, *85*, 177.
23 T. Maschmeyer, F. Rey, G. Sankar, J.M. Thomas, *Nature*, 1995, *378*, 159.
24 M.C. Klunduk, T. Maschmeyer, J.M. Thomas, B.F.G. Johnson, *Chem. Eur. J.*, 1999, *5(5)*, 1481.
25 T. Maschmeyer, M.C. Klunduk, C.M. Martin, D.S. Shephard, J.M. Thomas, B.F.G. Johnson, *Chem. Commun.*, 1997, 1847.
26 M. Crocker, R.H.M. Herold, A.G. Orpen, *Chem. Commun.*, 1997, 2411.
27 H.C.L. Abbenhuis, S. Krijnen, R.A. van Santen, *Chem. Commun.*, 1997, 331.
28 S. Krijnen, H.C.L. Abbenhuis, R.W.J.M. Hanssen, J.H.C. van Hooff, R.A. van Santen, *Angew. Chem., Int. Ed.*, 1998, *37*, 356.
29 L. Marchese, T. Maschmeyer, E. Gianotti, S. Coluccia, J.M. Thomas, *J. Phys. Chem. B*, 1997, *101*, 8836.
30 P.E. Sinclair, G. Sankar, C.R.A. Catlow, J.M. Thomas, T. Maschmeyer, *J. Phys. Chem. B*, 1997, *101*, 4232.
31 D. Tantanak, M.A. Vincent, I.H. Hiller, *Chem. Commun.*, 1998, 1031.
32 P.E. Sinclair, C.R.A. Catlow, *J. Phys. Chem. B*, 1999, *103*, 1084.
33 C.M. Barker, D. Gleeson, N. Kaltsoyannis, C.R.A. Catlow, G. Sankar, J.M. Thomas, *Phys. Chem. Chem. Phys.*, 2002, *4*, 1228
34 P.P. Pescarmona, T. Maschmeyer, *Aust. J. Chem.*, 2001, *54*, 583.

35 T. Maschmeyer, J.M. Thomas, A.F. Masters, *NATO ASI Ser., New Trends in Materials Chemistry*, 1997, *498*, 461.
36 H.C.L. Abbenhuis, *Chem. Eur. J.*, 2000, *6*, 25.
37 F.J. Feher, D.A. Newman, J.F. Walzer, *J. Am. Chem. Soc.*, 1989, *111*, 1741.
38 F.J. Feher, T.A. Budzichowski, R.L. Blanski, K.J. Weller, J.W. Ziller, *Organometallics*, 1991, *10*, 2526.
39 P.P. Pescarmona, J.C. van der Waal, I.E. Maxwell, T. Maschmeyer, *Angew. Chem., Int. Ed.*, 2001, *40*, 740.
40 T. Kudo, M.S. Gordon, *J. Am. Chem. Soc.*, 1998, *120*, 11432.
41 T. Kudo, M.S. Gordon, *J. Phys. Chem. A*, 2000, *104*, 4058.
42 K.R. Popper, *The Logic of Scientific Discovery*, Hutchinson of London, 1959.
43 K.R. Popper, *Conjectures and Refutations. The Growth of Scientific Knowledge*, 3rd edition, Routledge and Kegan Paul, 1969.
44 P.P. Pescarmona, J.J.T. Rops, J.C. van der Waal, J.C. Jansen, T. Maschmeyer, *J. Mol. Cat. A*, 2002, *182/183*, 319.
45 P.P. Pescarmona, J.C. van der Waal, T. Maschmeyer, *Chem. Eur. J.*, 2004, *10*, 1657.
46 P.P. Pescarmona, J.C. van der Waal, T. Maschmeyer, *Catal. Today*, 2003, *81*, 347.
47 J.F. Brown, L.H. Vogt, *J. Am. Chem. Soc.*, 1965, *87*, 4313.
48 P.P. Pescarmona, J.C. van der Waal, T. Maschmeyer, *Eur. J. Inorg. Chem.*, 2004, 978.
49 P.P. Pescarmona, M.E. Raimondi, J. Tetteh, B. McKay, T. Maschmeyer, *J. Phys. Chem. A*, 2003, *107*, 8885.
50 M.G. Voronkov, V.I. Lavrent'yev, *Top. Curr. Chem.*, 1982, *102*, 199.
51 P. Bussian, F. Sobott, B. Brutschy, W. Schrader, F. Schüth, *Angew. Chem., Int. Ed.*, 2000, *39*, 3901.
52 R. Bakhtiar, F.J. Feher, *Rapid Commun. Mass Spectrom.*, 1999, *13*, 687.
53 N.J. Harrick, *J. Phys. Chem.*, 1960, *64*, 1110.
54 J. Fahrenfort, *Spectrochim. Acta*, 1961, *17*, 698.
55 J.F. Brown, L.H. Vogt, *J. Am. Chem. Soc.*, 1965, *87*, 4313.
56 T.S. Haddad, B.M. Moore, S.H. Phillips, *Polym. Preprints*, 2001, *42*, 196.
57 F.J. Feher, R. Terroba, J.W. Ziller, *Chem. Commun.*, 1999, 2153.
58 N.J. Harrick, *Internal Reflection Spectroscopy*, John Wiley & Sons, New York, 1967.
59 R. Tauler, B. Kowalski, S. Fleming, *Anal. Chem.*, 1993, *65*, 2040.
60 E. Metcalfe, J. Tetteh, *Polym. Mater. Sci. Eng.*, 2000, *83*, 88.
61 J.E. Jackson, *A Users Guide to Principal Components*, John Wiley & Sons, New York, 1991.
62 P.D. Lickiss, S.A. Litster, A.D. Redhouse, C.J. Wisener, *J. Chem. Soc., Chem. Commun.*, 1991, 173.
63 H. van Bekkum, E.M. Flanigen, P.A. Jacobs, J.C. Jansen (eds.), *Introduction to Zeolite Science and Practice*, 2nd edition, Elsevier, Amsterdam, 2001.
64 K. Tanabe, W.F. Hölderich, *Appl. Catal. A*, 1999, *181*, 399.
65 R.J. Argauer, G.R. Landolt, *US* 3702886, 1972.
66 L.G.A. van de Water, J.C. van der Waal, J.C. Jansen, M. Cadoni, L. Marchese, T. Maschmeyer, *J. Phys. Chem. B*, 2003, *107*, 10423.
67 L.G.A. van de Water, J.C. van der Waal, J.C. Jansen, T. Maschmeyer, *J. Catal.*, 2004, *223*, 170.
68 M. Guisnet, P. Magnoux, *Catal. Today*, 1997, *36*, 477.
69 E.G. Derouane, C.J. Dillon, D. Bethell, S.B. Derouane-Abd Hamid, *J. Catal.*, 1999, *187*, 209.
70 E.G. Derouane, G. Crehan, C.J. Dillon, D. Bethell, H. He, S.B. Derouane-Abd Hamid, *J. Catal.*, 2000, *194*, 410.

10
Combinatorial Strategies for Speeding up Discovery and Optimization of Heterogeneous Catalysts on the Academic Laboratory Scale: A Case Study of Hydrogen Purification for Feeding PEM Fuel Cells

David Farrusseng and Claude Mirodatos

10.1
Introduction

Catalysis development and understanding are essential to most chemical synthesis advances and therefore require new materials able to steer the reactions towards identified target products. Because the topic of chemical synthesis is so broad and catalysis is so crucial to chemical synthesis, a significantly higher speed and efficiency in new materials discovery and process optimisation is required today to:

1. Shorten the market introduction delay of new clean catalytic technologies.
2. Reduce costs for process development.
3. Meet environmental regulations such as using cheaper and greener feed-stocks.

A combinatorial "discovery" approach for accelerating development of new catalytic chemical processes is expected to answer these challenges. Combinatorial catalysis is a new science which shows high potential for future progress. The first papers were published in the mid-1980s. Exponentially growing activities in this field have been observed ever since, fundamentally turning the conventional trial & errors research into a new strategy for chemical material discovery. Due to its strategic impact on numerous chemical processes, combinatorial catalysis is of immediate industrial interest, unlike most of the other new areas in fundamental research. As such, most large chemical and pharmaceutical companies have adopted combinatorial methodology. In addition, many new companies have been set up since the 1990s to develop and provide adapted technology in the catalysis domain [1–12].

Now, in parallel to the industrial implementation of this methodology, a growing interest of the scientific community for the combinatorial chemistry applied to heterogeneous catalysis is attested to by a continuous effort of dissemination (sustainable and recurrent international conferences on "COMBICAT" research [13], with specialized publications or dedicated issues in international journals [14], websites etc.), though still remaining a very controversial subject. Indeed, within universities as well as within public and private research centres, "combinatorial catalysis" may be understood as a simple random mixing of various chemicals, without chemistry rules and knowledge input, based on accelerated

High-Throughput Screening in Chemical Catalysis
Edited by A. Hagemeyer, P. Strasser, A. F. Volpe, Jr.

ISBN: 3-527-30814-8

search procedures by means of "High Throughput Experiments" (HTE). Actually, this "randomised search" is far from describing the combinatorial approach that embodies, in addition to conventional knowledge in catalysis, numerous other domains such as micro-mechanics, robotics, analytics, kinetics, modeling and informatics.

This chapter overviews some strategies that have been developed and applied at the academic level, by presenting a case study on the basis of our own experience: the search for new catalytic material and processes adapted to the production and purification of hydrogen for feeding PEM fuel cells.

The main application field for HTE deals with the iterative process of library design and testing to find well-performing catalysts. Each step of the so-called "combinatorial loop" (Fig. 10.1) will be presented, underlining the pros and cons that can arise from this new methodology.

1. *Design* of experiments and compounds: choice of the domain of research, identification of the targeted performance, which may be single or multi-objective (yield, cost, stability, safety, etc.), choice of the adapted strategies (for example, GA – genetic algorithms or DoE – design of experiments), the informatic tools, existing or to be developed, the partnership if necessary.

2. *Make:* implementation of automated catalysts synthesis by robots and the search for compromise by adapting the choice of the chemist to the effective possibilities of the available robots.

3. *Test:* evaluation of the catalytic performances, again involving the expertise of the available robots and the proper choice of operating conditions (tested reactions, stationary or non-stationary regimes, T&P, contact time, etc.).

4. *Model:* extraction of knowledge or data-mining (using for example artificial neural networks – ANN) providing trends or, better, precise relationship between materials design (composition, structure) and performances. On the basis of this information, the design of the next library is ordered by optimisation algorithms and the loop iterates again until the targeted performance is reached.

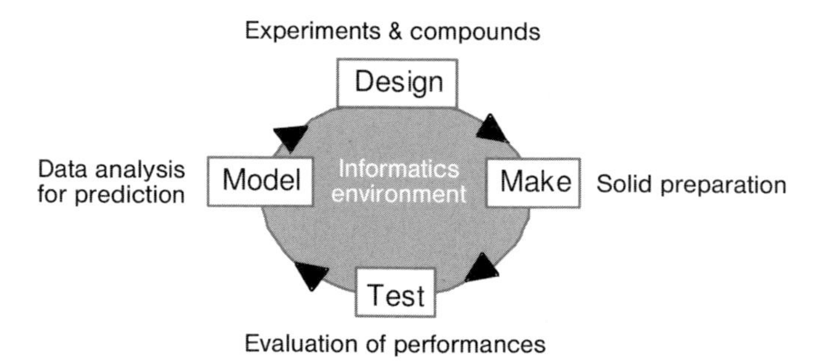

Fig. 10.1 Basic principle of the combinatorial approach applied to catalysis: iterative loop up to achievement of new product or process specification.

Another aspect of HT methodology, which may be considered as closer to the usual investigation mode at laboratory scale, aims to extract fundamental knowledge such as mechanistic and kinetic insights along the HTE process, which may in turn be used for further optimisation strategies.

5. *Informatic environment:* pivotal data management for each of the preceding steps, which involves capture of data, storage in a consistent way, query and also stream data analysis (data workflow).

10.2 Design: Targets and Strategies

10.2.1 Domain of Investigation

The first part of the combinatorial loop deals with the choice of the parameters to be investigated. Basically, all conventional or new chemistry fields for which suitable catalysts have not been found, which, despite decades of intensive research, are open for applying the combinatorial approach. This is the case for "dream reactions" such as propylene-to-propylene oxide, methane-to-methanol, alkane-to-alkene/oxygenates, etc. Conversely, due to the increasing constraints on releasing pollutants into the atmosphere or water, more efficient catalysts have to be discovered rapidly within the large field of environmental chemistry. As such, open fields have to be considered in the fine chemicals and pharma domain where non catalytic processes or homogeneously catalyzed systems could be replaced by more efficient and less-polluting heterogeneous systems enabling catalysts recovery and regeneration.

Here, the chosen domain for our case study is on-board hydrogen production to supply pure H_2 to a fuel cell in an electrical car. Among the sequential catalytic reactions that take place for H_2 production, the hydrogen purification units are located downstream, after the primary reforming of hydrocarbons into a $CO–H_2$ mixture or Syngas units. They consist of Reaction (1) the water-gas shift (WGS) reaction and Reaction (2), the selective or preferential oxidation of CO in the presence of hydrogen (Selox).

$$CO + H_2O = CO_2 + H_2 \qquad \Delta H = -41\ \mathrm{kJ\ mol^{-1}} \tag{1}$$

$$CO + H_2 + O_2 \rightarrow CO_2 + H_2O \qquad \Delta H = -524\ \mathrm{kJ\ mol^{-1}} \tag{2}$$

These are critical steps since they enhance the overall H_2 yield while minimizing the concentration of CO, which poisons the fuel cell.

Commercial WGS catalysts have been optimised for more than 50 years for the massive H_2 production in petrochemical plants. However, on-board production requires new catalytic properties such as short response in a dynamic regime, sulfur tolerance, low toxicity and safety (commercial catalysts generally contain Cr and are pyrophoric) and above all a higher efficiency to minimize the size of reactors. Immediately after the WGS units (high and low temperature), the Selox unit aims at lowering the CO concentration from about 1% to less than 10 ppm, by oxidizing CO into CO_2 after oxygen addition, but keeping the parallel hydrogen oxidation as low as possible. Again though considerable effort has been devoted to this critical step, which determines the lifetime of the fuel cell, much improvement in efficiency is required. Therefore, for these hydrogen purification steps, new miniaturized catalytic systems have to be developed and commercialised rapidly.

- *Pros:* The chosen domain implies an "old" chemistry to be revamped for "new" applications such as the electrical vehicle for the future. However, no existing proven process exists; therefore a significant breakthrough is expected, including the discovery of new eco-efficient materials and miniaturised technology that would allow car companies to develop light, compact vehicles. At the laboratory scale, the choice of these WGS/Selox reactions, which are closed and may require to be combined, appears favourable for a combinatorial approach since those two reactions can be catalyzed by many elements from the periodic table, which creates a vast space of parameters and therefore fits perfectly with the combinatorial strategy.
- *Cons:* Both reactions may involve transient operating conditions (e.g. light-off, shut-down) that may change considerably the state and performances of the catalysts (often mixed-oxide based, highly sensitive to the redox surrounding atmosphere). Therefore, ranking in a fast and single primary screening may lead to meaningless catalyst ranking and selection.

10.2.2
Choice of Objective Functions

The choice of target to be reached using the iterative loop represented in Fig. 10.1 or the "objective function" as defined by Wolf et al. [7, 15] is the second pivotal step before starting a combinatorial search. It should correspond to the studied case related to the envisioned application, e.g. the yield into a high value product, or only the selectivity in that product, if recycling is possible, or just the degree of conversion if selectivity is not a process issue. Any other characteristics of the catalytic system can also be considered as a target, like ageing resistance, possibility to regenerate, cost of materials, etc. Whatever the chosen target, one must ensure that the analytics of the combinatorial loop can rapidly and quantitatively evaluate the performance of each tested catalyst, to evaluate them through a criterion test (targeted performance reached or not), which will determine the decision to continue the iterative optimisation or to stop it. For instance, for testing the resistance to sulfur poisoning of reforming catalysts, the activated materials could be tested in parallel reactors in an accelerated way by increasing the concentration of H_2S or SOx in the gas feed. Similarly, the resistance to carbon deposition can be tested as well by favouring coke formation by lowering the temperature or decreasing the partial pressure of oxidant (water, carbon dioxide and/or oxygen) in the gas feed.

For the present case study of on-board cleaner hydrogen systems, the new generation of catalysts must perform at atmospheric pressure with maximised H_2 production as a goal. The conversion of CO both for WGS (1) and Selox (2) reactions can be chosen as the main objective function since it defines without ambiguity the efficiency of the catalysts to convert CO into CO_2. However, note that, for the WGS reaction, CO conversion can be limited by the thermodynamics equilibrium (Fig. 10.2), which means that two efficient catalysts having to reach the equilibrium (either at high or at low temperature) will not be distinguished by the single conversion criterion. High-throughput testing under higher space–time velocity will then be required.

For new applications such as on-board generation of hydrogen, the objective function is now to find a robust and highly active WGS catalyst, able to fulfil the requirements of a simplified and miniaturised reactor: robust, so that it can be used at all relevant temperatures (150–450 °C) that would allow the merger of the usual high and low WGS steps into a single one; highly active so that it can be used at the high space velocities required by miniaturised micro-structured systems (much higher than 4000 h^{-1}). In addition, new properties such as being non-pyrophoric (i.e. resistant to a sudden contact of the activated material to the open air) and non-toxic are required for these on-board domestic applications. Finally, the cost of the catalysts precursors may also be integrated within the combined objective function (e.g. by limiting a priori the use of noble metals).

For the Selox reaction, the oxidation mechanism may involve oxygen species that are stored in/over the catalyst support, such the ceria-based materials used in three-ways catalytic exhaust systems [16]. Consequently, the oxygen storage capacity (OSC) can also be considered as an objective function, if one assumes or dem-

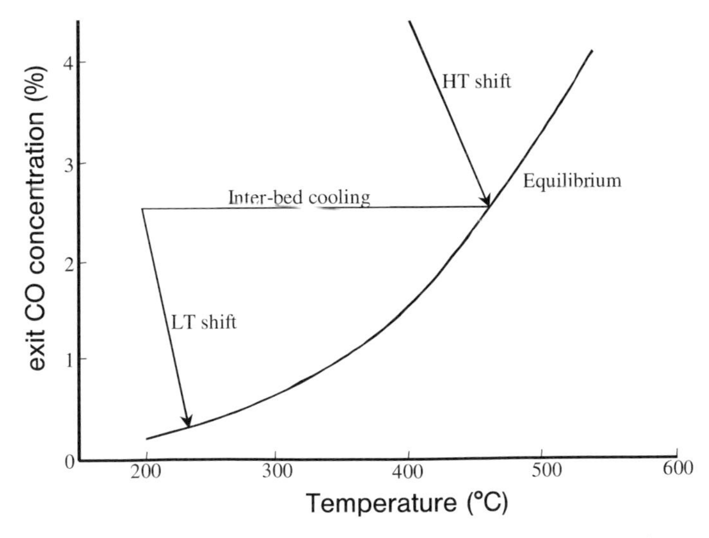

Fig. 10.2 Equilibrium CO concentrations in a WGS reaction showing how two (or more) water–gas shift reactors can be used in industrial applications

onstrates that OSC is directly linked to the CO conversion performance. Indeed, it will be necessary to check first that the HT analytics allow the operator to easily and rapidly measure the OSC within the iterative combi-loop. These technological issues will be detailed later.

- *Pros:* The choice of targets or objective functions is rather wide, depending on the reaction and application envisioned. Possibly, one easily accessible objective function such as CO conversion can be replaced by another more intrinsic parameter like the OSC for the Selox reaction, subject to a preliminary mechanistic investigation that clearly demonstrates the equivalence of the two parameters for evaluating performance. For the WGS reaction, selectivity is not a problem, except for methane and methanol side formation. For the Selox reaction, selectivity towards CO oxidation rather than towards H_2 oxidation becomes a priority objective function.
- *Cons:* This choice may be restricted by technical limitations and, also, by fundamental limitations such as thermodynamic equilibrium, which precludes any ranking of highly performing catalysts under single operating conditions. Other more difficult objective functions like catalyst stability should preferably be assessed at a later stage of catalyst development.

10.2.3 Strategies and Associated Algorithms

After the domain of investigation and the objective function(s) have been defined, the next step consists in the choice of the strategy to be implemented. Within the present domain of hydrogen purification, many strategies are potentially relevant. Three that are representative of a large panel of HT investigations have been selected and are illustrated here:

1. Black box discovery and optimisation of new catalysts.
2. Information-guided approach for discovery and optimisation of new catalysts.
3. Knowledge acquisition from data analysis: mechanistic and kinetic insights for a given reaction.

10.2.3.1 General Outlines

10.2.3.1.1 Black Box Discovery and Optimisation of New Catalysts

This first strategy, generally referred to as the “evolutionary strategy (ES)“, involves the full assistance of genetic algorithms [7, 11, 12, 15, 17]. This method deals with the iterative synthesis and screening of step-wise libraries of catalysts (typically 30 to 70 samples) or generations. The catalytic results of the preceding generation are used as feedback for the next one. The resulting combinatorial loop is schematised in Fig. 10.3.

In the first step, mixtures of compounds from a defined pool of elements and support materials, compiled on the basis of fundamental knowledge, are chosen

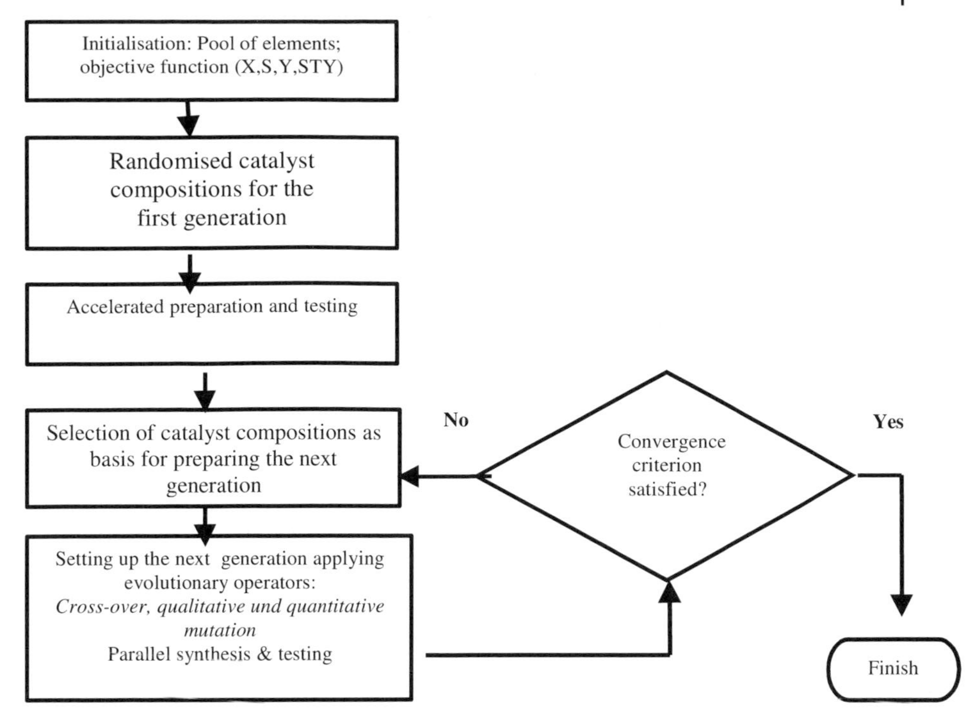

Fig. 10.3 Combinatorial loop for the evolutionary catalyst optimisation following Strategy WGS#1 [7, 11].

in a stochastic manner. On the basis of the objective function (e.g., selectivity, conversion, yield, etc.) evaluated for the individual catalytic materials of the first generation, the second generation of catalysts is designed by the software, applying mutation and crossover operators. The synthesis and testing of succeeding catalyst generations are repeated until the maximum of the objective function of the materials is achieved. This Darwinian approach of catalyst optimisation enables one to balance the exploration of vast experimental space with the final target of identifying classes of "promising" formulations. Indeed, the exploration of a wide range of combinations that is required in seeking entirely new catalysts is provided by the stochastic features of operators, while the optimisation process is guaranteed by the selection of best formulations at each generation.

This strategy will be illustrated below for the search of new WGS catalysts, being referred to as "Strategy WGS#1".

10.2.3.1.2 Information-guided Approach for Discovery and Optimisation of New Catalysts

This second approach is based on pre-existing knowledge of catalysis with the assistance of design of experiment (DoE) tools for library design and performance

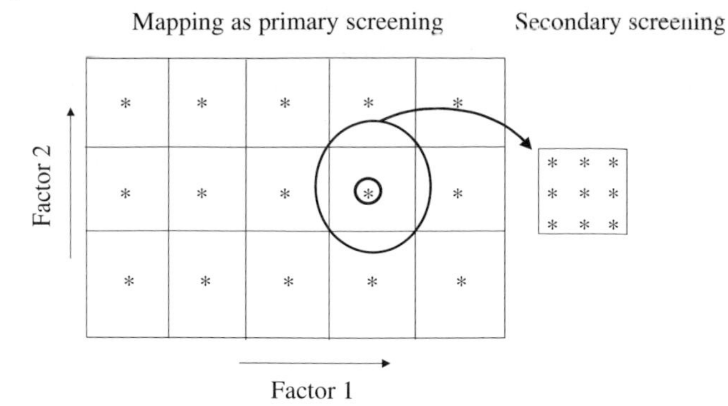

Fig. 10.4 Combinatorial loop for the evolutionary catalyst optimisation following Strategy WGS#2.

evaluation [12, 18]. This methodology resembles investigations usually conducted in laboratories because it mainly relies on the scientist's knowledge/intuition and results interpretation. In contrast to the ES, the main advantage of this approach is that parameters of synthesis and testing can be modified at each run, making the research much more flexible and therefore efficient. For example, if the presence of one element from the alkali series in a catalyst seems to be beneficial, other alkali can be introduced in subsequent libraries. In addition, DoE provides information on trends, e.g., main effects and interaction effects that can be easily visualised as a Pareto chart for example. In contrast to the ES, DoE does not provide classes of optimised formulations but only information on a mapping. Therefore the chemist has to carefully interpret the results in order to tune and to shrink the experimental space, run after run. This second strategy will be illustrated by the search for new Selox catalysts.

A slightly distinct strategy, referred to here as "Strategy WGS#2", consists of two steps (Fig. 10.4):

1. A first phase, which can be considered as a primary screening, aims at investigating a large number of diverse catalysts, so as to browse the parameter space. The main purpose is to cover, homogeneously, the search space by a portioning so as to maximise the chance of discovering synergetic effects between elements.

2. A second phase aims to optimise the hits arising from the first phase investigation. For this method, a novel algorithm combining genetic algorithm and datamining techniques on a database have been used [19].

10.2.3.1.3 Knowledge Acquisition from Data Analysis: Mechanistic and Kinetic Insights for a Given Set of Reactions

This third strategy aims at designing HT data acquisition procedure in such a way that more advanced information is collected, such as mechanistic and kinetic insights, as compared with the limited information obtained with the previous strategies. This newly acquired knowledge can be used in turn for further combinatorial strategies, as above described (ES or DoE) [12].

This third strategy will be illustrated by HT experiments carried out under WGS, reverse WGS *and* Selox conditions, varying operating parameters such as temperature, partial pressures, space velocity, under stationary and non-stationary conditions [20].

- *Pros:* The positive aspect in the choice of strategies is the widely open space for mixing a priori rules based on pre-existing knowledge and/or chemical intuition with the precise rules imposed by either the evolutionary or the DoE-based approaches. Such flexibility can be considered as an advantage for adapting the strategies and their parameters to the technical limitations existing in the laboratory (essentially the potential for HT synthesis and testing).
- *Cons:* A negative aspect of this step is the obvious lack of clear guidelines for designing the initial libraries in the infancy of this methodology, despite state-of-the-art data-mining [21].

10.2.4 Choice of Parameters for Libraries Design

10.2.4.1 WGS Catalysts

To compare the efficiency of the two evolutionary strategies WGS#1&2, the same initial parameter space has been considered, i.e., pool of elements, range of concentrations, testing operating conditions and objective function to evaluate the fitness of catalysts [22].

For the pool of elements considered for preparing catalysts, the choice was based on an arbitrary way of describing a WGS catalyst: an active phase formed by the combination of noble and non-noble metals from group VIII, a support formed from single or mixed oxides, eventually doped with promoters. Indeed, this arbitrary description does not mean that all the metal elements from the so-called "active phase" effectively participate in the WGS process and are under the metallic state under reacting conditions and, conversely, that the so-called "support" does not participate to the activation of the reacting molecules, in a tight interaction with the metal particles via "spill-over" processes. Tab. 10.1 reports the various elements and range of concentration considered in the initial pool used for automated catalysts synthesis.

The active phase metals were selected from the knowledge of elements able to activate both the CO molecule, which is a prerequisite for its oxidation, and the H_2O molecule into active oxygen and molecular hydrogen. The choice of supports was based on known properties for effectively dispersing the active phase, and the

Tab. 10.1 Initial pool of elements and selected synthesis parameters (synthesis method: impregnation).

Catalysts components	*Elements*	*Concentration range*	*Maximum number*
Noble metals	Au, Pt, Pd, Ru	0.5–1.0 wt%	1
Other metals	Fe, Cr, Cu, Zn, Ni, Mn, Co, Mo, La, Sm, Ce	20–80 wt%	4
Promoters	Ca, K, Li, Cs	0.5–1.0 wt%	2
Supports	α-Al_2O_3, γ-Al_2O_3, TiO_2, ZrO_2, α-Fe_2O_3, Ce(La)O_x, La_2O_3, MgO, "no support"	5–80 wt%	1

promoters belong to the alkali and earth alkali series known to tune CO adsorption on a metal via direct or indirect electron transfer effects.

The ranges of concentration were selected according to a general knowledge of metal efficiency for the WGS reaction. Thus, the activity of non-noble metals such as Cu, Fe or Zn is known to be effective for rather high concentration, such as in co-precipitated bulk phases. For noble metals, their higher efficiency, and economic considerations, leads us to restrict the concentration to 1 wt%.

For the present case, only wet impregnation of metal salts on supports was considered, to restrict the field of synthesis parameters and also since this process was found to be quite adapted to the step-by-step requirements of the automated synthesis.

10.2.4.2 **Selox Catalysts**

As indicated above, DoE tools were used for designing the Selox libraries [12, 18]. Tab. 10.2 reports the first library, which was designed to identify rapidly the elements (alkali/earth alkali, metal oxides, noble metals) or combination of elements that promote selective CO oxidation in a vast parameter space. The selection of element and the "combinatorial" rules for mixing them and generating multicomponent catalysts were done a priori, based on literature data and pre-existing knowledge. As such, the elements were classified into four groups (referred to later as classes), namely: "noble metals" (Pt, Pd, Ru, Rh, Au), "oxides" (transition metal oxides of Cr, Co, Mn, La, Sm, Mo), "dopants" (alkali or earth alkali Li, Cs, Ca) and "supports" (Al_2O_3, CeO_2, ZrO_2, ZnO, C). Among the a priori rules it was decided that all catalysts are composed of one support and two noble metals and, optionally, of one transition metal and one dopant. When present, the weight percentage of dopant, transition metal, and noble metal is fixed at 1, 20 and 0.5%, respectively. The choice of preparing catalysts with two distinct noble metals assumes that alloys may outperform the effects of respective single noble metals. The library was designed such that the main effects of supports, dopants and transition metals and each binary combination of noble metals could be computed. From these specifications, each class of elements can be considered as a qualitative factor and each element or binary com-

Tab. 10.2 DoE-based design for the 1st Selox library: each system is classified in the following classes of elements (or factors): "noble metal" (Pt, Pd, Ru, Rh, Au), "oxide" (transition metal oxides of Cr, Co, Mn, La, Sm, Mo), "dopant" (alkali and earth-alkali, Li, Cs, Ca) and "support" (Al_2O_3, CeO_2, ZrO_2, ZnO, C). Left: disjunctive coding. Right: factor coding used in DoE.

Samples	*Pt*	*Pd*	*Ru*	*Au*	*Rh*	*Cr*	*Co*	*Mn*	*La*	*Sm*	*Mo*	*Li*	*Cs*	*Ca*	*Al*	*Ce*	*Zr*	*Zn*	*C**	*Noble metal*	*Oxide*	*Dopant*	*Support*
1	1	1	0	0	0	0	0	0	0	0	0	0	0	0	1	0	0	0	0	1	1	1	1
2	0	1	1	0	0	0	0	0	0	0	0	0	0	0	0	1	0	0	0	3	1	1	2
3	0	0	1	1	0	0	0	0	0	0	0	0	0	0	0	0	1	0	0	6	1	1	3
4	0	0	0	1	1	0	0	0	0	0	0	0	0	0	0	0	0	1	0	10	1	1	4
5	1	0	1	0	0	0	0	0	0	0	0	0	0	0	0	0	0	0	1	2	1	1	5
6	0	1	0	1	0	0	0	0	0	0	0	1	0	0	1	0	0	0	0	5	1	2	1
7	0	0	1	0	1	0	0	0	0	0	0	0	1	0	0	1	0	0	0	9	1	3	2
8	1	0	0	1	0	0	0	0	0	0	0	0	0	1	0	0	1	0	0	4	1	4	3
9	0	1	0	0	1	0	0	0	0	0	0	1	0	0	0	0	0	1	0	8	1	2	4
10	1	0	0	0	1	0	0	0	0	0	0	0	1	0	0	0	0	0	1	7	1	3	5
11	0	1	1	0	0	1	0	0	0	0	0	0	0	0	1	0	0	0	0	3	2	1	1
12	0	0	1	1	0	0	1	0	0	0	0	0	0	0	0	1	0	0	0	6	3	1	2
13	0	0	0	1	1	0	0	1	0	0	0	0	0	0	0	0	1	0	0	10	4	1	3
14	1	0	1	0	0	0	0	0	1	0	0	0	0	0	0	0	0	1	0	2	5	1	4
15	0	1	0	1	0	0	0	0	0	1	0	0	0	0	0	0	0	0	1	5	6	1	5
16	0	0	1	0	1	0	0	0	0	0	1	0	0	0	1	0	0	0	0	9	7	1	1
17	1	0	0	1	0	1	0	0	0	0	0	0	0	0	0	1	0	0	0	4	2	1	2
18	0	1	0	0	1	0	1	0	0	0	0	0	0	0	0	0	1	0	0	8	3	1	3
19	1	0	0	0	1	0	0	1	0	0	0	0	0	0	0	0	0	1	0	7	4	1	4
20	1	1	0	0	0	0	0	0	1	0	0	0	0	0	0	0	0	0	1	1	5	1	5
21	0	0	1	1	0	0	0	0	0	1	0	1	0	0	1	0	0	0	0	6	6	2	1
22	0	0	0	1	1	0	0	0	0	0	1	1	0	0	1	1	0	0	0	10	7	2	2
23	1	0	1	0	0	1	0	0	0	0	0	1	0	0	0	0	1	0	0	2	2	2	3
24	0	1	0	1	0	0	1	0	0	0	0	1	0	0	0	0	0	1	0	5	3	2	4

Tab. 10.2 (cont.)

Samples	*Pt*	*Pd*	*Ru*	*Au*	*Rh*	*Cr*	*Co*	*Mn*	*La*	*Sm*	*Mo*	*Li*	*Cs*	*Ca*	*Al*	*Ce*	*Zr*	*Zn*	*C**	*Noble metal*	*Oxide*	*Dopant*	*Support*
25	0	0	1	0	1	0	0	1	0	0	0	1	0	0	0	0	0	0	1	9	4	2	5
26	1	0	0	1	0	0	0	0	1	0	0	1	0	0	1	0	0	0	0	4	5	2	1
27	0	1	0	0	1	0	0	0	0	1	0	1	0	0	0	1	0	0	0	8	6	2	2
28	1	0	0	0	1	0	0	0	0	0	1	1	0	0	0	0	1	0	0	7	7	2	3
29	1	1	0	0	0	1	0	0	0	0	0	1	0	0	0	0	0	1	0	1	2	2	4
30	0	1	1	0	0	0	1	0	0	0	0	0	1	0	0	0	0	0	1	3	3	3	5
31	0	0	0	1	1	0	0	1	0	0	0	0	1	0	1	0	0	0	0	10	4	3	1
32	1	0	1	0	0	0	0	0	1	0	0	0	1	0	0	1	0	0	0	2	5	3	2
33	0	1	0	1	0	0	0	0	0	1	0	0	1	0	0	0	1	0	0	5	6	3	3
34	0	0	1	0	1	0	0	0	0	0	1	0	1	0	0	0	0	1	0	9	7	3	4
35	1	0	0	1	0	1	0	0	0	0	0	0	1	0	0	0	0	0	1	4	2	3	5
36	0	1	0	0	1	0	1	0	0	0	0	0	1	0	1	0	0	0	0	8	3	3	1
37	1	0	0	0	1	0	0	1	0	0	0	0	1	0	0	1	0	0	0	7	4	3	2
38	1	1	0	0	0	0	0	0	1	0	0	0	1	0	0	0	1	0	0	1	5	3	3
39	0	1	1	0	0	0	0	0	0	1	0	0	0	1	0	0	0	1	0	3	6	4	4
40	0	0	1	1	0	0	0	0	0	0	1	0	0	1	0	0	0	0	1	6	7	4	5
41	1	0	0	1	0	1	0	0	0	0	0	0	0	1	1	0	0	0	0	4	2	4	1
42	0	1	0	0	1	0	1	0	0	0	0	0	0	1	0	1	0	0	0	8	3	4	2
43	1	0	0	0	1	0	0	1	0	0	0	0	0	1	0	0	1	0	0	7	4	4	3
44	1	1	0	0	0	0	0	0	1	0	0	0	0	1	0	0	0	1	0	1	5	4	4
45	0	0	0	1	1	0	0	0	0	1	0	0	0	1	0	0	0	0	1	10	6	4	5
46	1	0	1	0	0	0	0	0	0	0	1	0	0	1	1	0	0	0	0	2	7	4	1
47	0	1	0	1	0	1	0	0	0	0	0	0	0	1	0	1	0	0	0	5	2	4	2
48	0	0	1	0	1	0	1	0	0	0	0	0	0	1	0	0	1	0	0	9	3	4	3

bination in the case of noble metals corresponds to a distinct modality. Factor description and coding are given in Tab. 10.2. The Wilks model that can manage qualitative factors for linear regression (0 order) was used for that purpose. The model consists of 23 coefficients to be computed, which implies the carrying out of at least 24 uncorrelated experiments. D-optimal criteria were applied to select the best subset of experiments that provide coefficient estimation with the highest confidence intervals of a limited number of experiments. Computations were carried out with the commercial software Nemrodw. The total number of catalysts for this first library was fixed to 48 as this is a multiple of 16, which is the number of parallel runs that can be carried out in the high-throughput testing set-up available (see testing section). Inflation factors for each coefficient are around 2, indicating a fairly good parameter estimation.

The next libraries were designed step-wise on the basis of knowledge gained from previous ones. As such, the second, third and fourth libraries were based on platinum-ceria, metal oxide-ceria and copper-ceria formulas, respectively, each of them using a similar DoE model for their design [18].

10.3 Make: Automated Synthesis of Materials

For all the chosen strategies reported here, the catalysts were prepared using the same automated robot, which specially designed for solid-phase preparation (ZINSSER GmbH, Fig. 10.5), offering key functions such as heating, cooling, steering, reagent and solvent delivery, filtration and solid handling [8, 23, 24]. The elementary steps used for the automated synthesis of catalysts were chosen to be as close as possible to the conventional impregnation-processing method.

Catalytic materials following Strategy WGS#1 were prepared by sequential impregnation [25] of the different support materials with the individual nitrate precursor solutions, having a typical concentration of 1 mol l^{-1} on 1 g. After each impregnation step the samples were dried at 120 °C while shaking. Firstly, the elements of the transition metal group, then the alkaline promoters and finally the noble metals were added in the sequential impregnation procedure. The support materials were bulk oxides (particle size between 2.5 and 0.3 mm): γ-Al_2O_3 (Condea, 120 $m^2 g^{-1}$), α-Al_2O_3 (Condea, 5 $m^2 g^{-1}$), CuZnAl and ZrO_2.

Catalytic materials following Strategy WGS#2 were prepared by impregnating meshed α-Al_2O_3, and then by impregnating the non-noble metal or oxide-active phases and finally the noble metals and promoters [22]. For the primary screening step, more than 300 catalysts were prepared following the above described experimental planning. The element composition of the transition metals and rare earth metals was fixed to 20 wt.%, and the promoters to either 0.1 or 1 wt%. The active phase was deposited by sequential impregnation of precursor solutions in four steps (typical concentration 0.1 mol l^{-1}): a layer of Al, Zr, Mg or Ce was first synthesised on meshed (0.2 to 0.3 mm) γ-Al_2O_3 (100 mg of Condea α-Al_2O_3, 5 $m^2 g^{-1}$), then premixed solutions of nitrates of the main components were added, followed by the

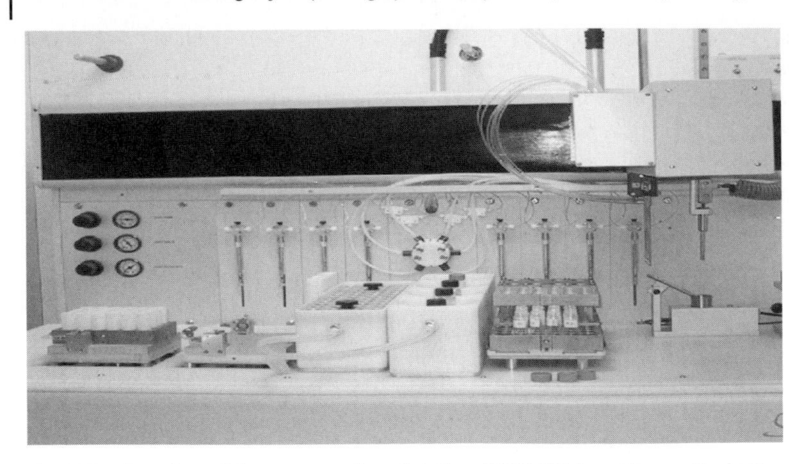

Fig. 10.5 Overview of the automated workstation (SOPHAS from Zinsser GmbH).

promoters and finally the noble metals. After each impregnation (0.4 ml), the suspension was evaporated under permanent shaking at 120 °C for 15 min. This procedure ensures the production of supported catalysts with the same amount of deposition (66 mg) on each catalyst, which can be easily transferred to small reactor vessels (no gel). In addition, the rather low quantity of prepared catalysts allows the robot to achieve the synthesis of 48 samples within few hours.

Similar automated preparation techniques were used for the Selox libraries. As an example, for the ceria-based materials (second library designed by DoE, see Tab. 10.2) the benchmark starting point for the catalyst search was a reference proprietary 4.75 wt% Pt/CeO_2 catalyst. Pt was kept as the active metal and the ceria support was modified by substituting it with 10 at% cations of various metals, by using the above-described robot synthesiser. Note that doping at such low fractions does not alter the cubic fluorite structure [20]. Dopant metals were chosen from groups IIIb–VIb and IIIa–Va, due to their ionic charge and/or size similar to Ce, with the additional requirement that they should be readily available and comparatively inexpensive. A set of 12 catalysts containing 2 wt% Pt on $Ce_{0.9}M_{0.1}O_x$ was prepared by impregnating pre-synthesized ceria powders (M = Bi, La, In, Mo, Pb, Sn, V, W, Y, and two combinations of Zr/Bi).

- *Pros:* The automated synthesis of heterogeneous catalysts, which is a prerequisite for effective high-throughput search, benefits from a mature technology, accessible for academic budgets and offers a rather wide range of technological solutions well adapted to most conventional preparation techniques (impregnation, ionic exchange, co-precipitation, sol-gel).

- *Cons:* Some technical limitations still exist, such as (1) the small amount of materials generally prepared for primary screenings may lead to difficulties in reproducing exactly a formula and (2) the lack of efficient solid handling from the preparation vessels to the parallel reactors, which may involve manual steps within the iterative loop. The main issue related to the automated synthesis lies in the fact that the preparation recipes are generally adapted to the robot possibilities, i.e. simplified and mostly based on sequential steps. This necessarily precludes any direct up-scaling and further commercialisation of the best formulas selected from these screening strategies. This is generally demonstrated by the fact that catalysts prepared by a synthesis robot with a composition similar to commercial systems never perform like the true commercial catalysts when tested under similar conditions. However, it has to be considered that trends in formula combination are searched for, rather than exact new formula.

10.4 Test: Automated Testing of Materials for Evaluation of Performances

Automated testing of heterogeneous catalysts by means of parallel reactors still requires laboratory-made systems at the academic scale, due to the prohibitive costs of existing commercial systems. In contrast, for liquid phase reactions, a domain closer to mature pharmaceutical research, reasonably cheap systems are available.

For the present case study, a 16-channel laboratory-made parallel reactor was used (Fig. 10.6). Other laboratory-scale reactors were also used with the same libraries to check the specificity of each testing system [13].

In each run, one channel was loaded with graphite or quartz and used as bypass reference, and one channel was loaded with reference catalyst (commercial or proprietary) for comparison. The temperature was varied from 100 to 250 °C, with a 5 min isotherm every 20 °C. Reaction products were monitored by fast gas chromatography (micro-GC) and on-line mass spectrometry (MS). The carbon mass-balance was calculated by monitoring the effluents from the bypass reactor and kept between 98 and 102%.

For the WGS reaction, the standard testing conditions were chosen as typical of the low-temperature shift inlet conditions imposed by conventional outlet high-temperature shift conditions, e.g., 200 °C; 3% CO, 37% H_2, 14% CO_2, 23% H_2O, Ar balance; GSHV = 3000.

For the Selox reaction, catalysts were tested both in the absence and presence of hydrogen and/or water, by admitting the following gas mixtures: 1% CO, 2% O_2 and N_2 to balance; 1% CO, 2% O_2, 20% H_2 and N_2 balance; 1% CO, 2% O_2, 20% H_2, 10% H_2O and N_2 to balance, keeping the same gas-space velocity. The oxygen storage capacity (OSC) was measured at various temperatures by measuring the H_2 consumption of each catalyst in transient mode after being oxidised under air flow, using the parallel reactor under isothermal conditions [20].

- *Pros:* Parallel testing has been rather widely described for decades, long before the appearance of the combinatorial strategy [26]. Therefore, even if the technology remains essentially home-made at the academic level, as stressed above, a rather satisfactory control of performances has been demonstrated simply by reproducibility tests or testing the same catalysts in all the channels of the reactor.

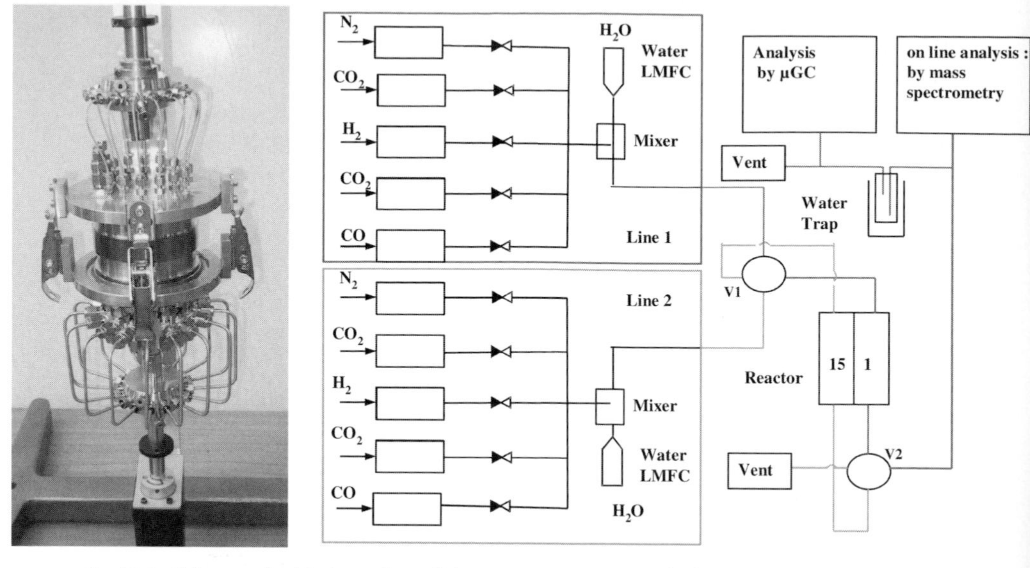

Fig. 10.6 Scheme of a 16-channel parallel prototype reactor jointly developed with AMTEC GmbH company, now commercialised as "SWITCH 16" reactor. See *http://www.amtec-chemnitz.de*.

- *Cons:* Perfectly isothermal conditions are not observed for simple prototypes such as described in Fig. 10.6, especially for highly exothermic reactions like partial or total oxidation, with observed temperature profiles and local hot spots (see results for the Selox reaction). However, these limitations do not prevent the observation of significant trends when large libraries are tested under various operating conditions. Commercial, improved systems adapted to the requirement of academia are now available on the market.

10.5 Model: Information Extraction & Modeling

This section reports the results obtained for the three above presented strategies:

1. Black box discovery and optimisation of new catalysts.
2. Information-guided approach for the discovery and optimisation of new catalysts.
3. Knowledge acquisition from data analysis: mechanistic and kinetic insights for a given reaction.

10.5.1
Black Box Discovery and Optimisation of New Catalysts Using an Evolutionary Strategy

Recently, the evolutionary strategy (ES) has been successfully applied for the iterative optimisation of catalysts in a series of reactions [7, 10, 15, 27–30]. For all the case studies reported, the optimisation process was found to converge rapidly after typically four to six generations.

For the present case study (Strategy WGS#1), seven generations of 72 catalytic materials each have been prepared and tested. CO conversion at standard conditions for the ten best compositions of each generation is shown in Fig. 10.7.

From the first to the seventh generation an improvement of the best CO conversion from 40% to nearly 70% is observed, though the improvement tends to stabilise for the last generations (Fig. 10.7). Under these operating conditions, the maximum CO conversion corresponding to the thermodynamic equilibrium would be 94% (calculated from free Gibbs energy minimisation). Therefore, the objective, to reach the thermodynamic equilibrium before stopping the optimisation procedure, was not achieved. However, due to the stabilisation trend observed by using this evolutionary algorithm, no further major improvement can be expected and other strategies should be tested for improving the best formulas.

From the chemical viewpoint, some essential elements and support materials were identified, such as Cu, Ce and Mn, as having the highest frequency in the catalytic materials (Fig. 10.8). More precisely, by checking synergistic and antago-

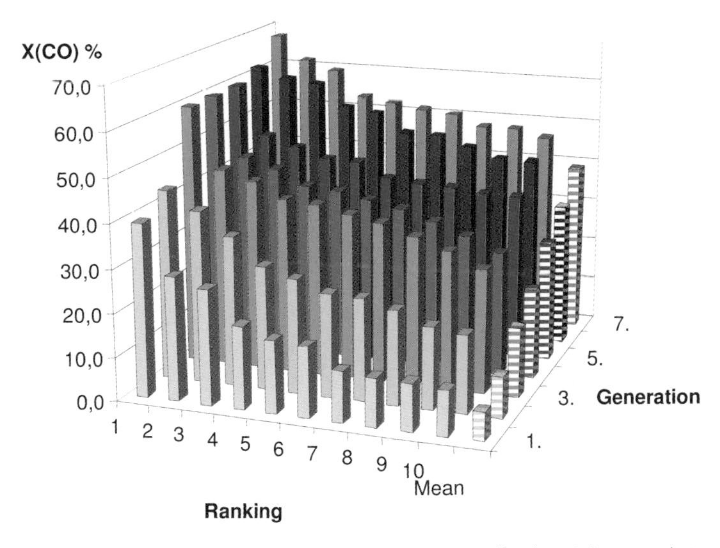

Fig. 10.7 CO conversion under standard conditions for the 10 best catalytic materials of each generation and the average CO conversion over all materials of each generation. (200 °C, 3% CO, 37% H_2, 14% CO_2, 23% H_2O, Ar balance, GSHV = 3000 h^{-1} (interpolated).)

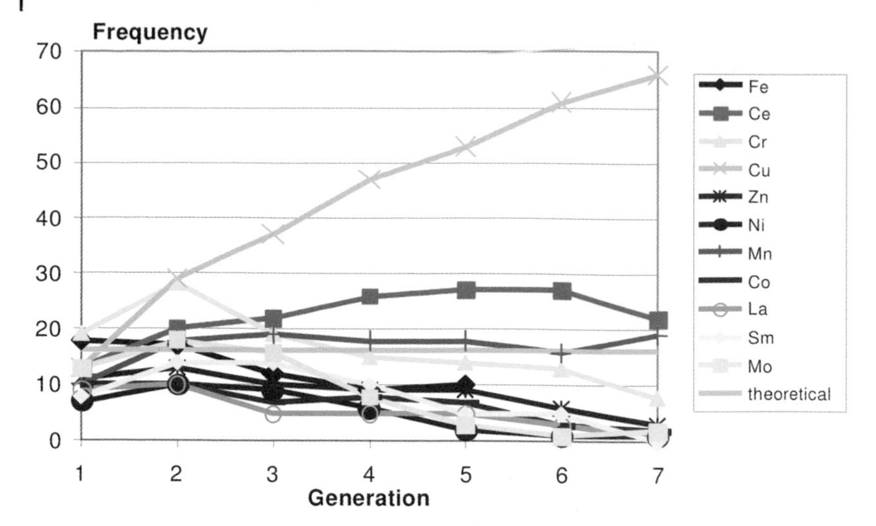

Fig. 10.8 Frequency of the base metals of the pool of elements with proceeding generations and theoretical value for a completely random distribution.

nistic effects among the elements as well as between elements and support materials, the following conclusions were drawn: For a Cu content > 10 wt% the performance of the material is positively influenced by the presence of Mn (< 6 wt%) and Ce, the latter when no alkali promoters are present; γ-Al_2O_3 was found to be the optimal support material. Other synergetic effects were also discovered, leading to new but lower-performing formulations.

This methodology was also applied in the same WGS domain, but with added new objective functions such as non-pyrophoricity of the systems, which was simply defined by excluding large amounts of copper. Again, new promising formulas were discovered, which are now in the process of being scaled-up [13].

10.5.2
Information-guided Approach for the Discovery and Optimisation of New Catalysts using DoE

10.5.2.1 Strategy WGS#2

As already presented in the section on "Design", for the strategy WGS#2, a primary library containing 490 catalysts was designed by partitioning the search space. In Fig. 10.9, the catalysts are ranked as a function of the CO conversion.

These results were further exploited by searching for interactions between the elements to guide the design of the next libraries [22, 31]. This data analysis was carried out using classification techniques [21]. Two techniques were used: (1) visualisation of contingency tables and (2) induction trees, aimed at predicting and identifying which catalysts are promising, to select a given formulation for a secondary screening.

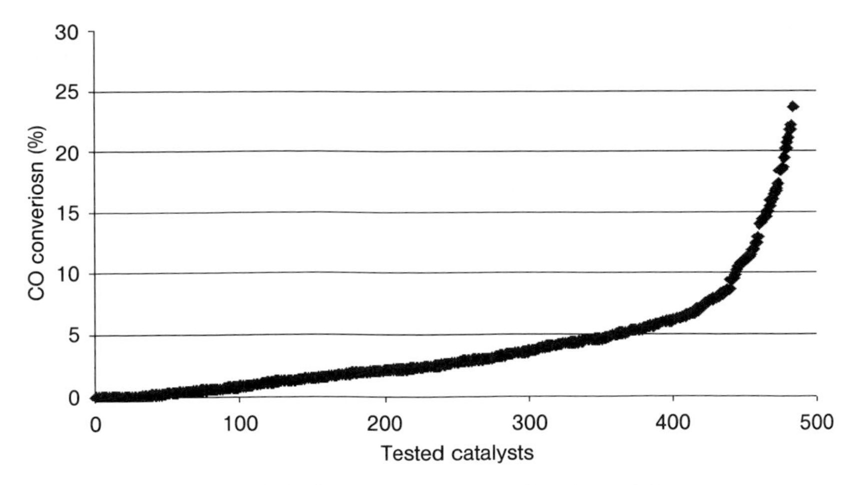

Fig. 10.9 CO conversion of the first screening step used in the WGS#2 strategy.

Contingency table analysis (Fig. 10.10, left), which can be considered as an unsupervised method, revealed, for example, the expected high correlation between Cu and Zn, and also an unexpected correlation between Mo and Sm. The same element combinations were highlighted by the induction tree results (Fig. 10.10, right).

From the output of this primary library analysis, a library of 64 catalysts was designed to focus on the experimental space that provides the highest probability to obtain improved catalysts. Unfortunately, due to still unclear reasons (probably linked to technological issues concerning the synthesis and testing steps, as already mentioned) the high data quality required for data exploitation was not achieved (e.g. many "outfliers" were mixed with relevant data) and the second stage of the strategy WGS#2 was not completed [22]. However, as for strategy WGS#1, interesting trends in the impact of elements and combination of elements have been used to prepare performing WGS catalysts.

10.5.2.2 **DoE Strategy for Selox Catalysts**

As a second example using the DoE assistant tool for designing libraries, the first library of 48 diverse Selox catalysts (Tab. 10.2, Section 10.2.4.2) was tested in both the presence and absence of hydrogen to evaluate the impact of operating conditions on Selox performance [18]. From the set of data giving the conversion of CO (not reported here), an analysis of the effect induced by the four selected factors (alkali addition, type of support, nature of the transition metals and of the noble metal binaries) was carried out following different models [18].

Accordingly, an exhaustive way to present the results was to compute the effects when moving from one factor to another one, everything else being kept equal and assuming a linear model. In other words, to indicate in a quantitative man-

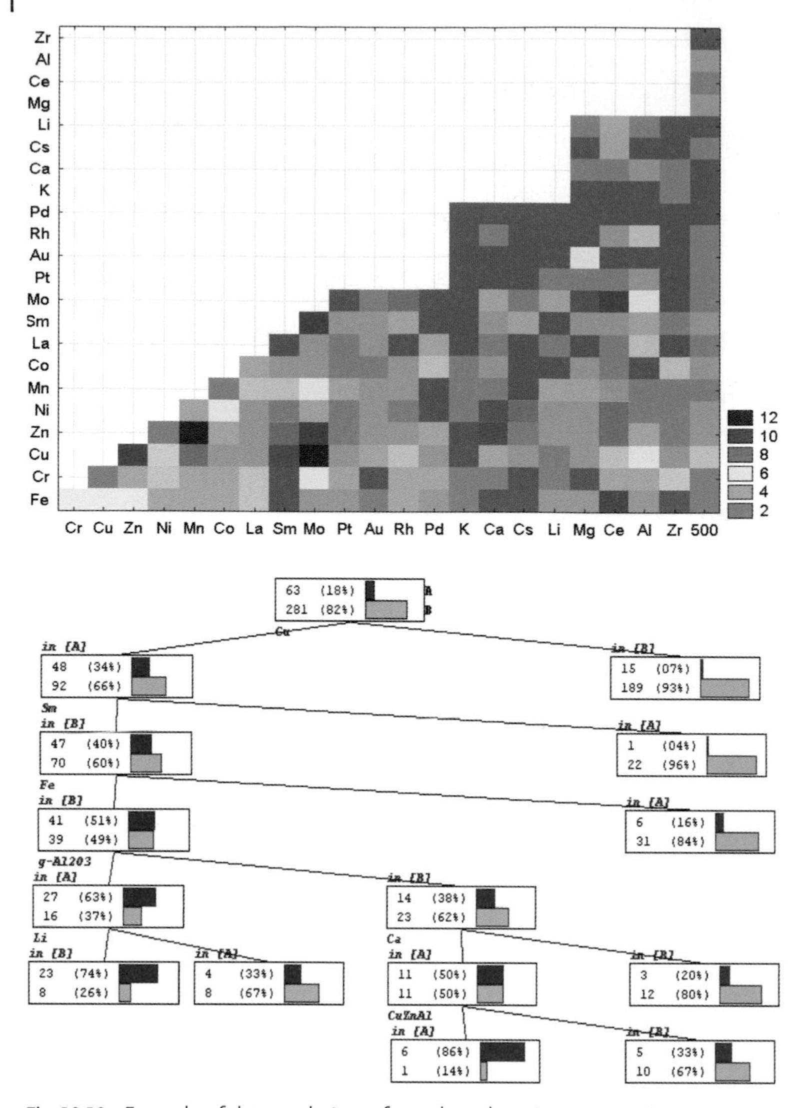

Fig. 10.10 Example of data analysis performed on the primary screening of 490 mixed oxides.

ner which change in CO conversion is induced by substituting one element by another one within the same class of elements. In addition, this indicates whether the effects are significant or not by deriving residual calculations. A representation of these effects is reported in Fig. 10.11.

For the two classes of alkali dopants and transition metal oxides, the effect of each of these elements was referred with respect to catalysts being free of dopant or transition metal oxide, respectively – this reference being level 1 (in blue). For the two

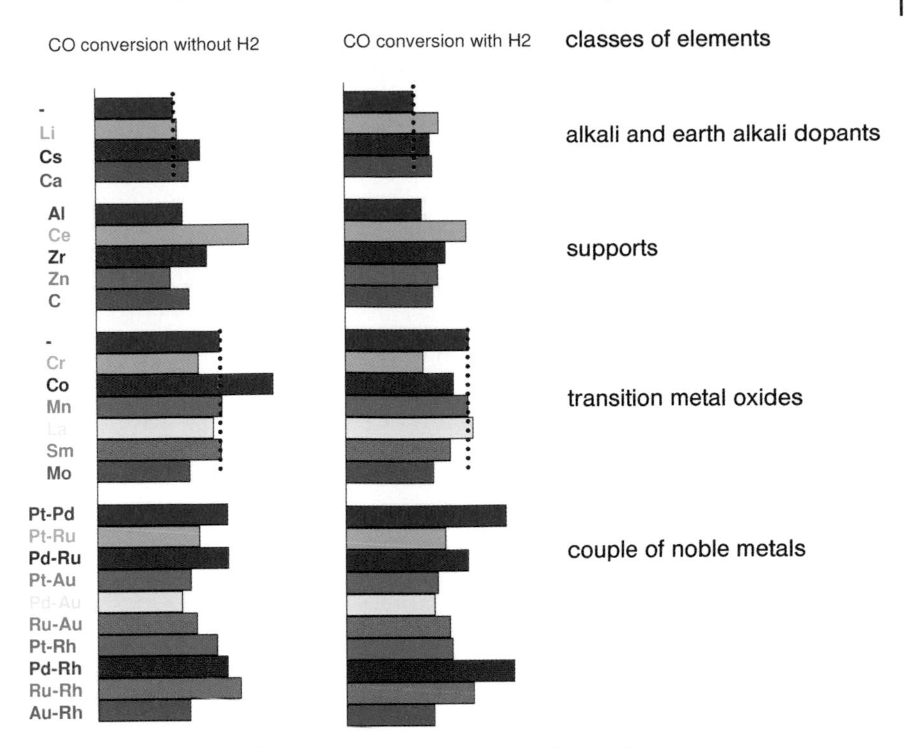

Fig. 10.11 Quantitative effects on CO conversion in the Selox reaction, both in the absence and presence of hydrogen, generated by the presence of one specific element (or couple of elements for the noble metals) for each of the four predetermined classes of elements.

other classes of supports and noble metals, no reference level for catalysts free from support or noble metal exists since it was decided initially that all catalysts are supported systems, including one couple of noble metals (Section 10.2.4). Fig. 10.11 clearly indicates that adding any alkali/earth alkali improves the CO conversion. In the absence of H_2, Cs doping has the largest effect on CO conversion (12% increase), whereas the presence of H_2 induces similar effects whatever the tested dopants. For the class of supports, CeO_2 clearly brings a higher beneficial impact than all other supports and, more specifically, in the absence of H_2.

In a summary, for H_2-free conditions, the combination of CeO_2 as support, Co as transition metal and the couple of noble metal Pt-Pd, Pd-Ru, Pd-Rh and Ru-Rh brings the highest positive effects on CO conversion. In the presence of H_2, the same trends are revealed except for Co which, surprisingly, exhibits detrimental effects compared with the metal-oxide-free reference. From this preliminary mapping, ceria was selected as the support to be used for designing the next libraries. Results obtained with the next libraries are reported in ref. [18].

10.5.3
Improvement of the Search Strategy by Means of Knowledge Extraction

While most combinatorial researches reported up to now involve the use of GA, using the traditional crossover and mutation operators (e.g. WGS#1), it has also been proposed to design new operators for each specific application, to improve search efficiency by means of knowledge extraction [32]. Hence, new methods that combine ES with a knowledge extraction engine have been reported recently within the field of heterogeneous catalysis, such as mining association rules [12, 18, 30, 33] and neural networks [19, 29, 34].

Basically, artificial neural networks (ANN) are designed for the following purpose: after being trained from a set of existing data (provided for instance by a tested primary library, like for the WGS#2 strategy, or by a collection of tested libraries, such as for the evolutionary WGS#1 strategy), they should be able to predict the performances of not-yet-tested catalysts that are ordered for the next generation (by DoE or GA). In turn, the experimentalist should be able to save experimental time by canceling trials for which a poor performance is predicted. This integration of ANN steps within the basis combinatorial loop is schematised in Fig. 10.12.

In the mid-1990s few papers related the use ANN for catalysts development by predicting performances from their elemental compositions [35–38] but without clear efficiency due to limited data sets. Today, hundreds of catalysts can typically be screened in a week, which opens up the prospect of using ANNs as pre-screening tools for speeding up the optimisation process. This concept was recently reported for modeling quantitatively the performances of ODHE catalysts as a function of elemental composition [39].

For the present case study, a first attempt to predict *quantitatively* performances of WGS catalysts by ANN regression technique led to a rather poor correlation between predicted and experimental CO conversion values (Fig. 10.13). This suggests that, in addition to noisy data, the used descriptors, which were restricted to the single elemental composition of the catalysts, do not contain per se sufficient

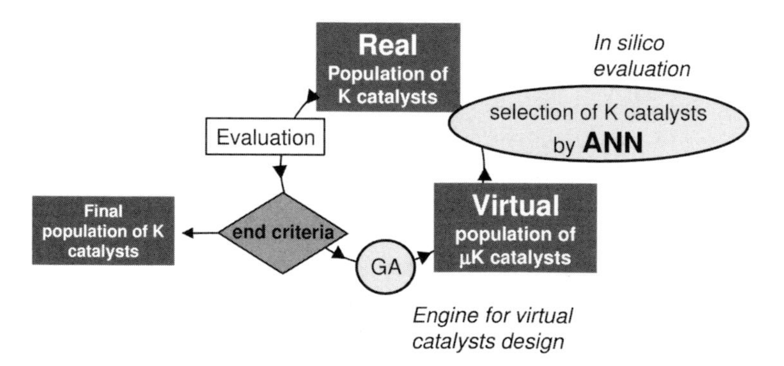

Fig. 10.12 Concept of pre-screening by data-mining as iterative screening proceeds [30].

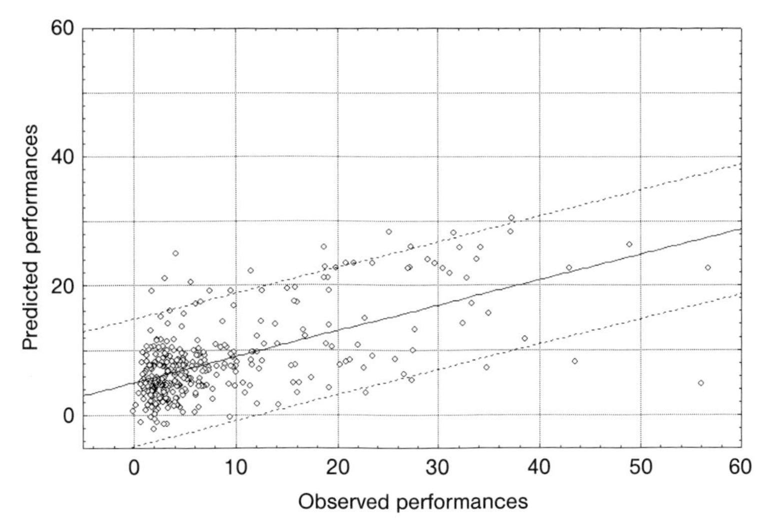

Fig. 10.13 Comparison of effective and predicted results for WGS catalysts [19].

information to correlate with performance. Probably, more relevant descriptors, such as redox potential, metal dispersion, etc., are required to reach a higher level of correlation like that expected for an expert system [19].

In contrast, using a simplified classification approach (dividing the data set into "good catalysts", which convert CO over a fixed threshold performance, and "bad catalysts", those converting CO under this threshold), it was possible to predict *qualitatively* the performances of a next generation, i.e. by finding that most of the effectively good catalysts were predicted as good with a high level of confidence, while most of the effectively bad ones were predicted as bad (Fig. 10.14). From these results, the various interactions between elements can now be predicted (instead of being deduced from experimental results, as shown in Fig. 10.10, left-hand side) for the virtual set containing all possible combinations (about 4500), as visualised in Fig. 10.15. Note that, although Fig. 10.10 (left-hand side) (from experiments) and Fig. 10.15 (from prediction) cannot be compared strictly, the main interactions, such as indeed the ternary Cu-Zn-Al, are revealed by the two tables, but new performing domains are proposed in the virtual library that were not detected from the DoE experimental library (for example the gold-based materials).

This simplified ANN methodology, which turns out to be efficient even when applied to noisy data sets, may therefore be used as a "first pass" filter for selecting the best performing systems, without testing the bad ones, during an optimisation process [12, 19].

- *Pros:* Converging conclusions were found between strategies applied for optimising WGS catalysts, such as the major role of some key elements or combination of elements. However, additional trends were revealed by the strictly ES

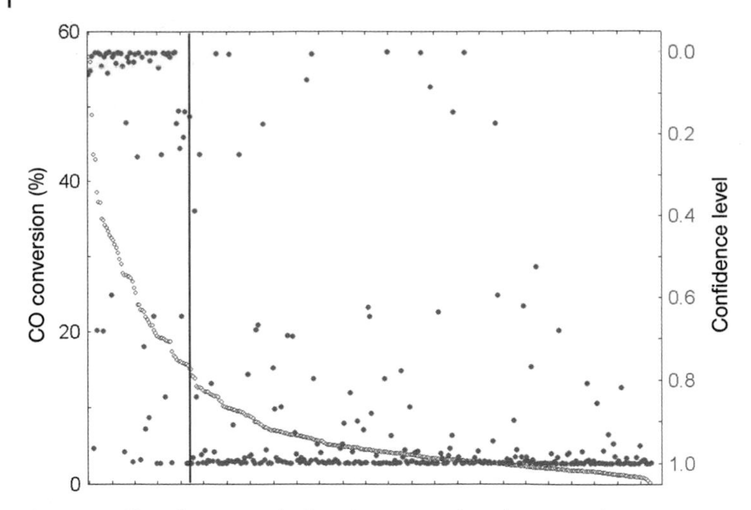

Fig. 10.14 Classification at the learning step and prediction on the next generation, as compared with the effective experimental results. Confidence level close to 0.0 = good catalysts, close to 1.0 = bad catalysts [19].

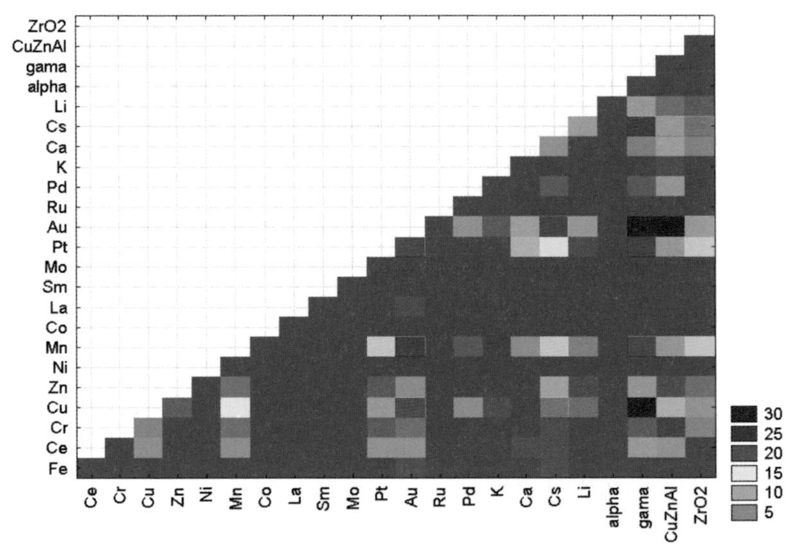

Fig. 10.15 Prediction of interactions between elements for WGS catalysts, deduced from calculated performances by ANN.

(WGS#1) while others were revealed when using a DoE step for the initial design of the first library and using statistical analyses of the data set (WGS#2) and vice versa. This stresses that the choice of evolutionary algorithms partially determines the type of discovered formulas.

The efficiency of integrating in the ES loop an additional step of prediction by ANN was also proved to be efficient, even for the worse case of noisy data.

- *Cons:* Data quality for HT synthesis and testing (reproducibility, homogeneity, scale-up ability), as shown for this case of metals/mixed oxides WGS and Selox catalysts, remains a drawback for efficient data-mining. As such, even if the trends for new catalytic formulas may still be detected, even in the presence of poorly reproducible systems or outliers, the ultimate target of predicting the quantitative performance of a catalyst after teaching an ANN with experimental data was found to require new catalysts descriptors that would contain more information than the simple elemental composition of the materials.

Therefore, the quality and diversity of data must be considered as a key objective for improving the efficiency of these new assistant tools.

10.5.4
Knowledge Acquisition from Data Analysis: Mechanistic and Kinetic Insights for a Set of Close Reactions

As mentioned in Section 10.2.3.1, explaining the general outlines of the search strategies, HT data acquisition may also be used to obtain more advanced fundamental information, such as mechanistic and kinetic insights, than the somewhat limited information found with the previously described combinatorial strategies (ES, DoE or ANN).

This strategy will be illustrated by HT experiments carried out not only under Selox ($CO + H_2 + O_2$) conditions but also under WGS ($CO + H_2O$) and reverse WGS ($CO_2 + H_2$) conditions, varying operating parameters such as temperature, partial pressures, space velocity, under stationary (CO conversion) and non-stationary (OSC measurements) conditions.

The library used was the one based on Pt-ceria-doped catalysts, i.e., the second one designed by DoE (Section 10.2.4). The results are reported in details elsewhere [18, 20].

Three major effects were revealed during this high-throughput study of this set of close reactions on ceria-based catalysts (Fig. 10.16):

1. The nature of dopants added to ceria strongly influenced the performances of Pt/ceria catalysts for CO oxidation in the absence of hydrogen (Fig. 10.16 a); these performances matched the OSC (Fig. 10.16 c).
2. Both catalyst performance and ranking for CO oxidation were strongly affected in the presence of hydrogen (Fig. 10.16 b).
3. Either experimental or simulated outlet gas composition indicated that, in any case, WGS/RWGS equilibria must be considered in addition to the oxidation processes.

In addition, the contact time or space velocity was found to have a major effect on Selox conversion for a given catalyst (Fig. 10.17), which is largely explained by the

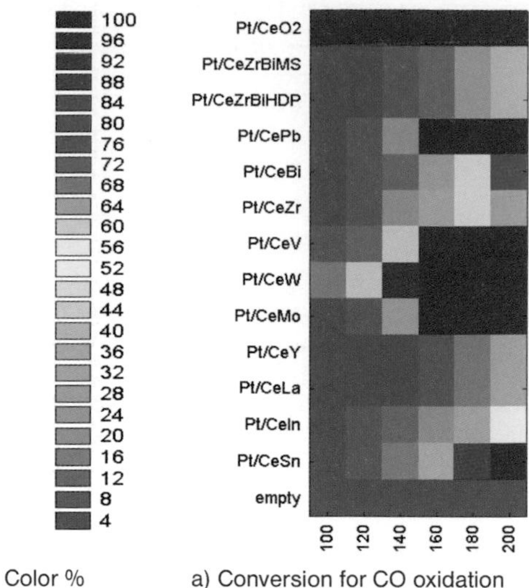

Color % scale of CO conversion

a) Conversion for CO oxidation in the absence of H_2 (%)

b) CO oxidation in the presence of H_2 (Selox)

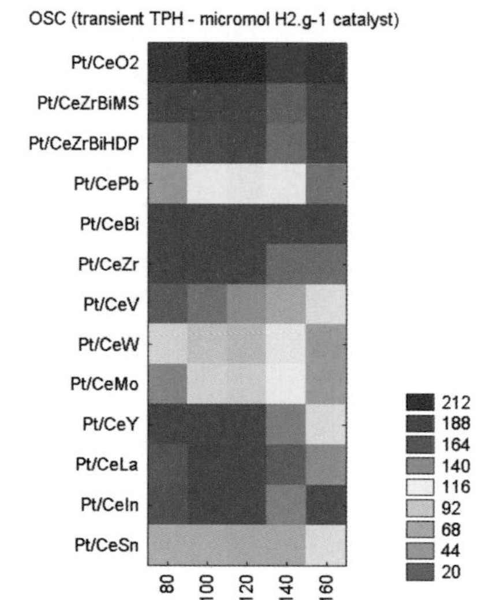

c): Oxygen storage capacity from transient TPH, mmol H_2g^{-1}catalyst

Fig. 10.16 Comparison of Pt-ceria-based catalysts performances (a, b) CO conversion, (c) OSC, as a function of temperature (°C) under various operating conditions.

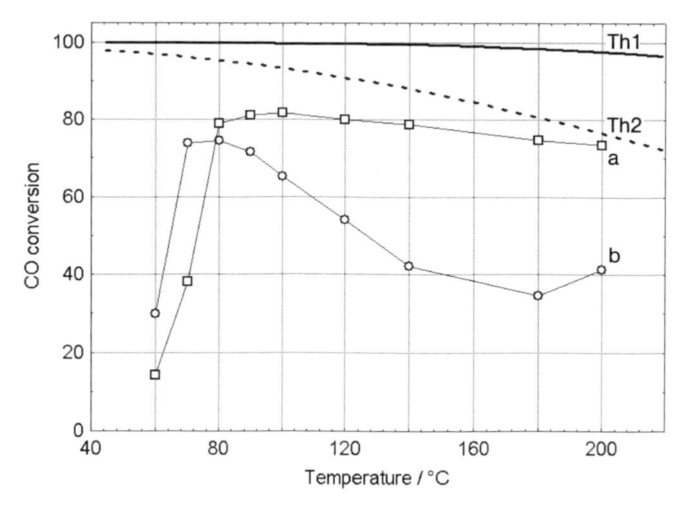

Fig. 10.17 Changes in CO conversion under Selox conditions as a function of temperature for two contact times (a) 170, (b) 48 ms. Th1: thermodynamic equilibrium for Selox inlet composition ($CO:O_2:H_2:N_2$ = 1:2:20:77 vol%), Th2: thermodynamic equilibrium for similar Selox inlet composition but assuming that the water formed by CO/H_2 oxidation is trapped by ceria.

occurrence of reverse water-gas shift after CO total oxidation (Fig. 10.18). Other parameters such as possibly non-isothermal conditions in the parallel reactor (hot-spots in the oxidation zone) were also considered to explain the observed kinetic trends [20].

- *Pros:* Methodologically, this approach of the Selox/WGS/RWGS reaction set, combining HT kinetics for under-stationary and non-stationary conditions, has provided key mechanistic insights, including the elementary steps coupling Selox and WGS reactions and the combined role of hydrogen and cations doping. This fundamental information will guide further optimisation steps [12, 18, 20].
- *Cons:* Throughout this investigation, the risks for data misinterpretation when HT experiments are carried out without satisfactory control of T&P conditions were pointed out, possibly combined with physico-chemical side effects like support-induced reaction products adsorption, which may change markedly the output gas concentration and, therefore, the final ranking of performances.

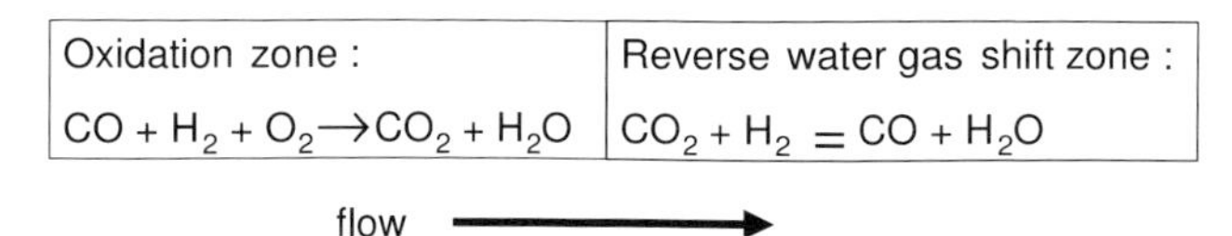

Fig. 10.18 Selox reactor axial mechanistic sequence.

10.6 Data Management

10.6.1 Database Development

The high-throughput production of data at each level of the combinatorial loop enforces an automated flow of data, which requires automated acquisition and retrieval. Since no database adapted to the specific case of heterogeneous catalysis is currently available (except those included in soft- and hardware packages provided by specialised companies, but at prices that academic groups cannot afford), the only solution was to develop an in-house database that fitted exactly the laboratory requirements.

"STOCAT™", a powerful database structure based on Oracle 9i has been developed within the frame of the European consortium "COMBICAT" [13], which allows the generation and integration of all the data related to synthesis, testing and performance of catalysts. All these data are saved via modules: Product & Reaction feeding, Synthesis configuration, Robot configuration, Chemical libraries design, Pre-mixed libraries design, Catalyst libraries, Heat treatment, Robot feeding and Test configuration. Data retrieval is carried out by a home-made query builder and Query Reporter that makes it easy to execute and report on the different queries.

"STOCAT™" manages chemicals and catalysts by providing unique bar-codes that can be printed and stuck onto vials. This is pivotal within the frame of research consortia, when large libraries are created in one place and then sent to other partners for further testing or when partners want to perform data analysis. Fig. 10.19 shows examples of the graphical user interface that enables data capture.

"STOCAT™" should soon be available in an academic version, i.e. flexible enough to be adapted to the specific requirements of each catalytic laboratory. Other close developments of in-house built databases will probably soon be on the market, offering different options in DB structure and flexibility, but essentially will answer the present lack of adapted instrumentation for data management at the academic laboratory scale.

For the present case study, only part of the data obtained for the WGS and Selox investigations were stored in the database since the quality of some experiments was estimated to be inadequate for storage (especially when technical artefacts were suspected in the case of catalyst composition or performance irreproducibility). This again points out the major issue of data management, which involves obtaining the highest data quality to avoid long-term pollution of the whole database, despite existing filters that can limit such damage.

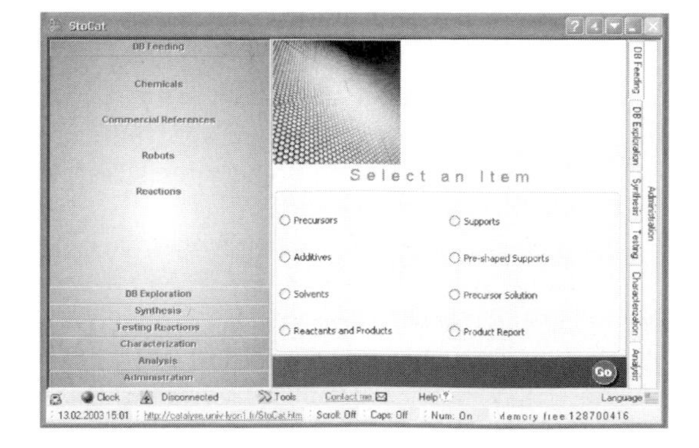

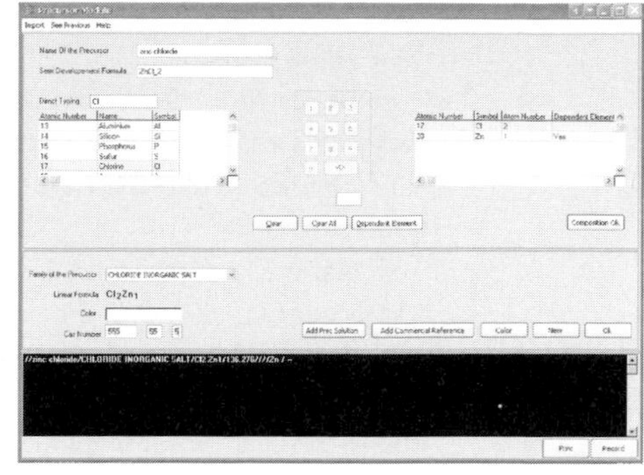

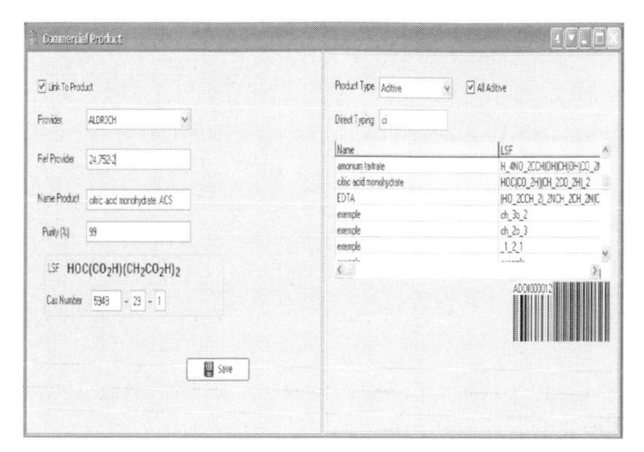

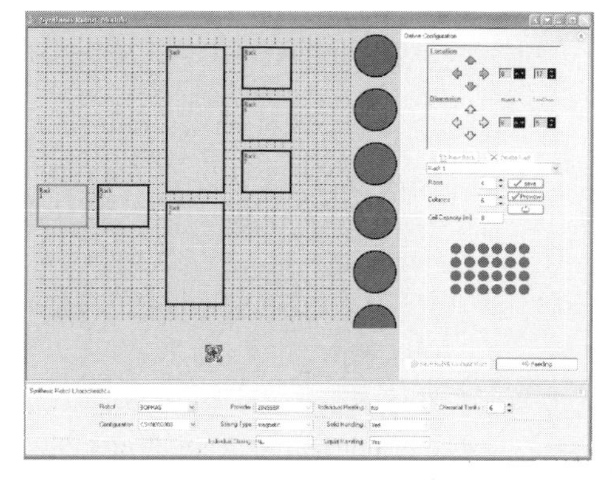

Fig. 10.19 Examples of graphical user interface that enable the capture of data in the STOCAT database.

10.6.2
e-Platforms and e-Languages Adapted to HTE

In parallel with database development, any laboratory developing an advanced HTE strategy needs a friendly and efficient way to access and integrate all the data and software applications, i.e., a powerful and flexible e-platform to mine, analyze and visualize all the diverse collected information.

Such an e-platform, "OPTICAT^{TM}", has been recently implemented in our laboratory using industry-standard Java and XML and links to data in a range of formats including any SQL-based database, such as Oracle. It allows the user to construct data processing treatments, from the simplest to the most complicated, by a drag & drop procedure, allowing available bricks, which contain treatments, to be linked into an entire complex data-process. Bricks can be database links, simple data processing tools like deleting of lines or columns or requests. The algorithms contained in the bricks can also be very powerful ones like data-mining tools, e.g. neural networks, knowledge-based systems or even parts of genetic algorithms. It is fully customisable in a few mouse clicks, which allows chemists without a deep knowledge of coding to design algorithms that match perfectly with their requirements.

Some comments should also be made about the e-languages that are used for these data management instruments and the consequences of the level and quality of communication between research groups not belonging to the same laboratory or institution.

To date, the numerous collaborations between industrial partners or industrial and academic partners in the catalysis domain have generally produced limited, easily manageable amounts of data, which did not require standardised formatting or databases to be exchanged. The development of HTE, however, will have major implications on data management, and urgent considerations on the form of the data and the means of exchanging them are due.

Up to now, no standards, e-languages and protocols exist. It may be foreseen that in the near future, large international programmes (US, EU, Japan) will work on common HTE facilities (technological and e-platforms) and will establish protocols. These platforms, by integrating standardised middleware interfaces, will contribute to the successful introduction and proliferation of High Throughput Technologies not only in academia but also in large companies and SMEs.

- *Pros:* Though not yet widespread, several databases adapted to heterogeneous catalysis have been developed recently at the laboratory scale, which mostly solves the crucial issue of data storage and retrieval.
- *Cons:* Special care must be taken in maximising the quality of the data to be stored, to avoid long-term pollution of the databases.

Most of the above-mentioned instruments still remain non-compatible and a huge effort in e-language standardisation and access protocols need to be carried out. International programmes should handle that challenge so as to allow HT laboratories to network and communicate easily.

10.7 Conclusions

This chapter has tried to follow, step-by-step for a given case study, the pros and cons that may be revealed at each stage of the so-called combinatorial loop, using academic hard- and software tools.

HT technology for catalysts-automated synthesis and testing appears to be reasonably adapted to date, but further improvements are expected for HT catalysts characterization, which is still restricted to costly and in general ex-situ spectroscopic techniques. These tools would provide the new catalyst descriptors needed to improve the ability to predict catalytic performances without testing.

The choice of combinatorial strategies, including the domain of investigation, the space of parameters, data management and processing techniques, appears as crucial. One option, as shown here, is to try and test sets of diverse catalyst formulations combined with real-time performance analysis. Robots are available for doing the experiments in a reasonable time scale, but the bottleneck now lies with data analysis and processing, with a possible integration of various new computer-based tools such GA, DoE, ANN, in respect of a continuous and scrupulous control on data quality and meaning. Optimising the methods for high-throughput data management remains therefore a pivotal objective for further improvement of the methodology.

Finally, the presented case study, which was for large part treated within the frame of the European consortium "COMBICAT", also revealed interesting features to be considered by teams willing to join the combi-domain: sharing research between several laboratories provides an obvious gain in time and allows partners to ascertain some conclusions due to a permanent double or triple check of the results. However, unexpected issues like the difficulties in information transfer and data management due to the enormous flow of circulating materials and data have to be solved. A huge effort in data handling, set-up and procedure standardisation, involving the implementation of recognised and normalised common e-languages and e-platforms, has to be carried out urgently as a compulsory step towards reaching the optimised integrated efficiency expected from this new shared research strategy.

10.8 Acknowledgments

The EU "COMBICAT" programme (GR5RD-CT1999-00022) is fully acknowledged for supporting part of the quoted work. Professor Manfred Baerns and his group at ACA (Berlin) and all IRC collaborators are fully acknowledged for their large joint collaboration within "COMBICAT".

10.9 References

1 Senkan, S.M. *Nature (London)* 1998, *394*, 350–353.
2 Senkan, S., Krantz, K., Ozturk, S., Zengin, V., Onal, I. *Angew. Chem., Int. Ed.* 1999, *38*, 2794–2799.
3 Weinberg, W.H., Jandeleit, B., Self, K., Turner, H. *Curr. Opin. Solid State Mater. Sci.* 1998, *3*, 104–110.
4 Weinberg, W.H., McFarland, E., Goldwasser, I., Boussie, T., Turner, H., Van Beek, J.A.M., Murphy, V., Powers, T., et al. In *PCT Int. Appl.* (Symyx Technologies USA, Wo, 1998, p 106
5 Maier, W.F. *Angew. Chem. Int. Ed.* 1999, *38*, 1216–1218.
6 Pescarmona, P.P., van der Waal, J.C., Maxwell, I.E., Maschmeyer, T. *Catal. Lett.* 1999, *63*, 1–11.
7 Wolf, D., Buyevskaya, O.V., Baerns, M. *Appl. Catal., A* 2000, *200*, 63–77.
8 Mirodatos, C. *Actual. Chim.* 2000, 35–39.
9 Reichenbach, H.M., McGinn, P.J. *J. Mater. Res.* 2001, *16*, 967–974.

10 Rodemerck, U., Wolf, D., Buyevskaya, O.V., Claus, P., Senkan, S., Baerns, M. *Chem. Eng. J.* 2001, *82*, 3–11.

11 Wolf, D., Gerlach, O., Baerns, M. in *Eur. Pat. Appl.* (Institut für Angewandte Chemie, Berlin-Adlershof e.V., Germany). Ep, 2002, p 13.

12 Farrusseng, D., Baumes, L., Mirodatos, C. in *High-Throughput Analysis: A Tool For Combinatorial Materials Science*, Potyrailo, R.A., Amis, E.J. (eds.), Kluwer Academic/Plenum Publishers, Dordrecht, 2003, pp 551–579.

13 Mirodatos, C. in *http://etip.cordis.lu/* and *www.ec-combicat.org*, 2003.

14 Perego, C., Mirodatos, C., Baerns, M. *Catal. Today* 2003, *81*, 307–317.

15 Wolf, D. In *Principles and Methods for Accelerated Catalyst Design and Testing*, Derouane, E.G., Parmon, V., Lemos, F., Ribeiro, F.R. (eds.), Kluwer Academic Publishers, Dordrecht, 2002, pp 125–133.

16 Descorme, C., Madier, Y., Duprez, D. *J. Catal.* 2000, *196*, 167.

17 Wolf, D., Buyevskaya, O., Baerns, M., Rodemerck, U., Claus, P. in *PCT Int. Appl.* (Institut für Angewandte Chemie, Berlin-Adlershof E.V., Germany) 2000, p 35

18 Farrusseng, D., Pereira, S.R.M., Hoffmann, C., submitted to *QSAR & Combi. Sci.*

19 Baumes, L., Farrusseng, D., Lengliz, M., Mirodatos, C. *QSAR Comb. Sci.* 2004, submitted.

20 Tibiletti, D., de Graaf, E.A.B., Rothenberg, G., Farrusseng, D., Mirodatos, C. *J. Catal.* 2004, in press.

21 http://www.statsoft.com/.

22 Vauthey, I., Baumes, L., Hayaud, C., Farrusseng, D., Mirodatos, C., Grubert, G., Kolf, S., Cholinska, L., Baerns, M., Pels, J.R. *Appl. Catal.* 2004, submitted.

23 Farrusseng, D., Baumes, L., Hayaud, C., Vauthey, I., Denton, P., Mirodatos, C., *Principles and Methods for Accelerated Catalyst Design*, E. Derouane et al. (eds.), NATO Science Series, Kluwer Academic Publishers, Dordrecht, 2002, pp 469–479.

24 Rodemerck, U., Ignaszewski, P., Lucas, M., Claus, P. *Chem. Eng. Technol.* 2000, *23*, 413–416.

25 Hahndorf, I., Buyevskaya, O., Langpape, M., Grubert, G., Kolf, S., Guillon, E., Baerns, M. *Chem. Eng. J.* 2002, *89*, 119–125.

26 Perez-Ramirez, J., Berger, R.J., Mul, G., Kapteijn, F., Moulijn, J.A. *Catal. Today* 2000, *60*, 93–109.

27 Buyevskaya, O.V., Bruckner, A., Kondratenko, E.V., Wolf, D., Baerns, M. *Catal. Today* 2001, *67*, 369–378.

28 Buyevskaya, O.V., Wolf, D., Baerns, M. *Catal. Today* 2000, *62*, 91–99.

29 Corma, A., Serra, J.M., Chica, A. *NATO Science Series, II: Mathematics, Physics and Chemistry* 2002, *69*, 153–172.

30 Serra, J.M., Corma, A., Farrusseng, D., Baumes, L., Mirodatos, C., Flego, C., Perego, C. *Catal. Today* 2003, *81*, 425–436.

31 Baumes, L., Jouve, P., Farrusseng, D., Lengliz, M., Nicoloyannis, N., Mirodatos, C. *7th Int. Conf. on Knowledge-Based Intelligent Information & Engineerging Systems (KES' 2003)*, 3–5 September 2003, University of Oxford, UK. In: Lecture Notes in AJ (LNCS/LNAI series), Eds: V. Palade, R.J. Howlett, L.C. Jain. Springer, Berlin.

32 Davis, L. *Handbook of Genetic Algorithms*, Van Nostrand Reinhold, New York, 1991.

33 Klanner, C., Farrusseng, D., Baumes, L., Mirodatos, C., Schüth, F. *QSAR Comb. Sci.* 2003, *22*, 729–736.

34 Serra, J.M., Corma, A., Chica, A., Argente, E., Botti, V. *Catal. Today* 2003, *81*, 393–403.

35 Cundari, T.R., Deng, J., Zhao, D. *Ind. Eng. Chem. Res.* 2001, *40*, 5475–5480.

36 Hou, Z.Y., Dai, Q., Wu, X.Q., Chen, G.T. *Appl. Catal. A: General* 1997, *161*, 183–190.

37 Kito, S., Hattori, T., Murakami, Y. *Appl. Catal.* 1989, *48*, 107–121.

38 Kito, S., Hattori, T., Murakami, Y. *Appl. Catal. A: General* 1994, *114*, 173–178.

39 Corma, A., Serra, J.M., Argente, E., Botti, V., Valero, S. *Chem. Phys. Chem.* 2002, *3*, 939–945.

11
Combinatorial Synthesis and High-Throughput Screening of Fuel Cell Electrocatalysts

Peter Strasser, Sasha Gorer, Qun Fan, Konstantinos Chondroudis, Keith Cendak, Daniel Giaquinta, and Martin Devenney

11.1
Introduction

Electrocatalysis is heterogeneous catalysis at an electrified electrode–solution interface [1]. The heterogeneously catalyzed chemical reaction, however, is distinguished from an electrocatalytic reaction in that it does not involve net charge transfer while an individual electrodic reaction *does* involve net charge transfer. The kinetics, therefore, are dependent on potential. This results in a powerful control parameter: varying the electrode potential directly translates into changes in the activation energy of an electrochemical process at the electrode surface.

The technological area currently most affected by advances in electrocatalysis is electrochemical energy conversion, especially fuel cells [2, 3]. A fuel cell converts chemical energy directly into electrical energy, without Carnot limitations. Environmental concerns and new regulations have given renewed stimuli to the development of fuel cell-powered devices and vehicles. Fuel cell technology promises low to zero emission energy for power plants, backup generators, and the transportation sector. Fuel cells also offer high fuel efficiency on the order of 50% or more. Low-temperature polymer electrolyte membrane fuel cell (PEMFC) technology, in combination with high-energy density fuels, is an especially attractive alternative to battery power for today's portable electronics applications.

With the increasing interest in fuel cells as alternative power sources for stationary, portable and automotive applications, there is a growing need to develop cheaper and more active anode and cathode electrocatalysts. The cost of current precious metal-based electrocatalysts is a considerable hurdle for the successful commercialization of fuel cells. Currently, two areas of PEMFC catalyst research are receiving the most attention in the scientific literature [4]: decreasing the Pt content in cathodes and increasing the activity of direct fuel cell anodes. The cathodic power losses of a hydrogen-operated PEMFC can amount to as much as 40% of the total power loss. Hence, low-Pt cathode catalysts are needed for the reduction of oxygen [4, 5]. Likewise, more active anode electrocatalysts for fuel cells operating on small organic molecule feeds such as methanol (direct methanol fuel cells, DMFC) are required to achieve improved Pt-based power densities. DMFCs are the most commonly studied among the direct fuel cells [6, 7].

High-Throughput Screening in Chemical Catalysis
Edited by A. Hagemeyer, P. Strasser, A. F. Volpe, Jr.

ISBN: 3-527-30814-8

The conventional method of discovering and developing fuel cell catalysts is manually intensive and time-consuming. The traditional starting point is often an heuristic model based on which bulk catalysts are synthesized and characterized one at a time. Catalyst inks are then formulated by the suspension of catalyst powders in a solvent for membrane electrode assembly (MEA) preparation. MEAs are tested in single-cell fuel cell test stations, often in duplicate or triplicate. An advantage of single-cell testing is that catalysts prepared using industrially relevant methods are tested under commercially relevant fuel cell conditions. Under these circumstances, however, the testing of 10 catalyst powders may require over half a year [8]. In this situation, the development of reliable, high-speed discovery and testing alternatives is desirable [9].

Electrocatalysis presents a special opportunity but also a challenge for combinatorial and high-throughput catalyst research. The opportunities lie in the huge parameter space encompassed by all possible compositional, electrical and process variables that are relevant to the development of a new fuel cell catalyst. As a result of this complexity, traditional research methods are unable to address the full variable space. Instead much of today's fuel cell electrocatalyst research is concentrated on optimizing the interfacial contact and utilization of the catalyst within a complex material system. Herein lies the challenge for combinatorial electrocatalyst research. The combinatorial research program must be designed in such a way that all parallel and/or high-throughput synthetic and characterization methodologies provide relevant experimental data.

Combinatorial electrocatalysis (CE) – for the purpose of improving the intrinsic catalyst activity – does not eliminate the need for slow, single-cell testing, but simply precedes it, allowing the number of slow, single-cell measurements to be minimized. Additionally, CE does not attempt to eliminate the use of heuristic modeling. Instead, CE is intended provide a straightforward testing mechanism for modeling techniques. In fact, experimental results from a combinatorial study may serve to support a given model, as shown later in this chapter.

This chapter presents the design and application of a two-stage combinatorial and high-throughput screening electrochemical workflow for the development of new fuel cell electrocatalysts. First, a brief description of combinatorial methodologies in electrocatalysis is presented. Then, the primary and secondary electrochemical workflows are described in detail. Finally, a case study on ternary methanol oxidation catalysts for DMFC anodes illustrates the application of the workflow to fuel cell research.

11.2 Combinatorial Methods for Electrocatalysis

An ideal combinatorial and high-throughput workflow consists of a synthesis and a screening step of comparable throughput.

Three techniques have been described in the literature to prepare combinatorial libraries of fuel cell electrocatalysts: solution-based methods [8, 10–14], electrodeposition methods [15–17] and thin film, vacuum deposition methods [18–21]. Vacuum deposition methods were chosen herein for electrocatalyst libraries in order to focus on the intrinsic activity of the materials, e.g., for ordered or disordered single-phase, metal alloys.

Two general techniques have been described for the high-throughput screening of electrocatalyst libraries: optical screening and electrochemical screening.

Optical screening of electrochemical half-cell reactions is an indirect detection method based on the idea that each half-cell reaction generates or consumes ions. Half-cell reactions relevant to fuel cells include the anodic oxidation of methanol or hydrogen and the cathodic reduction of oxygen in acidic environments where protons are generated and consumed, respectively. In the presence of a suitable indicator molecule, e.g. a pH indicator, an optical fluorescence or absorbance signal can be observed in the vicinity of an active electrocatalyst. Optical screening of an array of diverse electrocatalysts supported on a conductive substrate (catalyst library) indicates where on the array the half-cell reaction is occurring, and to what degree. Reddington et al. demonstrated parallel, high-throughput optical screening of electrocatalyst libraries using a fluorescence-based method [10, 11, 22, 23]. Fig. 11.1 illustrates the optical screening of an array of Pt-Rh-Os DMFC anode catalysts supported on a Toray paper electrode. Quinine, an acid-sensitive fluorescence indicator, was used as 'chemosensor' to map the local proton concentration near each individual catalyst composition. The simplicity of this truly parallel, high-throughput screening method is intriguing, requiring only a suitable indicator and a light (UV–Vis) source. Direct measurement of current flowing through individual electrocatalysts is not performed, however, and the method requires conditions compatible with the performance of the indicator, conditions that are generally inconsistent with the conditions of conventional fuel cell testing.

High-throughput electrochemical screening methods are direct measurements in the sense that the current and potential of each individual catalyst is accurately controlled and monitored. Electrochemical cell arrays for electrochemical screening have been reported by Warren et al. [16], Gorer [24], Strasser et al. [18–21], Sullivan et al. [25], Jiang and Chu [26], Liu and Smotkin [27] and Guerin et al. [28]. While Warren electrodeposited Pt-Ru on an array of Pt electrodes to show that Pt50Ru50 is the most active methanol oxidation catalyst, Sullivan et al. studied alkanethiol-modified gold array electrodes. Jiang and Chu developed an array method with a movable probe and an electrolyte link that can represent the setup of a simple, single fuel cell measurement. Gorer studied electrodeposited ternary methanol oxidation catalysts in multi-electrode arrays. Liu and Smotkin developed a fuel cell array with 25 individually controllable electrodes.

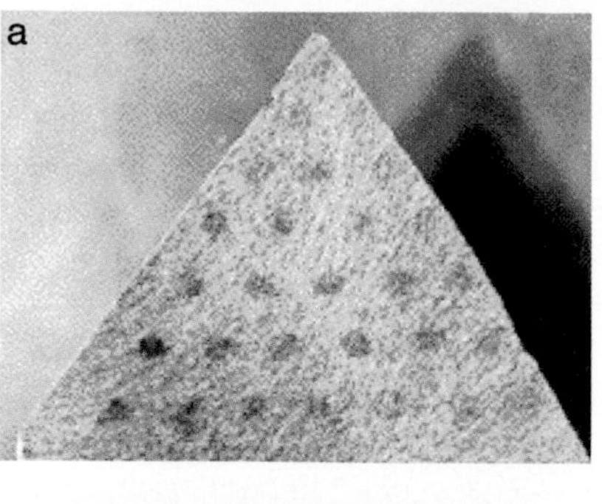

Fig. 11.1 An array of 28 Pt-Rh-Os catalyst spots in 6 M aqueous methanol, pH 6, quinine indicator. (a) Image in ambient light; (b) fluorescence image at low overpotential; (c) fluorescence image at high overpotential where methanol oxidation occurs at every spot in the array. (Reprinted with permission by Reddington et al., *Science* 1998, 280, 1785. Copyright 1998 AAAS.)

A third screening method for arrays of electrocatalysts was recently introduced by Hillier and coworkers [15, 29, 30]. Using a scanning electrochemical microscope (SECM), a microelectrode tip is moved over an electrocatalyst array. The resulting electrochemical feedback currents are measured and used to generate an activity map of the electrocatalyst library. This method does not require individual electronic addressability for each electrocatalyst.

11.3
An Integrated, High-Throughput Screening Workflow for Electrocatalysis

This section introduces a two-stage, fully automated combinatorial synthesis and screening workflow for electrocatalyst research developed by scientists at Symyx Technologies [18, 19, 21]. The primary synthesis and screening workflow includes vacuum deposition of thin film electrocatalysts, high-throughput structural characterization and parallel electrochemical array screening. The secondary synthesis and screening workflow includes the preparation of high surface area, supported electrocatalysts using a freeze-drying technique, followed by a parallel rotating disk electrode (PRDE) screening method for powder catalysts.

11.3.1
Electrode Array for Primary Synthesis and Screening

Libraries of electrocatalytic materials were synthesized directly on an 8×8 electrode array consisting of 64 individually addressable electrodes (Fig. 11.2). The array was prepared on an insulating 3″ diameter quartz wafer using photolithographic methods [16]. Suitable electrode materials were chosen to eliminate electrode-based Faradaic activity in the potential range of interest. Typical examples include Au or Ti. Each of the 64 electrodes was approximately 1.5 mm in diameter and electrical contact was made with each electrode via a conductive contact pad on the edge of the wafer. The edge pads and the center electrodes were connected by fine address lines of the same material. The address lines were covered by an insulating material to prevent them from interfering in the electrochemical electrode processes.

11.3.2
Design of Combinatorial Thin Film Libraries

The design of a combinatorial catalyst library refers to the act of composing a sequence of synthesis steps such that the desired portion of a compositional or process parameter space will be mapped onto the final materials (catalyst) library. The design of libraries is generally preceded by the selection of some rational catalyst concept, yet may include some degree of serendipity. Library Studio® [31] served as the principal library design tool. To design a thin film library, the substrate or destination for the materials array is selected, in this case an 8×8 electrode array, as described above. The set of chemical sources (e.g., metals or metal

Fig. 11.2 A 64-element, individually addressable, Ti electrode array on a 3″ diameter quartz wafer for the synthesis of electrocatalyst libraries.

compounds) to be employed in the synthesis process is then selected. Using the graphical interface of Library Studio®, the sources (metals) are mapped to the appropriate destinations (individually addressable electrodes) such that the compositional space of interest is created by appropriately varying the thicknesses of different metal sources. Mappings may include the deposition of layers of constant thickness or of varying thicknesses. Mapping information such as metal source, deposition location and layer thickness is communicated to the vacuum deposition instrument using a Library Studio output file. A graphical representation of the final library design, for example, displaying the molar fractions of each library element together with the thickness information of each individual metal layer, can be stored in a database for later reference and retrieval [32].

11.3.3 Combinatorial Thin Film Synthesis

The development of new and improved electrocatalysts begins with the discovery of materials displaying improved *intrinsic* electrochemical activity. Intrinsic activity is best observed and compared in a well-controlled catalyst environment where variables that may disguise the intrinsic activity trends are minimized or absent. Particle size, particle size distribution, surface area, catalyst utilization and the distribution of crystallographic phases are parameters that are typically difficult to control. Vapor deposition of unsupported thin film electrocatalysts eliminates many of these variables. This method provides a controlled synthetic route to smooth, single-phase or multi-phase, ordered or disordered metal alloy phases depending on deposition and processing conditions.

Parallel synthesis of the 64-element metal alloy electrocatalyst library was achieved using automated rf-magnetron vacuum sputtering. The individual constituents of each catalyst composition were sputter-deposited onto the 64-element electrode array using a combination contact-mask/moving-shutter approach (Fig. 11.3 a) [33]. The proprietary moving-shutter technique allowed the controlled deposition of thin film compositional gradients ranging from a few ångstroms to several micrometers in thickness. Compositional variations in the thin films were achieved either by sputtering parallel and orthogonal thickness gradients (Fig. 11.3 b) or else by sequential sputtering of metal slabs of constant thickness but varying orientation across the wafer. Typical individual sputtering steps ranged from a thickness of about 5 to 15 Å (Fig. 11.3 b, left-hand side). Sputtering of superlattice structures (Fig. 11.3 b, right-hand side) led to deposition of thin film thicknesses of up to several thousand Ångstroms. The thicknesses of the deposited layers are in the range of the diffusion length of the metal atoms (e.g., about 10 to about 30 Å), which allows in-situ alloying of the metals. Post-deposition processing of the libraries included thermal annealing in an appropriate atmosphere.

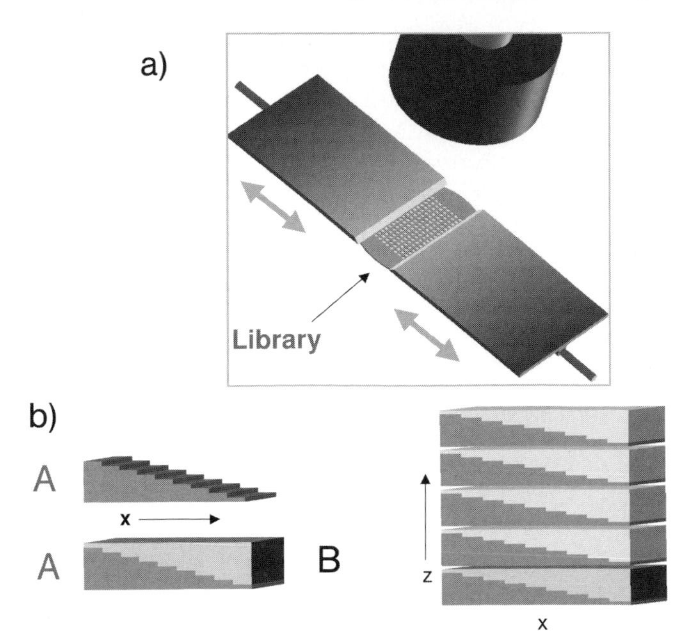

Fig. 11.3 (a) An rf-magnetron sputter technique with automated target selection; automated moving shutters and physical shadow masking is employed to deposit the electrocatalyst libraries. (b) Sequential gradient sputtering of very thin material slabs results in in situ thin film alloy formation on the electrodes. The width of the slabs on the left extends over the entire width of the electrode array. Superlattice deposition in the *z* direction results in the deposition of thick films of the desired stoichiometric gradient.

11.3.4
High-Throughput Characterization of Electrocatalyst Libraries

Structural and compositional characterization of individual elements of a combinatorial library can be important for the initial validation of a particular combinatorial synthesis method. Many earlier reports on combinatorial synthesis and screening of electrocatalysts fall short of reporting the complete structural and compositional characterization of individual library elements of interest. The workflow described here includes catalyst characterization before and after screening, thereby establishing an activity–composition–structure–stability relationship for electrocatalysts. This can be relevant in light of the extreme conditions present in a conventional fuel cell environment.

The present workflow includes two complementary high-throughput characterization methods: X-ray diffraction (XRD) for structural information and scanning electron microscope/energy dispersive X-ray (SEM/EDX) analysis for quantitative compositional information. Automated rapid-serial XRD and EDX analyses were performed using programmable *xy*-stages in combination with suitable software.

Thin film electrocatalysts were characterized before and after screening to monitor structural as well as compositional changes due to the electrochemical testing environment. Changes in the peak intensity of the EDX spectra provided information on base metal leaching, while shifts in XRD peak position or intensity revealed structural changes. The combination of this information enables a more detailed and quantitative evaluation of corrosion processes occurring during electrochemical testing.

To demonstrate the viability of the synthesis and characterization approach in the present workflow, two binary Pt-Fe alloy libraries were designed, synthesized and characterized by XRD (Figs. 11.4 to 11.6) [19]. One library (Fig. 11.5) was characterized as synthesized, while the other (Fig. 11.6) was annealed at 400 °C for 12 h in a hydrogen/argon atmosphere. Pt-Fe is a well-known binary alloy system, exhibiting both substitutional solid solution compositional ranges and intermetallic compounds.

Figs. 11.5 and 11.6 show eight XRD patterns as measured along row C. In Fig. 11.5, the dominant (111) diffraction peak for 100% Pt appears at around $2\theta = 40°$, as expected. The absence of diffraction peaks of pure Pt or Fe in the multi-metal XRD patterns is indicative of complete alloying of the multi-metal thin films by interlayer diffusion.

As the amount of Fe is increased, the (111) peak shifts to smaller *d*-spacings, reflecting a contraction of the lattice. The (111) peak positions in Fig. 11.5 show a continuous shift from pure Pt to pure Fe. The Pt-Fe XRD patterns are consistent with a single-phase, substitutional solid solution (disordered alloy) over the entire compositional range. In contrast, Fig. 11.6 clearly displays diffraction from intermetallic compounds of lower symmetry. Post-deposition annealing has resulted in an ordering of the Pt and Fe atoms, the effect of which is the crystallization of an ordered metal alloy of lower symmetry than 100% Pt. In essence, the applied vacuum deposition method is ideally suited for the preparation of multi-component,

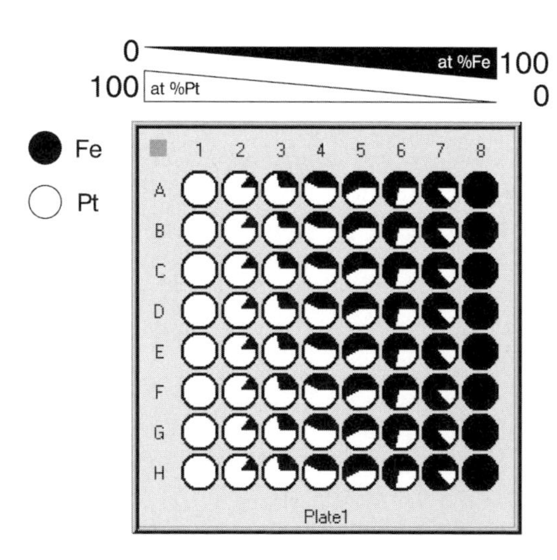

Fig. 11.4 Library design of a 64-element electrocatalyst library of Pt-Fe binary alloys. The square (Plate 1) and the 64 round spots represent the wafer substrate and the location of individual electrocatalyst alloys, respectively. The pie-chart character of each catalyst represents its chemical composition, ranging, from left to right, from 100% Pt to 100% Fe. Each row, A–H, is identical. During synthesis, this library design will be deposited onto the electrode array. The design was created using Library Studio® [31].

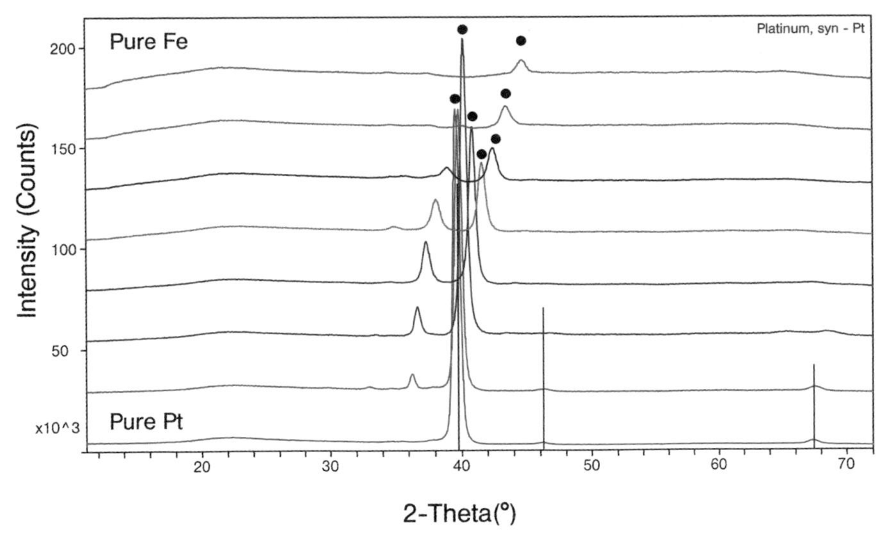

Fig. 11.5 Eight XRD patterns taken along row C of the library in Fig. 11.4 with no post-deposition annealing. The XRD patterns of the synthesized alloys are consistent with a substitutional solid solution with a face-centered cubic (fcc) structure. The black dots indicate diffraction from the fcc (111) plane. The thin vertical lines indicate the expected positions of the diffraction peaks of 100% Pt.

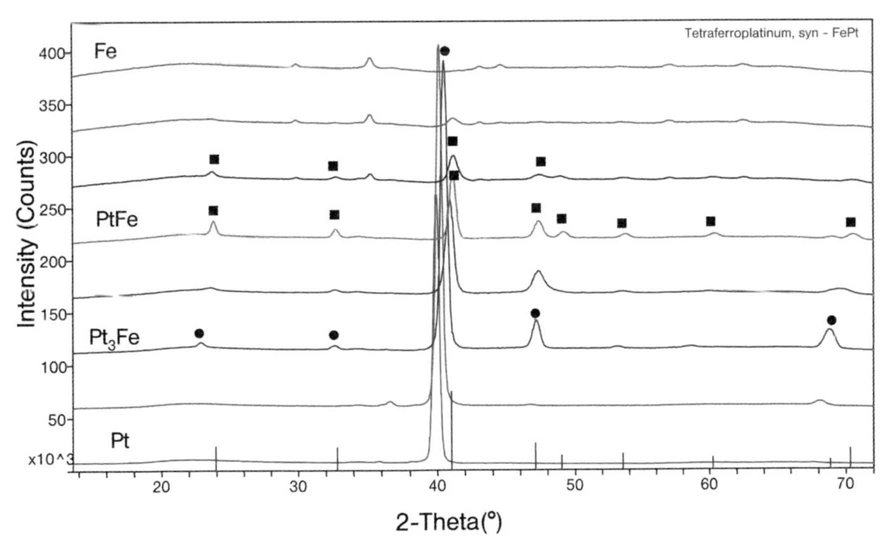

Fig. 11.6 Eight XRD patterns taken along row C of the library in Fig. 11.4 *after* post-deposition annealing (400 °C, 12 h, H_2/Ar) in a flow-through tube furnace. Intermetallic phase formation is observed. The black dots indicate the formation of the primitive cubic Pt_3Fe phase, while black squares indicate the formation of tetragonal PtFe. The thin vertical lines indicate the expected positions of the diffraction peaks of tetragonal PtFe.

single-phase materials, and enables the study of both ordered and disordered metal alloy phases.

11.3.5 High-Throughput Electrochemical Screening of Electrocatalyst Libraries

For high-throughput electrochemical screening, the electrocatalyst library wafer was combined with a cylindrical cell body such that all 64 electrocatalysts were facing upwards and were exposed to the electrolyte. All contact pads at the edge of the wafer were isolated from the electrolyte. A Pt mesh was placed in parallel to the quartz wafer and served as the common counter electrode. An Hg/Hg_2SO_4 electrode was used as reference electrode and was located in a glass compartment with a Luggin-Haber capillary, filled with supporting electrolyte. The capillary tip of the reference electrode compartment was placed between the working electrode array and the counter electrode mesh. The distances between the capillary tip of the reference compartment and the working electrodes at the edge and at the center of the array were chosen to be large compared with the distance between individual working electrodes in order to minimize differences in the uncompensated ohmic resistance between catalysts. Parallel electrochemical measurements were performed using a 64-channel multi-potentiostat (Arbin), avoiding complications and limitations associated with pseudo-parallel multiplexing techniques.

Fig. 11.7 demonstrates a typical multi-channel screening result using the experimental apparatus described above. Shown are 64 voltammetric scans of the electroreduction of oxygen measured in parallel under identical conditions. The catalyst library consisted of 64 identical pure Pt thin film catalysts. Since the catalyst array is stationary, the hydrodynamics near each electrode interface are not well defined, resulting in variations in the mass-transport limited currents. In the kinetically controlled regime (near +0.2 $V/Hg/Hg_2SO_4$), however, the 64 scans show good reproducibility and low experimental errors of about 10%. This is where intrinsic activity trends are evaluated. The inset of Fig. 11.7 shows 64 cyclic voltammetric measurements performed in supporting electrolyte prior to oxygen reduction screening. The 64 thin film electrodes exhibit voltammetric features typical for smooth polycrystalline Pt electrodes.

The primary screening data is intended to provide trends in the intrinsic activity of large numbers of catalyst materials.

11.3.6 Closing the Primary Workflow Loop

The results of primary screening data are used to determine the design of a more focused electrocatalyst library, thereby beginning the next cycle of the primary workflow. In each subsequent cycle, the examined experimental variables become more and more limited as the experimental results move toward an optimized set of process parameters or an optimized catalyst composition. There are several strategies as to how best design the next cycle of catalyst libraries to maximize the

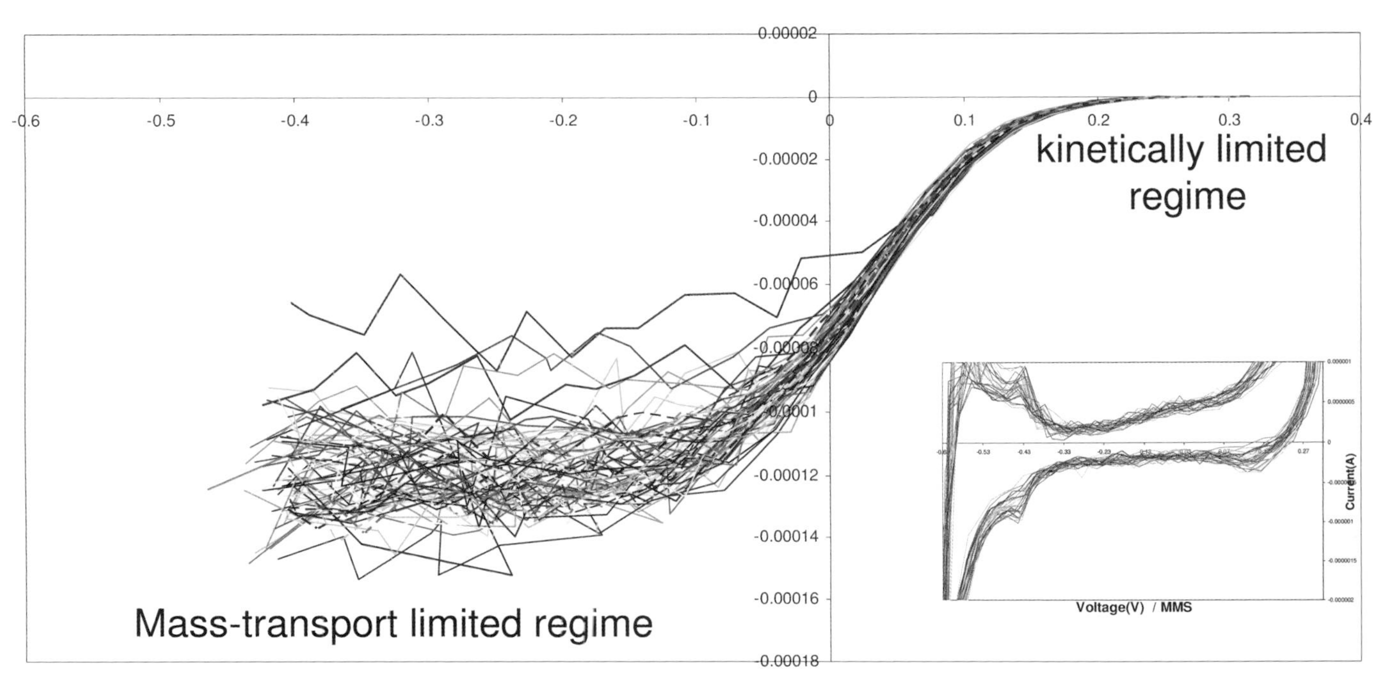

Fig. 11.7 Parallel voltammetric screening of 64 Pt thin film catalysts for the electroreduction of oxygen in acidic solution. In the kinetically controlled region, the activity of all 64 catalysts shows good reproducibility. Conditions: 0.5 M H_2SO_4, oxygen saturated; 20 mV s^{-1} anodic scan rate, 60 °C, electrolyte stirring, potentials are plotted on the mercury/mercury sulfate electrode scale. Inset: voltammogram of 64 Pt thin film catalysts in oxygen-free sulfuric acid, 20 mV s^{-1}, prior to oxygen screening.

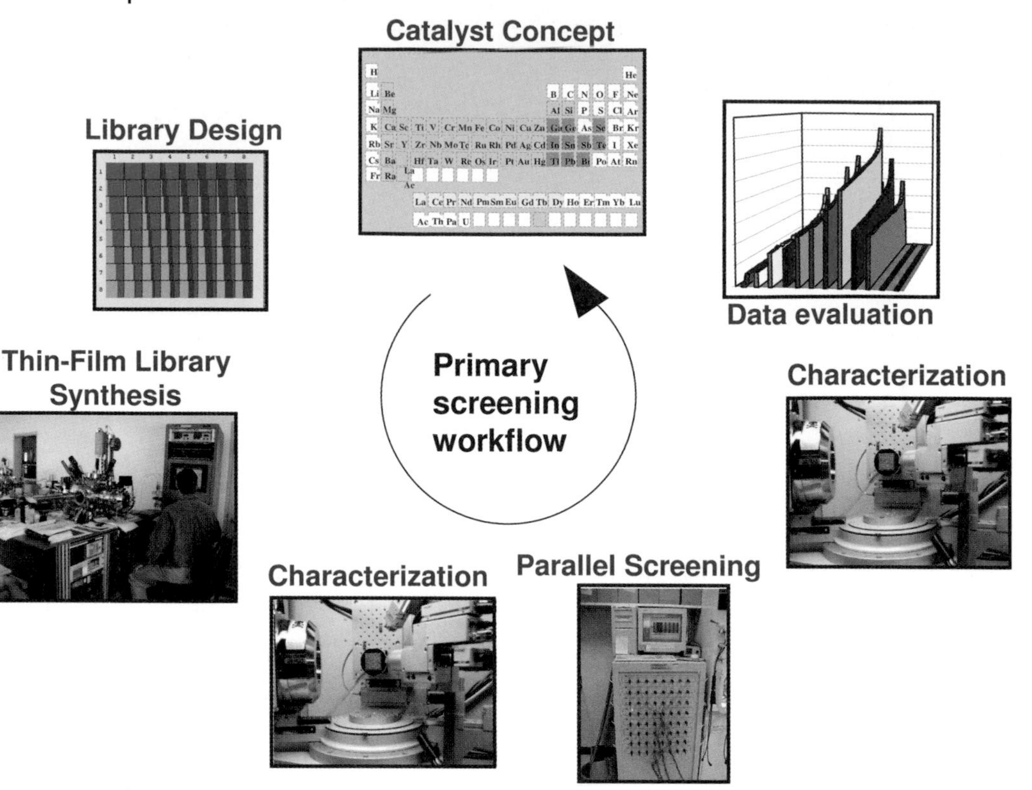

Fig. 11.8 Schematic of the automated primary high-throughput electrochemical workflow employed at Symyx Technologies for the combinatorial development of new fuel cell catalysts. Individual steps of the workflow include: choice of catalyst concept, design of appropriate materials library using Library Studio® [31], synthesis of electrocatalyst library on electrode array wafer, XRD and EDX characterization of individual electrocatalysts before screening, high-throughput parallel electrochemical screening of library, XRD and EDX characterization of catalysts after screening, data processing and evaluation.

rate of convergence toward an optimum in the performance hyperspace. For a more detailed description of such strategies, the reader is referred to earlier chapters in this book.

The complete primary screening workflow for the discovery of new fuel cell electrocatalysts is shown in Fig. 11.8. The individual components of this workflow are designed such that no bottleneck occurs.

11.3.7
Secondary Screening of High Surface Area Electrocatalysts

Primary thin film 'hits', i.e., optimized catalyst compositions determined during one or more primary screening cycles as represented in Fig. 11.8, are examined in a secondary combinatorial screening workflow. The secondary workflow again consists of an automated combinatorial synthesis method and high-throughput screening tools of similar throughput.

We have developed a combinatorial solution-based synthesis workflow that has allowed the rapid-serial preparation of high surface-area carbon-supported electrocatalysts using robotic liquid-dispensing capabilities and a novel freeze-drying methodology. Combinatorial powder catalyst libraries are designed using Library Studio® [31]. Impressionist® [34] software controls the solution-dispensing robots. Libraries of electrocatalyst powders are prepared in arrays of quartz vials. Precursor solutions are deposited onto automatically weighed samples of high surface-area carbon blacks. All powders, in parallel, are suspended in solvent, frozen in liquid nitrogen, and freeze-dried. Dried powders are annealed and reduced in flow furnaces.

Parallel high-throughput screening of electrocatalyst powders is performed using a modification of a well-known thin film powder testing method [35–38]. The RDE method requires the preparation of electrocatalyst/Nafion ink that is deposited onto a polished glassy carbon electrode as a thin film. The electrocatalyst film is dried and attached to a rotation shaft. The powder thin film method allows a very accurate determination of the kinetic characteristics of the electrocatalyst powders and has been well established in the electrocatalytic community. The RDE method also provides accurate guidance as to the performance of an electrocatalyst powder in a single fuel cell measurement at comparable catalyst utilization.

Fig. 11.9 Photograph of a 16-channel parallel rotating disk electrode (PRDE) test station for high-throughput screening of high surface-area carbon-supported electrocatalysts. The PRDE station is part of the secondary screening workflow for electrocatalysts.

Scientists and engineers at Symyx have designed and developed an automated, parallel RDE apparatus for the screening of high surface area powder electrocatalysts. A 16-channel version of the new screening tool is shown in Fig. 11.9. Sixteen identical RDE systems with stepping motors, rotating shafts, moving cell stages, gas supply, electrode contacts, and a 16-channel multi-potentiostat are controlled and operated in parallel. Library design information, including metal alloy composition and metal loading information, is automatically linked with the electrochemical results to generate data plots such as voltammograms, chronoamperometric plots, and mass transport-corrected Tafel plots.

Structural and compositional characterization of high surface-area catalysts is crucial for evaluating whether the secondary synthesis of a particular electrocatalyst was successful. Similar to the primary screening workflow, XRD is used for the structural characterization of catalyst powders, while SEM/EDX is employed for the compositional characterization of electrocatalyst powders before and after electrochemical screening.

11.4 New Ternary Fuel Cell Catalysts for DMFC Anodes

The primary and secondary electrochemical workflows presented above have been successfully validated and applied to the development of new compositions for fuel cell catalysts, specifically to the search for more active ternary and higher-order catalyst compositions for the electrochemical oxidation of methanol in acidic solutions [18, 19]. Some results of this study are now illustrated.

Over the past 35 years, much has been learned about the electrooxidation of methanol on the surface of noble metals and metal alloys, in particular platinum and ruthenium [2, 4, 6, 7]. Significant overpotential losses occur in the reaction due to poisoning of the alloy catalyst surface by carbon monoxide. Yet, Pt-based metal alloys are still the most popular catalyst materials in the development of new fuel cell electrocatalysts, based on the expectation that a more CO-tolerant methanol catalyst will be developed. The vast ternary composition space beyond Pt-Ru catalysts has not been adequately explored. This section demonstrates how the ternary space can be explored using the high-throughput, electrocatalyst workflow described above.

11.4.1 Primary Screening of Diverse Catalyst Compositions

A diverse composition space of ternary transition metal–Pt alloys was chosen as the starting point for the combinatorial investigation. Four transition metals, W, Co, Ni, and Ru, were selected to be base metal components of Pt-based metal alloys [18, 19]. The ternary permutation of these metals results in a total of six Pt-based ternary alloys, namely Pt-Ru-W, Pt-Ru-Co, Pt-Ru-Ni, Pt-Co-Ni, Pt-Co-W and Pt-Ni-W. Each ternary system was sampled broadly; six stoichiometries were chosen:

$Pt_{20}M'_{20}M''_{60}$, $Pt_{40}M'_{40}M''_{20}$ and all permutations thereof. Consequently, a total of 36 ternary compositions were synthesized. In addition, 100% Pt and Pt-Ru binaries were included in the design of the library to provide internal standards. A stochastic optimization algorithm [39] was used to optimize a sequence of thin film slab depositions such that all the desired binary and ternary compositions could be accommodated on an 8×8 wafer array. The final solution contained the 36 desired ternary compositions, several Pt-Ru binary catalysts, 100% Pt, and 24 additional binary, ternary, and quaternary compositions that were not in the original design. These latter compositions are a by-product of the synthesis design optimization. The library was synthesized by vapor deposition techniques and electrochemically screened in a methanol-containing electrolyte using the primary synthesis and screening tools described above.

Fig. 11.10 shows the results and conditions of a chronoamperometric screening at 450 mV/RHE. The Pt internal standard and two Pt-Ru binary alloys are indicated. As expected, alloying Pt with Ru results in a significant increase in activity. It is also apparent from the data, however, that several of the ternary thin film alloy catalyst display higher activity than Pt or Pt-Ru. Fig. 11.11 shows the geometric current density of all 64 channels after 5 min testing time. Base metal-rich alloys, with a base metal content of about 80 at%, exhibit extremely high activity. Inspection of both the cyclic voltammograms in supporting electrolyte and EDX data after screening revealed that the base metal-rich catalysts suffered significant corrosion, thereby increasing their surface area and consequently their geometric

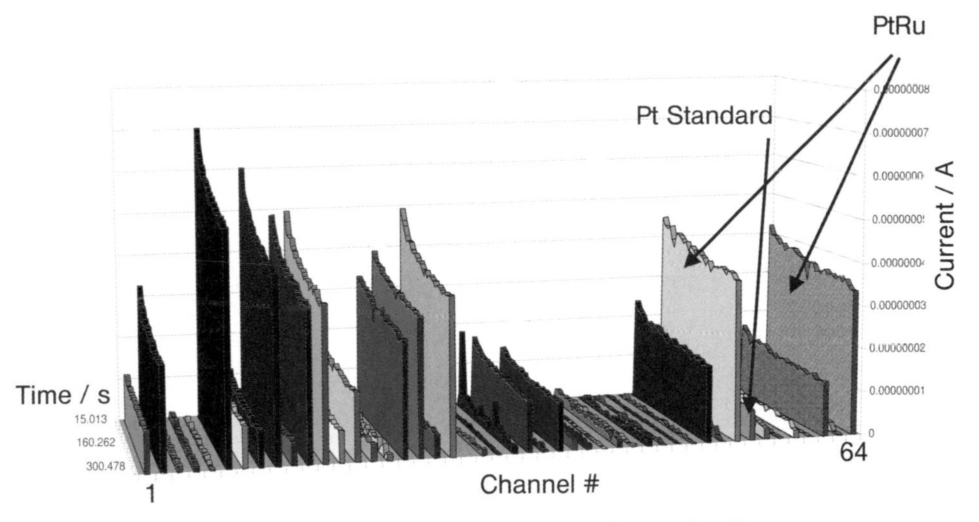

Fig. 11.10 Parallel chronoamperometric screening of a 64-element, thin film electrocatalyst library for the oxidation of methanol. The library contained a diverse set of binary, ternary and quaternary electrocatalyst compositions consisting of Pt in combination with W, Ni, Co and Ru. The graph plots current vs. time and channel number. Conditions: 1 M methanol, 0.5 M H_2SO_4, room temperature, $E=+450$ mV/RHE, test time=5 min. For clarity, channel numbers 2–4, 10, 12, 19, 20, 23, 26–29, 42, 45 and 57 are omitted. (Reproduced from [18]).

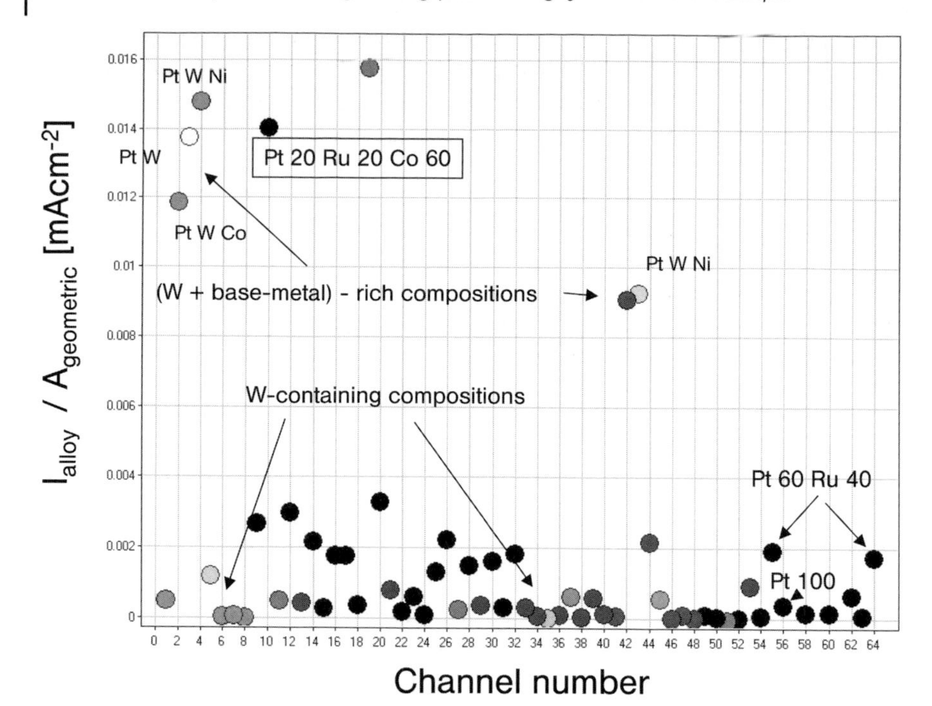

Fig. 11.11 Geometric current density vs. channel number for the chronoamperometric screen in Fig. 11.10. Dots indicate the final chronoamperometric geometric current density of each channel. The grey coloring of the dots encodes the W content of the electrocatalysts (white – high, black – low). The composition of a few active channels is indicated.

current density. To correct for corrosion, surface-area-normalized activity data were plotted (Fig. 11.12). When surface-area-normalized activity is compared, all Ru-free compositions display low activities, while Ru-containing ternaries and quaternaries display high activities. Fig. 11.13 shows the observed experimental activity trends of this broad discovery library relative to 100% Pt.

11.4.2
Comparison with Model Calculations of CO Tolerance of Ternary Electrocatalysts

Several heuristic models have been put forward to describe and predict the oxidation of methanol on pure noble metal and noble metal alloy surfaces. It is generally accepted that methanol undergoes adsorption followed by a number of hydrogen stripping steps to form adsorbed carbon monoxide (CO). During the H stripping steps, various short-lived intermediates [4, 40–44] have been proposed. While the initial adsorption and C–H bond breaking is thought to be rate determining at high overpotentials, the poisoning of the surface by CO is believed to impede the oxidation process at low, fuel cell-relevant overpotentials. According to the bifunctional model [44–47], the rate of CO removal at low overpotentials will be en-

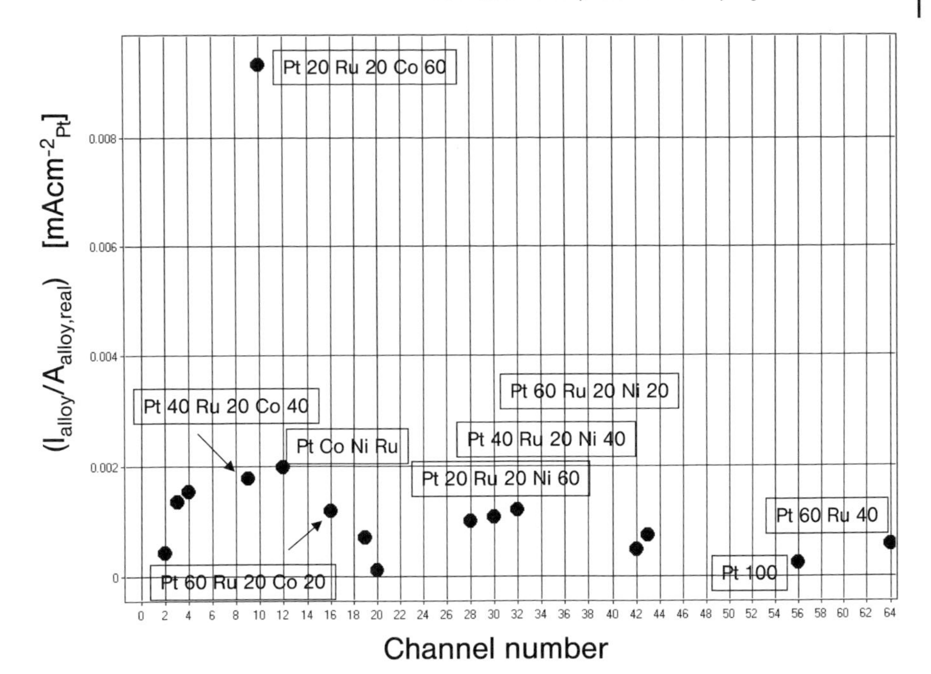

Fig. 11.12 Electrochemical activity (area-specific activity) normalized by the surface area of Pt for the most active alloys. The Pt surface area was determined electrochemically using the hydrogen adsorption integral of each catalyst.

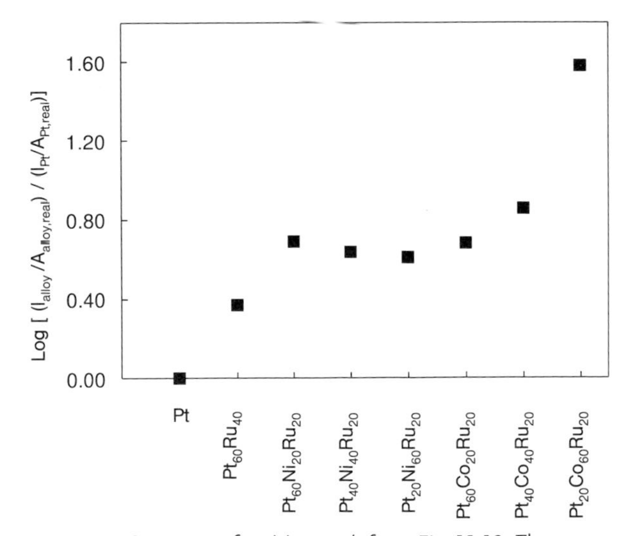

Fig. 11.13 Summary of *activity trends* from Fig. 11.12. The area-specific activities are plotted after normalization with repect to Pt. The logarithmic value represents an activation energy difference (relative kinetic improvement) and follows the DFT-calculations discussed in the text. The gains in activity on going from pure Pt to Pt-Ru, Pt-Ru-Ni and Pt-Ru-Co compositions are obvious. Quaternary compositions offer additional potential for improvement.

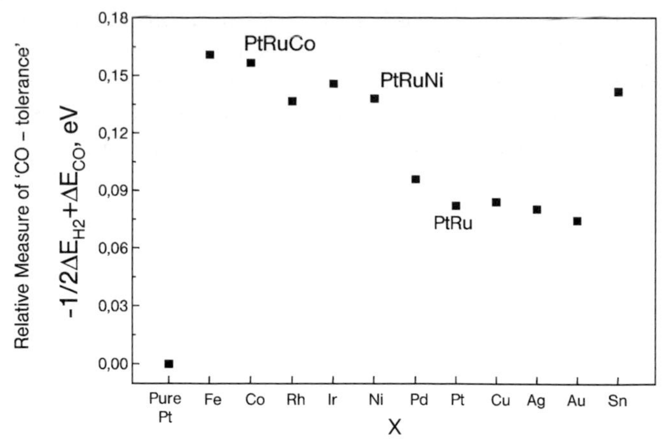

Fig. 11.14 Calculated CO tolerance of various ternary alloys, Pt-Ru-X. A mixed-metal top layer of Pt and metal sits on two hexagonal Ru metal layers. The top layer contains 2/3 monolayers Pt and 1/3 monolayers "M" (MPt_2/Ru). The alloyed atoms, M, on the surface, are respectively listed along the horizontal axis. "Pure Pt" = 100% Pt; "Pt" = Pt binary alloy Pt-Ru with a Pt top layer on two Ru layers. The observed activity trend from Pt to Pt-Ru to Pt-Ru-Ni to Pt-Ru-Co is similar to the trends observed in the experimental screening results. (Reproduced from [18]).

hanced by the introduction of oxophilic alloy components. These oxophilic metals provide oxygen-containing species (H_2O, OH, etc.) that can interact with and remove adsorbed CO. Conversely, the ligand model [48–52] states that alloying Pt may change the chemical properties of the Pt atoms on the surface of the alloy. If alloyed Pt bonds CO more weakly than pure Pt, the CO coverage should become smaller and hence the CO poisoning effect weaker. Recently, Nørskov and Liu [48, 49] presented a density functional theory (DFT) and microkinetic computational study of the CO and H_2/CO oxidation reaction on alloy surfaces. In this study, they found new evidence for a ligand effect involved in the electrocatalytic oxidation of hydrogen in the presence of CO on bimetallic surfaces. The authors demonstrated that the CO tolerance of Pt-Ru surfaces with respect to H_2/CO oxidation is not so much dependent on additional oxophilic Ru sites, but rather that Ru alters the electronic properties of Pt such that CO adsorption becomes less favorable. As a result, CO coverage of Pt-Ru alloys is reduced compared with pure Pt. This allows more hydrogen to be oxidized, and, hence, results in higher electrocatalytic activity [48].

Since the formation of strongly bonded surface CO constitutes the major kinetic hurdle for the oxidation of methanol at low overpotentials, model calculations of the CO tolerance should also give guidance in the development of ternary methanol oxidation catalysts. In fact, model calculations of the CO tolerance of ternary Pt-Ru-X alloys have been performed (Fig. 11.14) [18] revealing activity trends similar to those observed in the experimental combinatorial methanol oxidation study (Fig. 11.13): Figs 11.13 and 11.14 identify Pt-Ru-Co ternary compositions as more active than Pt-Ru-Ni compositions, followed by Pt-Ru binary and pure Pt electrocatalysts. The similarities in the observed activity trends of ternary alloys for methanol oxidation and H_2/CO oxidation may indicate that a common mechanism is operating in both cases. For H_2/CO, the calculations suggest that the reduced coverage of strongly bonded CO creates a larger number of bare Pt surface sites and, hence, allows for more hydrogen to be adsorbed and oxidized. Similarly, for methanol oxidation on ternary alloys, an analogous mechanism would predict that the rate of formation of strongly bonded surface CO in the final H-stripping step would be suppressed in favor of an oxidation reaction channel of a preceding H-containing intermediate (e.g., a formyl species being oxidized to a formate species followed by the decomposition to CO_2). The latter hypothesis would correspond to a shift in the relative rate between the methanol oxidation channel via CO and a 'direct' methanol oxidation channel involving an H-containing intermediate.

In conclusion, the computational study of ternary Pt-Ru-X alloys suggests that future strategies toward more active electrocatalysts for the oxidation of methanol should be based on a modification of the CO adsorption energy of Pt (ligand effect), rather than on the enhancement of the oxophilic properties of alloy components (enhanced bifunctional effect).

11.4.3 Screening of a Pt-Ru-Co Focus Library

The diverse combinatorial screening described in the previous sections revealed active Pt-Ru-Co methanol oxidation catalysts, especially near $Pt_{20}Co_{60}Ru_{20}$. A second primary screening cycle (Fig. 11.8) focused on a more limited compositional space to isolate the active region in more detail. Figs. 11.15 and 11.16 show the design and the sampled compositional space of a Pt-Ru-Co ternary library. The design contains a pure Pt catalyst, Pt-Co binaries (row A), Pt-Ru binaries (column 1)

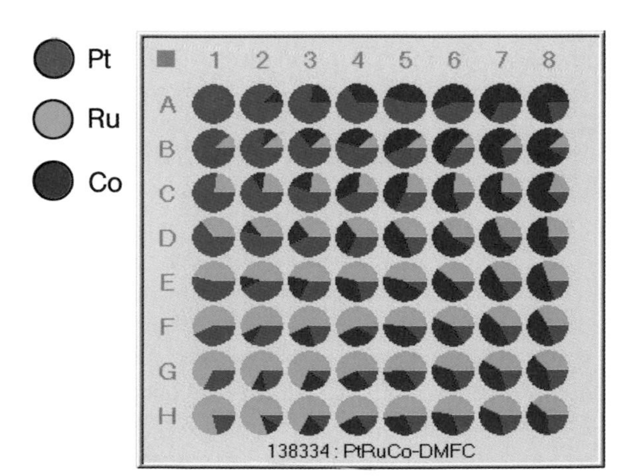

Fig. 11.15 Design of an electrocatalyst library of Pt-Ru-Co alloys for a more focused examination of the ternary composition space. The pie-chart character of each catalyst represents its chemical composition, with pure Pt in the upper left corner. The design was created using Library Studio® [31].

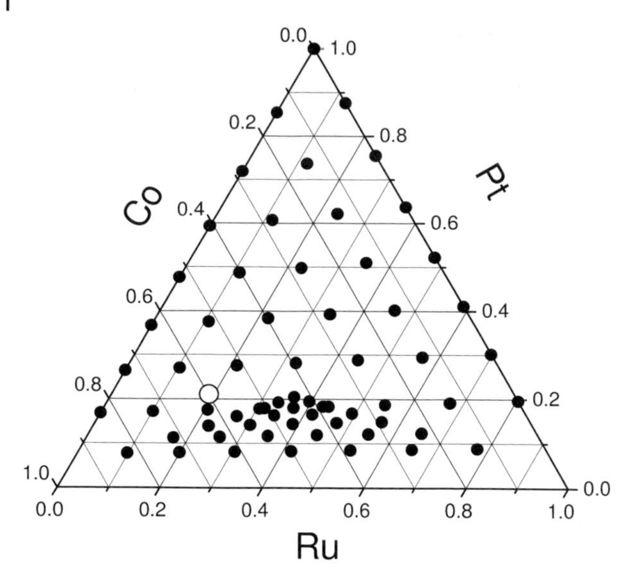

Fig. 11.16 Ternary composition diagram of the Pt-Ru-Co associated with the library design described in Fig. 11.15. Black dots indicate the individual binary and ternary compositions synthesized and screened. The larger, lighter dot indicates the stoichiometry of the most active material identified through the original primary screening protocol (Figs. 11.10 to 11.13).

as well as 49 ternary compositions. Detailed characterization of the thin film catalysts was performed both before and after screening.

High-throughput screening of the Pt-Ru-Co ternary library resulted in the activity-composition map of Fig. 11.17. Activity is represented by the surface-area-normalized current using the hydrogen adsorption integral derived from cyclic voltammograms (CVs) in methanol-free electrolyte. The CVs were measured before and after screening for methanol oxidation, to capture surface changes during screening.

As shown in Fig. 11.17, a line connects Pt-Ru binary compositions. Interestingly, under the experimental conditions (room temperature), a Pt-Ru binary composition of $Pt_{87}Ru_{13}$ shows the highest activity of all Pt-Ru binaries. This is in substantial agreement with earlier findings by Gasteiger et al. [44] who, at room temperature, reported a higher activity for a smooth $Pt_{90}Ru_{10}$ alloy surface than for a $Pt_{50}Ru_{50}$ surface. The authors argue that the activity enhancement is based on an ensemble effect: at low temperatures methanol adsorbs preferentially on Pt, three adjacent Pt atoms in the immediate neighborhood of a Ru atom should provide an ideal methanol adsorption and oxidation site. A statistical model for a (111) surface geometry showed that this configuration is most likely with a Ru content of 10–14 at% Ru in the binary alloy. Additional Ru (beyond about 50 at%) appears to lower the methanol oxidation activity significantly, in agreement with a recent study by Tripkovic et al. [53].

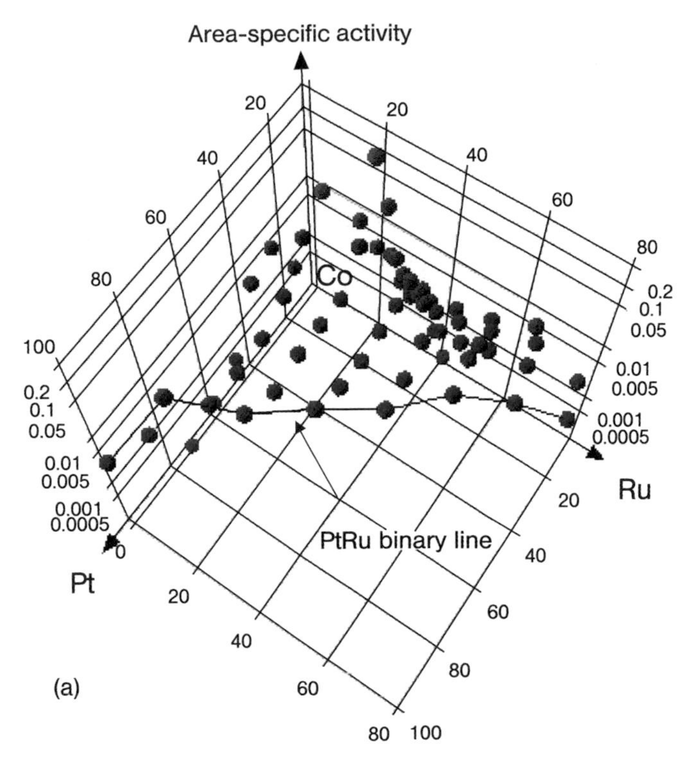

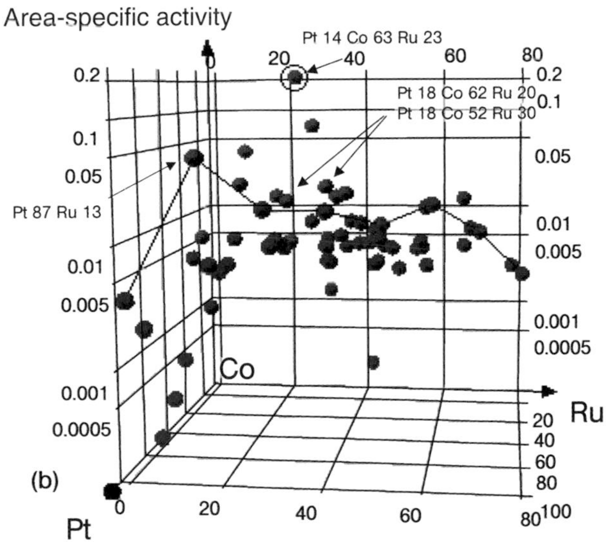

Fig. 11.17 Chronoamperometric screening results from the ternary catalyst library described in Figs. 11.15 and 11.16. Surface-area-normalized activity values of each individual composition are plotted as a function of composition. Color-coding indicates activity: red = high, blue = low. The Pt-Ru binary compositions are connected by a solid line to underscore the activity trends observed in this binary system. Conditions: 1 M methanol, 0.5 M H_2SO_4, 550 mV/RHE, 5 min.

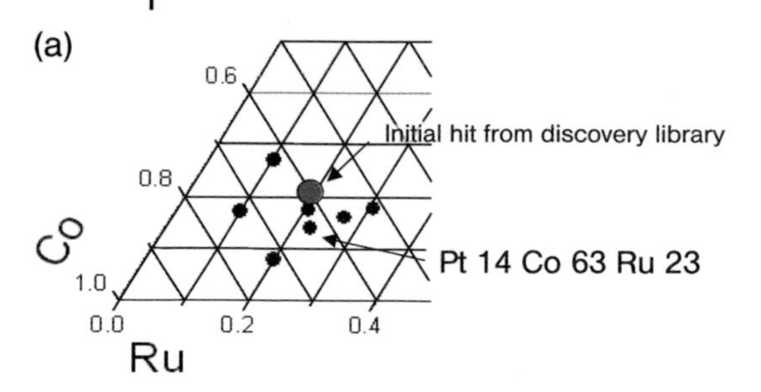

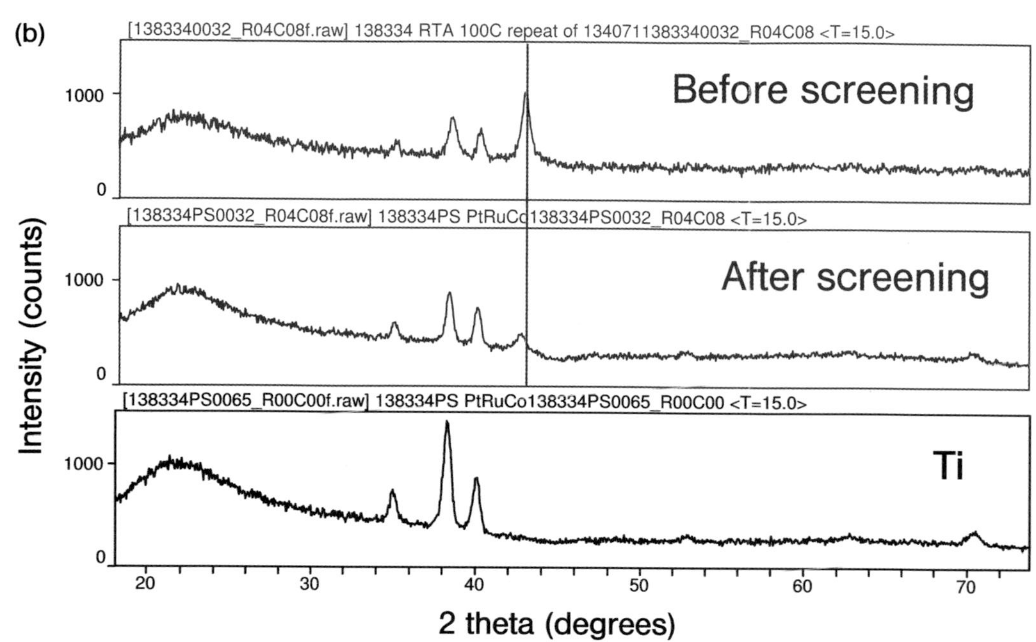

Fig. 11.18 Stability analysis of the most active ternary composition, $Pt_{14}Co_{63}Ru_{23}$, shown in Fig. 11.17. (a) Location of the electrocatalyst within the ternary composition map. (b) Comparison of the XRD profile of the electrocatalyst before and after screening. The dominant diffraction peak shifts slightly to larger lattice parameters, indicating leaching of cobalt. Significant intensity degradation (relative to the Ti electrode) has occurred after screening. Diffraction of a bare Ti electrode is shown for comparison.

Fig. 11.19 Stability analysis of an active but stable ternary composition, $Pt_{18}Co_{62}Ru_{20}$. (a) Location of the electrocatalyst within the ternary composition map. (b) Comparison of the XRD profile of the electrocatalyst before and after screening. The dominant alloy diffraction peak remains essentially unchanged, indicating structural stability of the alloy. Diffraction of a bare Ti electrode is shown for comparison. (c) EDX spectra of the electrocatalyst before (red) and after (yellow) electrochemical screening suggest compositional stability.

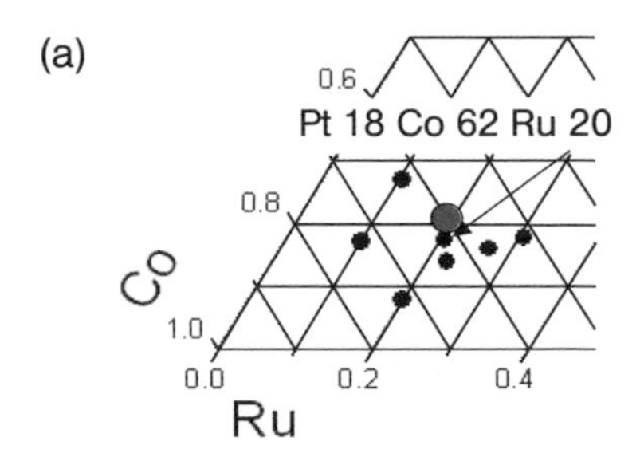

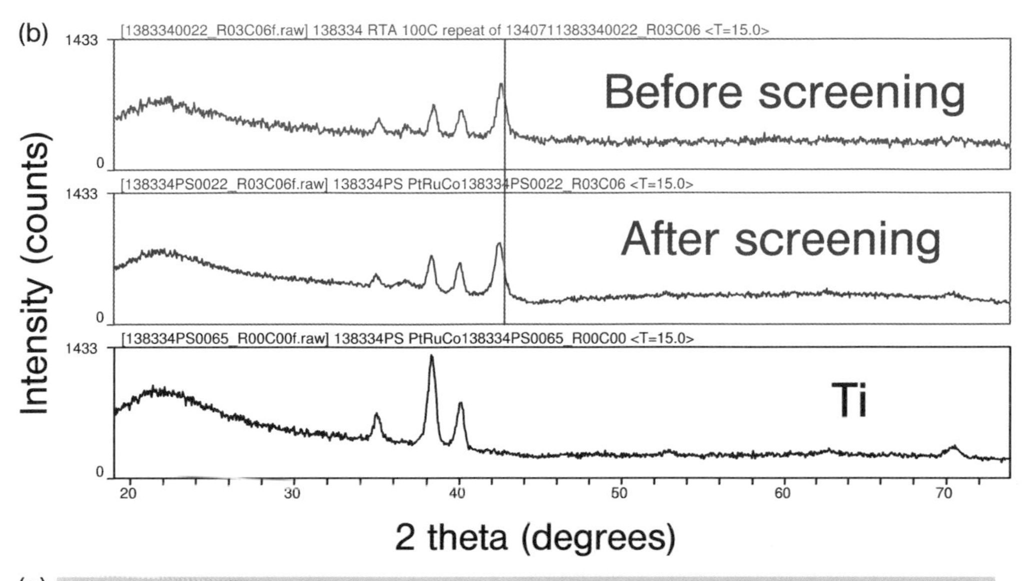

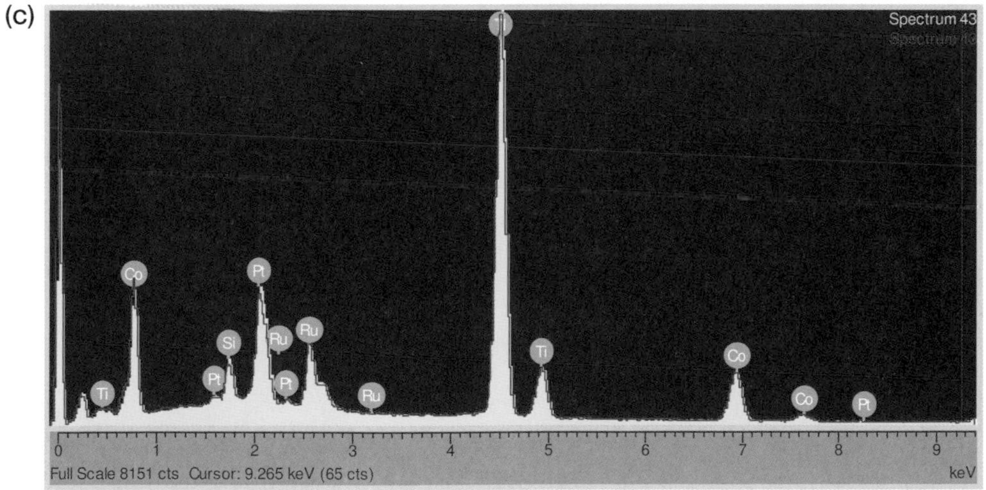

Fig. 11.19 (legend see page 292).

High methanol oxidation activity is observed for Co-rich compositions that are similar to stoichiometries identified in the broad screening study (Fig. 11.17). Particular examples are $Pt_{14}Co_{63}Ru_{23}$ or $Pt_{18}Co_{62}Ru_{20}$.

In addition to activity, the primary screening workflow yields information regarding the structural and compositional stability of the thin film catalysts by XRD and EDX. Fig. 11.18 depicts the stability evaluation of the most active composition $Pt_{14}Co_{63}Ru_{23}$. Fig. 11.18 (a) shows the position of the composition within the Co-rich corner of the Pt-Ru-Co ternary composition space in relation to other compositions sampled in this library. Although activity shows a maximum at this stoichiometry, a Co content greater than 63 at.% appears to be detrimental to stability. Diffraction of this composition (Fig. 11.18 b) demonstrates that $Pt_{14}Co_{63}Ru_{23}$ is unstable under the experimental conditions: The (111) alloy peak at around $2\theta=43^{\circ}$ shifts slightly to larger *d*-spacings after screening, indicating an expansion of the lattice due to loss of alloying components, i.e. Co. Diffraction intensity is also significantly decreased. This is corroborated by EDX results indicating a loss of Co after screening compared with the as-synthesized composition.

In contrast, $Pt_{18}Co_{62}Ru_{20}$ is an active composition on the Pt-rich side of the activity maximum. A similar stability evaluation reveals relatively stable behavior, as shown in Fig. 11.19. This composition is similar to the original catalyst identified during the broad screening study. Diffraction after screening is substantially similar to that before screening (Fig. 11.19 b), and EDX data before and after screening indicate stability (Fig. 11.19 c).

Fig. 11.20 summarizes experimentally observed trends in the area-normalized activity (black bars) and the Pt weight-fraction-normalized activity (white bars) of some electrocatalysts in Fig. 11.17. It can be seen that a small difference in area-

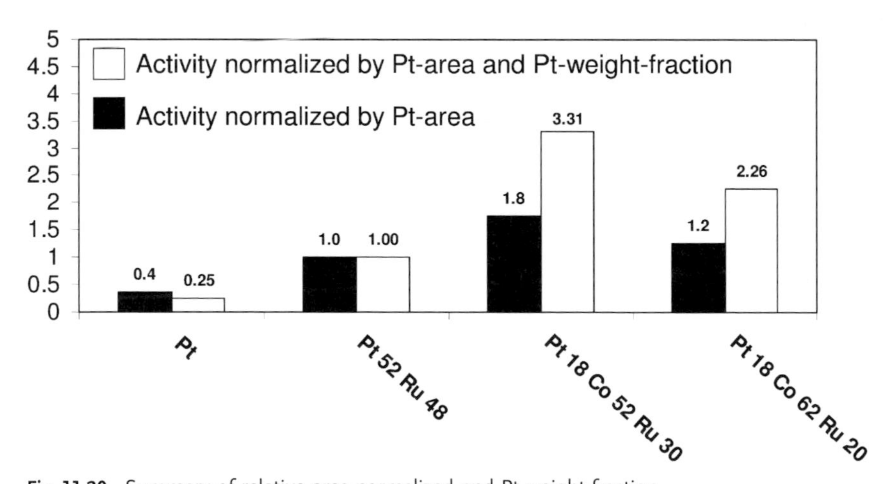

Fig. 11.20 Summary of relative area-normalized and Pt weight-fraction-normalized activities of pure Pt, $Pt_{52}Ru_{48}$, and two Pt-Ru-Co compositions at 550 mV/RHE in acidic solution. Hydrogen adsorption integrals were measured before electrochemical screening for methanol.

normalized activity between a catalyst and the PtRu standard may translate into a significant activity advantage of the ternary composition, once the Pt content is factored in on a weight basis. $Pt_{18}Co_{52}Ru_{30}$, for instance, contains only 36% Pt by weight, compared to 68% for $Pt_{52}Ru_{48}$. This additional enhancement becomes relevant when evaluating Pt-mass-based activities of high surface-area powder electrocatalysts.

11.5 Conclusions

Primary and secondary high-throughput synthesis and screening workflows have been developed and applied to the search for improved DMFC anode catalysts.

The described high-throughput electrochemical screening method allows for the rapid, accurate and direct measurement of potential–current–time activity data in the primary synthesis and characterization step.

The secondary screening workflow builds on the information gained in the primary screening step and demonstrates the development of a combinatorial, supported, high surface-area catalyst synthesis methodology for the preparation of identified active catalyst phases. Robotics facilitate parallel weighing of carbon powders, liquid dispensing of stock solutions, ultrasonication, impregnation, drying and annealing. High-throughput screening of catalyst powders is performed with a novel parallel rotating disk electrode (PRDE) tool wherein up to 16 RDE experiments can be performed and evaluated in parallel. This combination of primary and secondary high-throughput electrochemical workflows represents a significant advance in the search for high activity electrocatalysts. In contrast to other combinatorial and high-throughput electrochemical approaches, the presented workflow enables a detailed characterization of individual library elements, including structure, composition and stability in electrochemical environments. The presented high-throughput electrochemical strategies offer the opportunity to advance the understanding of fuel cell electrocatalysts, and contribute to our understanding of electrocatalytic processes at an unprecedented speed.

11.6 Acknowledgment

The authors are indebted to S. Nguyen for engineering support, K. Gordeeva for software support and S. Klein for careful assembly of bibliographic references.

11.7 References

1 Bockris, J.O.M., Reddy, A.K.N. *Modern Electrochemistry*, 3rd ed., Plenum/Rosetta, New York, 1977, Vol. 2.

2 *Handbook of Fuel Cells – Fundamentals, Technology, and Application*, Vielstich, W., Lamm, A., Gasteiger, H. (eds.), Wiley, New York, 2003.

3 Larminie, J., Dicks, A. *Fuel Cell Systems Explained*, Wiley, New York, 2000.

4 *Electrocatalysis*, Lipkowski, J., Ross, P.N. (eds.), Wiley-VCH, New York, 1998.

5 Thompsett, D., Pt alloys as oxygen reduction catalysts. In: *Handbook of Fuel Cells – Fundamentals, Technology and Applications*, Vol. 3, Vielstich, W., Gasteiger, H. A., Lamm, A. (eds.), Wiley, New York, 2003.

6 Parsons, R., VanderNoot, T. *J. Electroanal. Chem.* 1988, *257*, 9.

7 Beden, B., Leger, J.-M., Lamy, C. In: *Modern Aspects of Electrochemistry*, Vol. 22, Bockris, J.O.M., Conway, B.E., White, R.E. (eds.), Plenum Press, New York, 1992, p 97.

8 Smotkin, E.S., Diaz-Morales, R.R. *Annu. Rev. Mater. Res.* 2003, *33*, 557.

9 Archibald, B., Bruemmer, O., Devenney, M., Gorer, A., Jandeleit, B., Uno, T., Weinberg, W.H., Weskamp, T., Combinatorial methods in catalysis. In: *Handbook of Combinatorial Chemistry – Drugs, Catalyst, Materials*, Nicolaou, K.C., Hangko, R., Hartwig, W. (eds.), Wiley, New York, 2002.

10 Reddington, E., Yu, J.-S., Chan, B.C., Sapienza, A., Chen, G., Mallouk, T.E., Gurau, B., Viswanathan, R., Liu, R., Smotkin, E.S., Sarangapani, S., Combinatorial discovery and optimization of electrocatalysts. In: *Combinatorial Chemistry*, Fenniri, H. (ed.), Oxford University Press, Oxford, 2000, p 401.

11 Reddington, E., Sapienza, A., Gurau, B., Viswanathan, R., Sarangapani, S., Smotkin, E.S., Mallouk, T.E. *Science* 1998, *280*, 1735.

12 Mallouk, T.E., Reddington, E., Pu, C., Ley, K.L., Smotkin, E.S., Discovery of methanol electro-oxidation catalysts by combinatorial analysis. In: *Fuel Cell Seminar Extended Abstracts*, Orlando, USA, 1996, p 686.

13 Mallouk, T.E., Smotkin, E.S., Combinatorial catalyst development methods. In: *Handbook of Fuel Cells – Fundamentals, Technology and Applications*, Vielstich, W., Lamm, A., Gasteiger, H. (eds.), John Wiley & Sons, New York, 2003, *2* (Part 3), 334.

14 Choi, W.C., Kim, Y.J., Woo, S.I., Hong, W.H. *Studies in surface science and catalysis* 2003, *145* (Science and Technology in Catalysis 2002), 395.

15 Jayaraman, S., Hillier, A.C. *J. Phys. Chem. B* 2003, *107*, 5221.

16 US 6,187,164, 2001.

17 Baeck, S.H., McFarland, E.W. *Korean J. Chem. Eng.* 2002, *19*, 593.

18 Strasser, P., Fan, Q., Devenney, M., Weinberg, W.H., Liu, P., Nørskov, J.K. *J. Phys. Chem. B* 2003, *107*, 11013.

19 Strasser, P., Fan, Q., Devenney, M., Weinberg, W.H., Combinatorial Exploration of ternary fuel cell electrocatalysts for DMFC anodes – a comparative study of PtRuCo, PtRuNi and PtRuW systems, AIChE fall meeting, 2003, San Francisco.

20 Strasser, P., Gorer, S., Devenney, M., Combinatorial electrochemical strategies for the discovery of new fuel-cell electrode materials. In: *Proceedings Volume of the International Symposium on Fuel Cells for Vehicles – 41st Battery Symposium*, Yamamoto, O. (ed.), The Electrochemical Society of Japan, Nagoya, 2000, p 153.

21 Strasser, P., Gorer, S., Devenney, M., Electrochemical techniques for the discovery of new fuel-cell cathode materials. In: *Direct Methanol Fuel Cells*, Vol. 4, Narayanan, S.G., Zawodzinski, T. (eds.), The Electrochemical Society, Washington, 2001, p 191.

22 Reddington, E., Yu, J.-S., Sapienza, A., Chan, B.C., Gurau, B., Viswanathan, R., Liu, R., Smotkin, E.S., Sarangapani, S., Mallouk, T. E. *MRS Symp. Proc.* 1999, *549*, 231.

23 Reddington, E., Yu, J.-S., Sapienza, A., Chan, B.C., Gurau, B., Viswanathan, R., Liu, R., Smotkin, E.S., Sarangapani, S., Mallouk, T.E., Advanced

catalytic materials. In: *MRS Symp. Proc.*, LEDNOR, P.W., NAKAGI, D.A., THOMPSON, L.T. (eds.), 1999, *549*, 231.

24 US 6682837, 2004; 6498121 2003; 6517965, 2003.

25 SULLIVAN, M.G., UTOMO, H., FAGAN, P. J., WARD, M.D. *Anal. Chem.* 1999, *71*, 4369.

26 JIANG, R., CHU, D. *J. Electroanal. Chem.* 2002, *527*, 137.

27 LIU, R., SMOTKIN, E.S. *J. Electroanal. Chem.* 2002, *535*, 49.

28 GUERIN, S., HAYDEN, B.E., LEE, C.E., MORMICHE, C., OWEN, J.R., RUSSELL, A.E., THEOBALD, B., THOMPSETT, D. *J. Comb. Chem.* 2004, *6*, 149.

29 JAMBUNATHAN, K., SHAH, B.C., HUDSON, J.L., HILLIER, A.C. *J. Electroanal. Chem.* 2001, *500*, 279.

30 JAYARAMAN, S., HILLIER, A.C. *Langmuir* 2001, *17*, 7857.

31 EP 1080435, 2002.

32 US 6658429, 2003.

33 US 6045671, 2000; 6364956, 2002; 5985356, 1999.

34 US 6507945, 2003, EP 1175645, 2003.

35 MARKOVIC, N.M., SCHMIDT, T.J., STAMENKOVIC, V., ROSS, P.N. *Fuel Cells* 2001, *1*, 105.

36 SCHMIDT, T.J., GASTEIGER, H.A., STAEB, G.D., URBAN, P.M., KOLB, D.M., BEHM, R.J. *J. Electrochem. Soc.* 1998, *145*, 2354.

37 PAULUS, U.A., WOKAUN, A., SCHERER, G.G., SCHMIDT, T.J., STAMENKOVIC, V., MARKOVIC, N.M., ROSS, P.N. *Electrochim. Acta* 2002, *47*, 3787.

38 GLOAGUEN, F., NAPPORN, T., CROISSANT, M.-J., BETHELOT, S., LEGER, J.M., LAMY, C., SRINIVASAN, S., Electrocatalysis of fuel cell reactions – a RDE investigation using high surface area dispersed electrocatalysts dispersed in recast proton conductive membrane. In: *Electrode Materials and Processes for Energy Conversion and Storage IV*, Vol. 13, MCBREEN, J., MUKERJEE, S., SRINIVASAN, S. (eds.), The Electrochemical Society, Pennington, 1997, Vol. 97-13, p 131.

39 EP 1350214, 2003.

40 KAURANEN, P.S., SKOU, E., MUNK, J. *J. Electroanal. Chem.* 1996, *404*, 1.

41 HAMNETT, A. *Catal. Today* 1997, *38*, 445.

42 ANDERSON, A.B., GRANTSCHAROVA, E., SEONG, S. *J. Electrochem. Soc.* 1996, *143*, 2075.

43 GURAU, B., VISWANATHAN, R., LIU, R., LAFRENZ, T.J., LEY, K.L., SMOTKIN, E.S. *J. Phys. Chem. B* 1998, *102*, 9997.

44 GASTEIGER, H., MARKOVIC, N., ROSS, P.N., CAIRNS, E.J. *J. Phys. Chem.* 1993, *97*, 12020.

45 WATANABE, M., MOTOO, S. *J. Electroanal. Chem.* 1975, *60*, 275.

46 WATANABE, M., SHIBATA, M., MOTOO, S. *J. Electroanal. Chem.* 1986, *206*, 197.

47 GASTEIGER, H., MARKOVIC, N., ROSS, P.N. *J. Phys. Chem.* 1995, 99, 8290.

48 LIU, P., LOGADÓTTIR, Á., NØRSKOV, J.K. *Electrochim. Acta* 2003, *48*, 3731.

49 LIU, P., NØRSKOV, J.K. *Fuel Cells* 2001, *1*, 192.

50 IGARASHI, H., FUJINO, T., ZHU, Y.M., UCHIDA, H., WATANABE, M. *Phys. Chem. Chem. Phys.* 2001, *3*, 306.

51 WATANABE, M., ZHU, Y.M., IGARASHI, H., UCHIDA, H. *Electrochemistry* 2000, *68*, 244.

52 WOLOHAN, P., MITCHELL, P. C. H., THOMPSETT, D., COOPER, S.AJ. *J. Mol. Catal. A* 1997, *119*, 223.

53 TRIPKOVIC, A.V., POPOVIC, K.D., GRGUR, B.N., BLIZANAC, B., ROSS, P.N., MARKOVIC, N.M. *Electrochim. Acta* 2002, *47*, 3707.

12
High-Throughput Approaches in Olefin Polymerization Catalysis

Vince Murphy

12.1
Introduction

The purpose of this contribution is to discuss recent advances in high-throughput techniques for the discovery of new olefin polymerization catalysts and new polymer products. Particular attention is given to the development of novel reactor designs, new rapid screening techniques, and catalyst discoveries through the application of high-throughput approaches.

12.2
Background

Combinatorial chemistry and high-throughput screening were first developed within the pharmaceutical industry, where long development times and high research costs forced the development of a new strategy to accelerate the drug discovery process. The new strategy is based upon the parallel synthesis and high-throughput screening of a targeted group or library of molecules, far surpassing the more traditional "make one and screen one" approach. Combinatorial approaches have now become mainstream within the pharmaceutical industry, and companies without such technology typically seek other ways to distinguish themselves. High-throughput synthesis and screening has since been extended to materials science and catalysis, and not unlike the situation during the early stages of development within the pharmaceutical industry, such approaches have initially met with healthy scientific skepticism. Skepticism is of course understandable, but the benefits of the high-throughout approach will ultimately ensure its acceptance. Increased rates of innovation, cost effectiveness, improved intellectual property, reduced time to market, and an improved probability of success are some of the attractive features that can benefit the user.

Within the polyolefin industry, the ongoing discovery and development of new single-site catalysts is a highly competitive research activity driven by the lucrative benefits of a product polymer possessing superior properties and low production costs. Single-site catalysts are organometallic compounds possessing stabilizing

High-Throughput Screening in Chemical Catalysis
Edited by A. Hagemeyer, P. Strasser, A. F. Volpe, Jr.

ISBN: 3-527-30814-8

ligands that tune the catalytic properties of the metal center and present identical catalytic centers to the incoming monomers. Unlike the original Ziegler-Natta catalysts, single-site catalysts offer polymer products with narrow molecular weight distributions, greater control over comonomer incorporation, and uniform interchain comonomer distributions. Since their introduction some 20 years ago, impressive advances have been made in the development of single-site catalysts. During this period, research has mostly been centered on metallocene complexes composed of high-valent group (IV) metal centers supported by cyclopentadienyl-based ligands [1]. Structural modifications of the cyclopentadienyl ligand framework have led to substantial improvements in catalyst performance, and the emergence of commercial processes employing metallocene complexes for the preparation of a wide range of commodity and specialty polymer products [1, 2]. More recent research has uncovered single-site catalyst classes based upon new non-cyclopentadienyl ligands, and mid and late transition metals [3].

The polyolefin industry continues to seek improvements in catalyst performance features such as productivity, temperature stability, comonomer incorporation capability, copolymer sequence distribution control, molecular weight capability, and the ability to copolymerize new monomer combinations. However, the discovery and development of new olefin polymerization catalysts is impeded by a limited understanding of the complex relationships between molecular structure and performance. For example, it is currently not possible to predict which new metal–ligand combinations will lead to active catalyst classes, and which metal–ligand combinations will have low or no activity. Furthermore, upon discovering a novel catalytically active metal–ligand combination, catalyst optimization is always desirable, and yet for any new catalyst class it is extremely difficult to predict the structural features necessary to impart the desired improvements in catalyst performance. Minor structural modifications can have profound effects upon catalyst performance, and thus, even for a well-studied catalyst class, it is not possible to accurately predict the effects of a simple ligand modification. Thus, the conventional approach of designing, making and screening catalysts one at a time is a perilously slow trial and error process that offers no guarantee of success. Such problems have contributed to many years of development time prior to the commercialization of metallocenes.

The high-throughput approach is particularly suited to problems where the parameter space is too large to be addressed efficiently using conventional approaches, and where the outcome is the result of an unpredictable interdependence among the variables of the parameter set. With a seemingly limitless ligand landscape, and unpredictable structure–performance relationships, single-site olefin polymerization catalysts fit this description perfectly. Conceptually, a high-throughput screen can rapidly identify catalytically active systems and, importantly, identify and reject inactive systems. Additionally, the ability to rapidly generate meaningful catalyst performance data can lead to rapid optimization, and provide comprehensive and ultimately predictive structure–property relationships.

It is certainly necessary to mention the level of difficulty associated with setting up a sustainable high-throughput screening program for olefin polymerization catalysts. The more conventional methods of catalyst discovery research involve complex ligand syntheses, and challenging organometallic chemistry. The catalysts are often extremely air and moisture sensitive, require careful handling, and their performance is highly sensitive to the polymerization conditions. Additionally, the screening is usually performed in carefully engineered reaction vessels with controlled stirring and sophisticated gas-flow controllers to ensure that the product from the polymerization represents the capabilities of the catalyst rather than limitations of the process conditions. However, despite these technical difficulties, the last few years have witnessed the emergence of the first high-throughput synthesis and screening approaches in olefin polymerization catalysis. The first advances, which signify a systematic departure from the "make one and screen one" approach, are discussed in the following sections.

12.3 High-Throughput Screen: What is Required?

The application of high-throughput approaches to olefin polymerization catalysis requires the integration of several components: (1) the ability to express large numbers of ligands with sufficient steric and electronic diversity is likely to be critical to the success of any high-throughput program directed towards catalyst discovery; (2) efficient synthetic routes to the catalyst precursors and activated catalysts; (3) an adaptation of catalyst synthesis and screening to smaller scales, necessary to sustain high throughput and avoid time-consuming scale-up synthesis; and (4) suitably rapid methods to perform the experiments and assess the results. Additionally, sustainable high-throughput programs will require sophisticated software to aid experimental design, screening, product characterization, and the analysis of copious amounts of data. With regard to (1) it is not the purpose of this chapter to review strategies to express small molecule diversity, since this is a comprehensive subject that has been reviewed recently [4]. Instead, we focus on (2) to (4), with particular emphasis on recent developments in new reactor designs, new rapid screening techniques, and olefin polymerization catalyst discoveries through the application of high-throughput approaches.

12.4 Reactor Designs

Microreactor designs can range from simple arrays of small glass vials to sophisticated assemblies of computer-controlled continuous flow reactors. For certain applications, an array of vials may represent a perfectly suitable design. Typically, however, olefin polymerization involves the use of gaseous monomers, elevated temperatures, efficient stirring to prevent mass transport limitations, and the ability to inject components into the reactor at temperature and pressure. In spite of these demanding conditions, several groups have reported impressive designs of

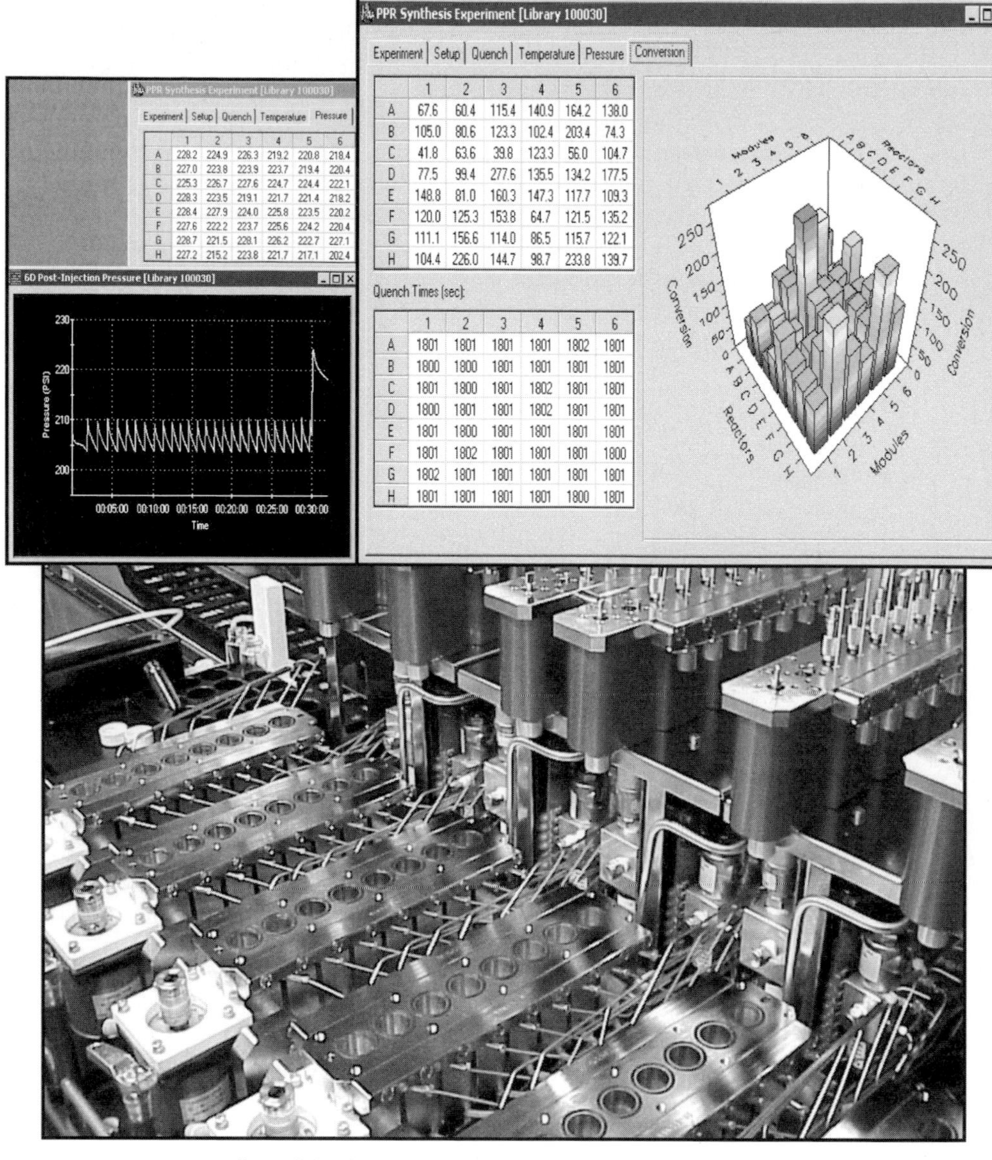

Fig. 12.1 48-Cell Parallel Polymerization Reactor with associated software.

parallel and rapid serial reactors. Coughlin et al. have recently described a convenient gas manifold and associated reactor design that accommodates different reactor vessel sizes (70–300 mL) in addition to multi-well arrays (20 mL) [5]. The simple design represents an alternative to highly sophisticated systems that can be prohibitively expensive for many laboratories. The multi-well design can be used to perform seven simultaneous polymerizations. The reactor utility was demonstrated using literature catalysts for ethylene and propylene polymerizations. A team from the Dow Chemical company has recently described a rapid serial "electrothermal microreactor" designed to assess polyolefin catalyst performance at high temperature (175 °C) and pressure (400 psig) within a tubular reactor containing a series of temperature sensors [6]. Measurable exotherms were correlated to the differences in the nascent activity of two well-studied polyolefin catalysts, and the authors report a potential throughput of ten catalyst runs per hour. Researchers at Symyx have recently described a parallel pressure reactor [7]. The instrument consists of 48 individual high-pressure batch reactors where each reactor possesses its own pressure and temperature control and gaseous feed line. Thus, 48 individual reaction events can be monitored in real time under conditions that provide meaningful information about the performance capabilities of each catalyst at high temperatures (200 °C) and pressures (500 psig). More recently, this instrument has been further developed to accommodate supported catalysts, and semicontinuous reagent additions at high pressures (Fig. 12.1) [8–10]. Examples of catalysts and new polymer architectures discovered using this technology are presented in the following section.

Mülhaupt et al. [11] have recently reported a fully automated single polymerization reactor with a series of carousels used to deliver catalyst solutions and collect resultant polymer products. The authors claim that initial scoping polymerizations are performed in a commercially available Chemspeed parallel reactor [12], a process that provides appropriate candidates for the larger scale automated single reactor (300 mL). The throughput is specified at 120 polymerizations per day, provided that the unit can be operated continuously over a 24 h period. Additionally, the authors provide a convenient list of the ever increasing number of commercially available parallel reactors. There is no doubt that the advances in reactor hardware begin to address the challenges associated with high-throughput screening of olefin polymerization catalysts.

12.5 Rapid Screening Techniques

Rapid screening approaches have also advanced significantly since the introduction of the high-throughput concept. A clever technique has been described by Sutherland, in which an array of thermistors is immersed into reaction media to monitor temperature changes as small as 100 μK [13] (Fig. 12.2). The thermistor technique is well suited to high-throughput catalyst screening, although thus far it has not been reported for olefin polymerization catalysis.

Willson and Symyx have introduced an alternative infrared thermal imaging technique used for the screening of olefin polymerization catalysts. As with the thermistor technique, the technique is readily adapted to screen catalyst arrays, and exothermic events such as an olefin polymerization can be followed with exquisite sensitivity [14]. Thales Technologies have recently introduced a mecha-

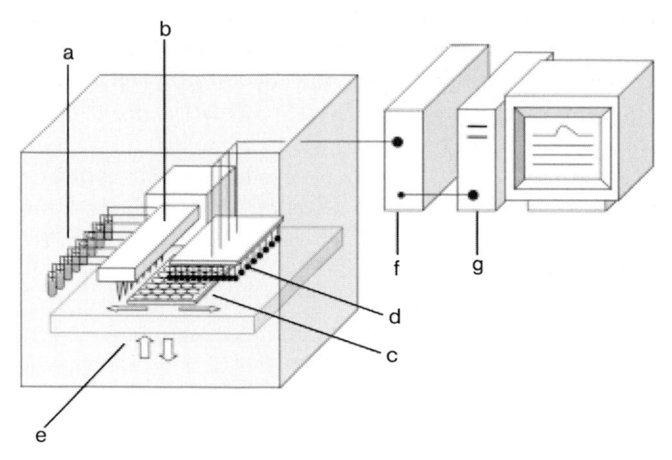

Fig. 12.2 Thermistor array apparatus consisting of (a) eight reagent/substrate reservoirs, (b) an 8-channel dispensing head, (c) a 96-well plate and holder (movable in the *x*-direction), (d) an 8×12 thermistor array, (e) a base plate (movable in the *z*-direction), (f) a multiplexing apparatus, (g) a microcomputer.

nism-based approach to high-throughput screening of olefin polymerization catalysts [15]. The technique uses an electrospray ionization tandem mass spectrometer to assay trapped organometallic intermediates during an olefin polymerization reaction. An olefin polymerization catalyst can be activated in the presence of monomer, quenched with a trapping agent, and then introduced through an electrospray into the mass spectrometer. Kinetic analysis of the intermediates can reveal rates of initiation, chain propagation and termination for a given catalyst. The authors claim that the monomer uptake rate and resultant polymer molecular weight can be predicted for each catalyst for a given set of polymerization conditions. Additionally, the technique is reported to be suitable for rapid serial and pooled approaches to catalyst screening. A group from Avantium has recently discussed strategies for catalyst discovery using high-throughput approaches, but no screening details were presented [16]. To enhance their high-throughput initiative in olefin polymerization catalysis, researchers at Symyx have reported a suite of rapid polymer characterization techniques for downstream analysis of polymer arrays. The techniques include a rapid GPC that can provide molecular weight and conversion information at rates up to 90 s per sample [17], an FTIR method that provides polymer composition analysis at a rate of about 2 min per sample [18], a rapid technique to determine phase transition temperatures [19], a parallel device for the measurement of dynamic viscosity or rheology [20], and a parallel dynamic thermomechanical measurement device [21].

Mülhaupt et al. has recently attempted to correlate the thermal, rheological and mechanical properties of polymers with spectroscopic fingerprints using multivariate analyses [11]. This approach is driven by the fact that polymer properties and processing issues are the ultimate drivers for olefin polymerization catalyst research, and conventional analyses directed at these properties are slow and require large amounts of sample. In an initial study, the Freiburg team were able to use partial least-squares analysis to correlate near-infrared signatures with melting temperature, molecular weight, and degree of branching for a series of branched polyethylenes produced from late metal catalysts. Encouraged by these first results, they are now in the process of examining more ambitious correlations such as tensile strength, hardness, and density. Initial results were presented for a training set of around sixty commercial ethylene-α-olefin copolymers. Clearly the approach is a worthy one, but the authors do stress that the scope of the approach is heavily dependent upon the availability of model polymers for the training set.

12.6 Catalyst Discoveries

This final section concerns the discovery of new catalysts through the application of high-throughput approaches. Not surprisingly, given the advances in reactor design and screening techniques, there are now illustrations of catalyst discovery experiments that would have been extremely difficult to perform using more conventional approaches. An early example was reported by Symyx Technologies, who developed a parallel synthesis and screening protocol for a polystyrene-bound 96-member library of 1,2-diimine nickel and palladium complexes. The key intermediate, a polystyrene-bound 1,2-diketone was converted into a 48-membered 1,2-diimine library through a condensation reaction with 48 electronically and sterically diverse anilines. The ligand library was then divided and converted into 48 nickel and palladium complexes through reactions with [DME]$NiBr_2$ and [COD]PdMeCl [22, 23]. The report also demonstrates the application of a chemical encoding strategy in which polystyrene-bound Ni(II) and Pd(II) complexes were encoded with cleavable tertiary amine tags. After a pooled polymerization, the tags were cleaved from the polymer products, and detected using HPLC to reveal the identity of the catalyst. Although no new significant catalyst discoveries emerged, the high-throughput approach to screening olefin polymerization catalysts was demonstrated. Müllen and co-workers at Max Planck Institute have also reported a screening approach in which catalysts were encoded on solid supports [24]. The Max Planck group employed fluorescent dyes encoded onto silica supports. Firstly, various metallocene–MAO mixtures were supported on silica in the usual manner, exposed to the appropriate dye, and then dried to produce the encoded supported catalysts. The dye used in the study had been previously shown to have no affect upon the performance of the metallocene catalysts. After pooled polymerizations the resultant polymer particles can be read by UV spectroscopy to identify the fluorescent tag, and hence the metallocene. Kaminsky has recently prepared and screened a library of sixteen 1,2-diimine palladium catalysts for ethylene–norbornene copolymerizations [25]. The catalysts were screened by a simple mix and screen approach in which the ligands were mixed together with palladium acetate and activators in the presence of the monomers. Polymer compositions, molecular weights and sequence distributions were determined using conventional analytical techniques and found to be strongly dependent upon the catalyst structure. To validate the screening results, two isolated palladium catalysts were synthesized and studied in detail. Bazan and collaborators at Symyx have reported the use of the parallel polymerization reactor and the rapid polymer characterization techniques previously described to perform a tandem catalysis experiment in which the action of three well-defined catalysts were coordinated to produce a wide range of branched polyethylenes from a single monomer (ethylene) (Fig. 12.3) [26]. The high-throughput screening infrastructure consisting of the parallel reactor, liquid-handling robots and rapid polymer characterization tools was essential for optimizing reaction conditions and probing the effects of catalyst compositions on the polymer properties.

Gibson and workers [27] at BP have reported a high-throughput approach directed towards the optimization of chromium polyethylene catalysts. The study aimed to improve the performance of a previously discovered chromium catalyst. In this study, the ca. 200-member hemi-salen ligand library was complexed in situ to a soluble chromium precursor [*p*-tolylCrCl$_2$(thf)$_3$], followed by addition of 180 equiv. of MAO, and exposure to 1 atm of ethylene for 15 min. Relative activities were determined from the polyethylene yields, and a new highly active chromium catalyst was uncovered. The Coates group has reported a clever approach to the

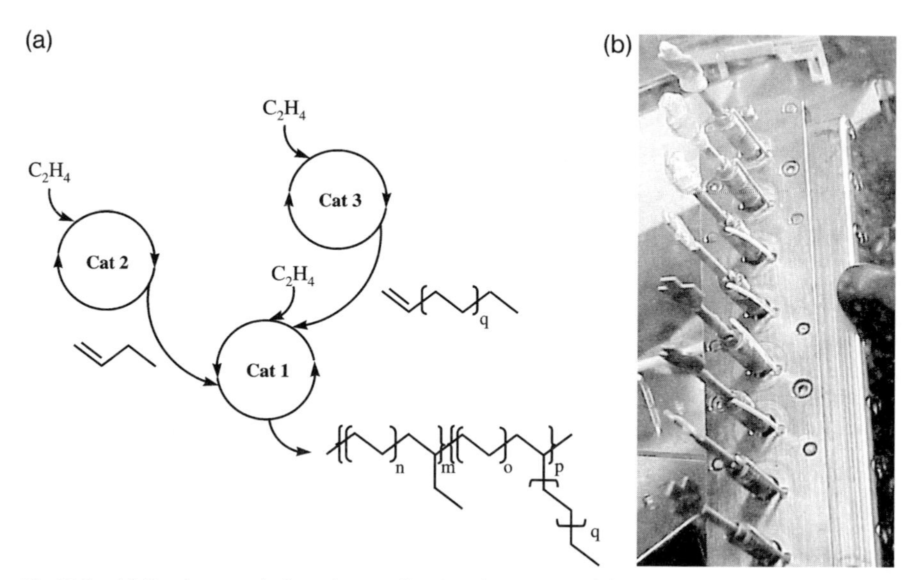

Fig. 12.3 (a) Tandem catalysis cycle coordinating the actions of three well-defined catalysts. (b) A photograph of the stirrers from a parallel reactor, showing the different polymers produced as the percentage branching increases due to changes in catalyst composition. High-throughput techniques were essential for optimizing reactions conditions and probing the effects of catalyst compositions on the polymer properties.

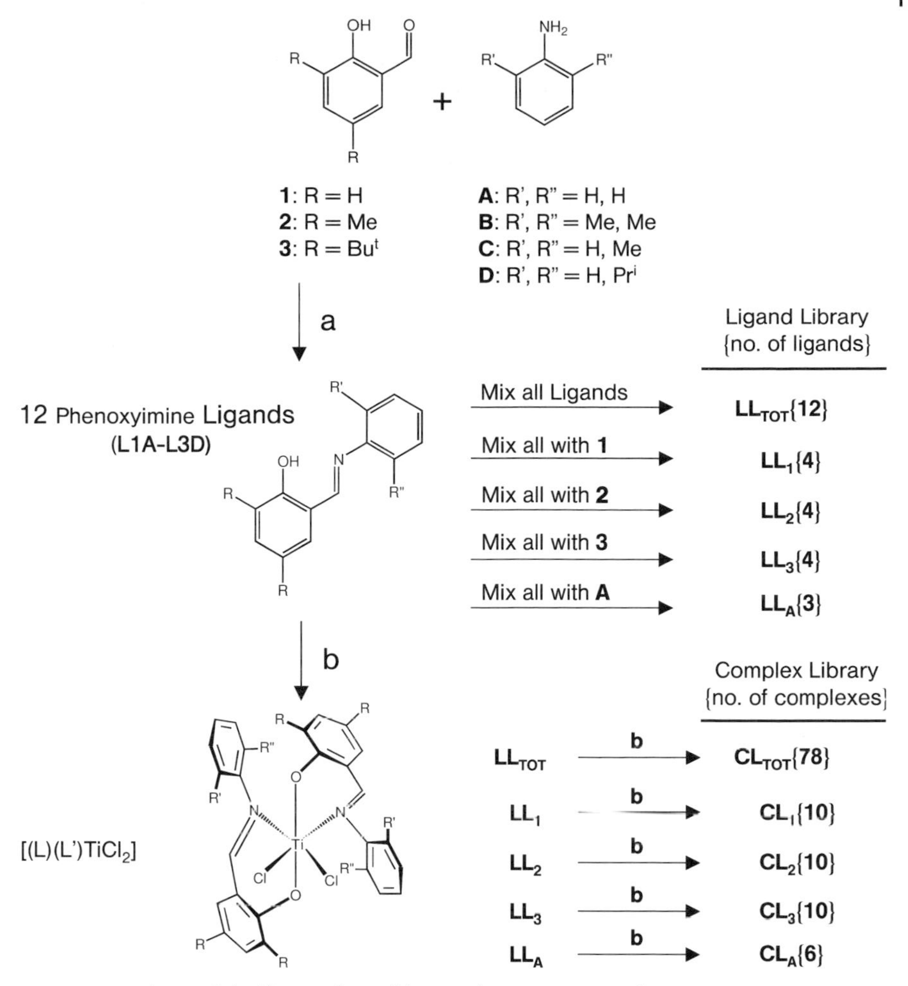

Fig. 12.4 Synthesis of the library of mixed hemi-salen titanium complexes. (a) Parallel synthesis: MeOH, reflux, 12 h, >95%; (b) pooled synthesis: n-BuLi, –60 °C to 20 °C, 4 h (1 equiv.); $TiCl_4$, –60 °C to 20 °C, 16 h (0.5 equiv.), >90%.

discovery of stereospecific polypropylene catalysts [28]. A small collection of bidentate hemi-salen ligands was pooled in solution, deprotonated and reacted with 0.5 equiv. $TiCl_4$, producing a complex library with a total of 78 possible metal–ligand complexes (CL_{TOT} in Fig. 12.4). The pooled complex library was then activated with MAO, and screened at room temperature under 2.7 atm of propylene for 6 h. The authors reasoned that the complex mixture of polymers produced would serve as a stereochemical recording of the events of each polymerization catalyst. Since atactic polymers are amorphous and highly soluble, deconvolution through selective extraction procedures should reveal the identity of any catalyst exhibiting stereocontrol. Ninety percent of the polymers produced were determined to be atactic through diethyl ether extraction experiments. The remaining 10% were found to contain syndiotactic polypropylene. Small subsets of the complex libraries were then resynthesized and screened to ascertain the metal–ligand combination responsible for the stereocontrol (CL_1–CL_A in Fig. 12.4). The authors concluded the study by isolating and screening a new syndiospecific propylene polymerization catalyst.

Symyx have recently reported the first fully integrated methodology for the discovery and optimization of olefin polymerization catalysts using high-throughput approaches [29]. The Symyx approach utilizes high-throughput primary and secondary screening techniques supported by rapid polymer characterization methods. In their paper, the Symyx team introduced micro-scale 1-octene primary screening polymerization experiments combining arrays of ligands with reactive metal complexes $M(CH_2Ph)_4$ (M = Zr, Hf) and multiple activation conditions as a new high-throughput technique for discovering novel group IV polymerization catalysts. In this case the primary screen was performed in around 250 μL of solvent in arrays of 1 mL glass vials. Despite the simplicity of the 1-octene primary screen, it can be used to predict the performance of catalysts for subsequent ethylene–1-octene copolymerizations. For instance, catalysts capable of polymerizing 1-octene to high conversions and high molecular weights in the primary screen would be predicted to possess

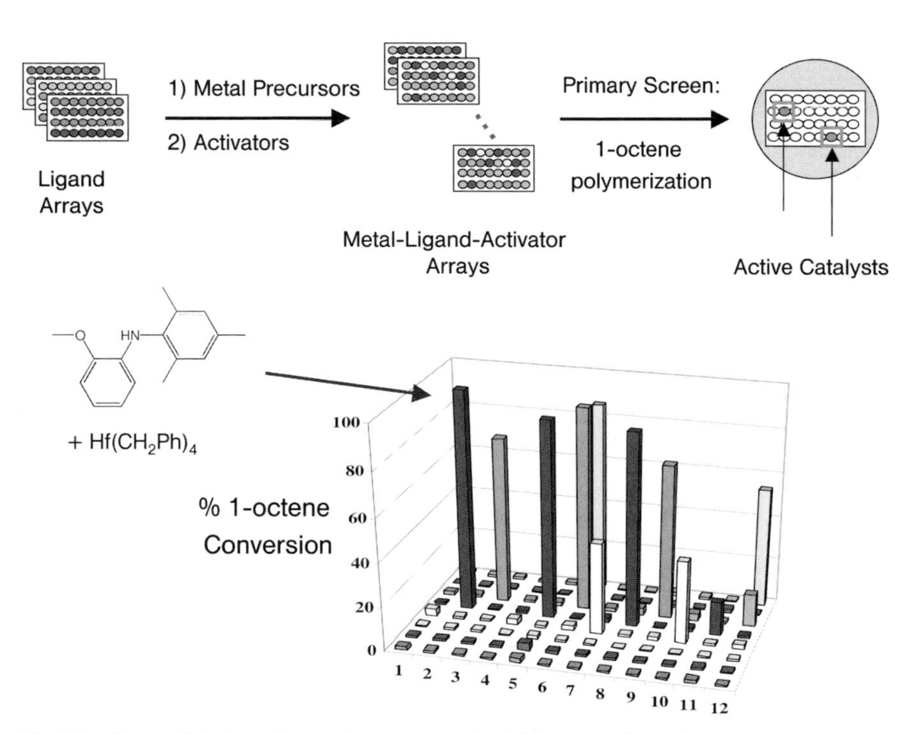

Fig. 12.5 Symyx high-throughput primary screen, in which arrays of metal-ligand combinations are rapidly surveyed for olefin polymerization activity. In this example, a 1-octene primary screen was used to discover a new amide ether-based hafnium catalyst.

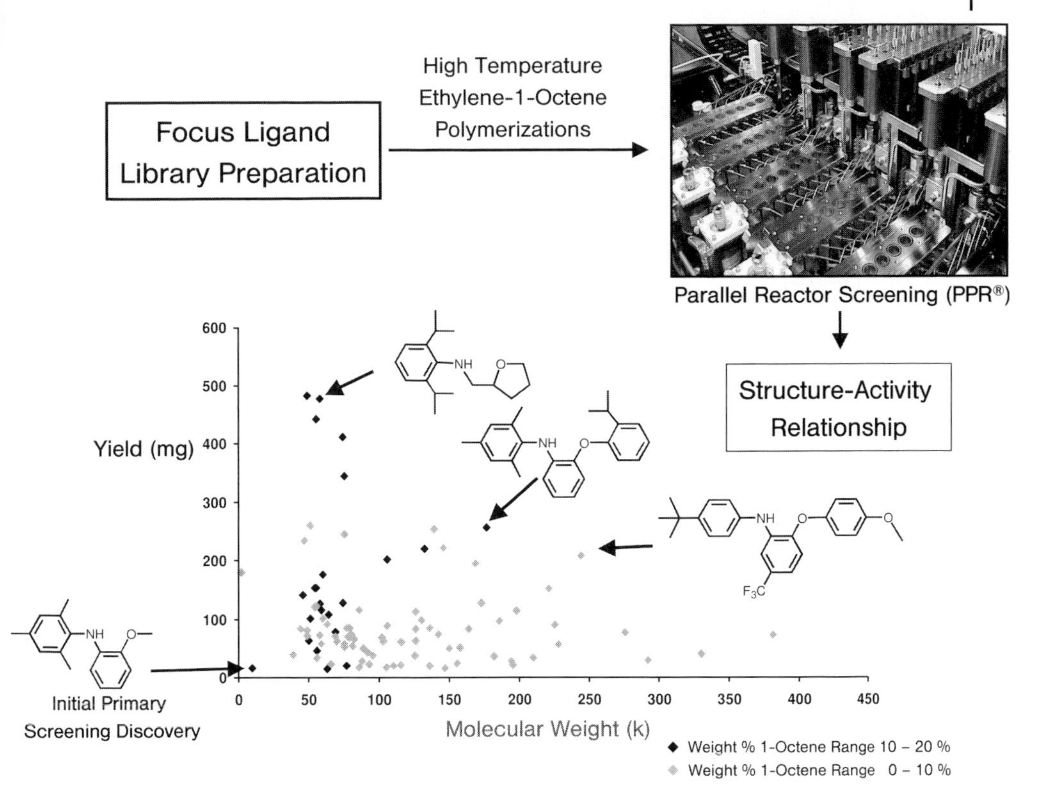

Fig. 12.6 Symyx secondary screen, which is used to rank primary screening discoveries and to optimize the most promising catalyst leads. In this example, high-temperature ethylene-1-octene copolymerization experiments were performed on a focused 96-member amine-ether library.

high alpha-olefin incorporation capability, and produce high molecular weight LLDPE in a subsequent ethylene–1-octene screen [30]. The primary screening methods were first validated using a commercially relevant polyolefin catalyst, and implemented rapidly to discover a new amide ether-based hafnium catalyst $[\eta^2\text{-(N,O)-(2-MeO-}C_6H_4)(2,4,6\text{-}Me_3C_6H_2)N]Hf(CH_2Ph)_3$ (Fig. 12.5).

Larger scale secondary screening experiments performed on a focused 96-member amine-ether library demonstrated high temperature ethylene-1-octene copolymerization capabilities for this catalyst class, and led to catalysts with significant performance improvements over the initial primary screening discovery (Fig. 12.6).

Importantly, conventional one-gallon batch reactor copolymerizations performed using selected amide-ether hafnium compounds confirmed the performance features of this new catalyst class, thus validating the overall approach and demonstrating the utility of small-scale polymerizations. The work is significant because it is the first example of a new catalyst family discovered *and* optimized using high-throughput approaches supported by rapid synthesis, liquid handling ro-

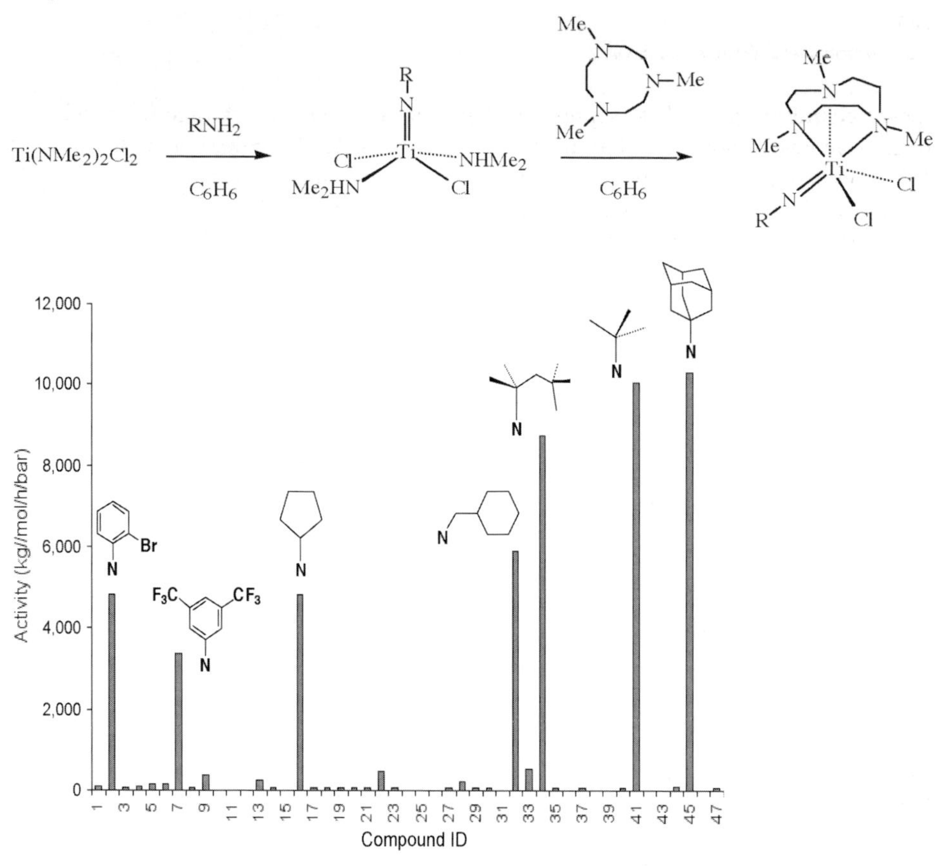

Fig. 12.7 Synthesis and screening of 47 imidotitanium compounds of the formula $[Ti(NR)(Me_3[9]aneN_3)Cl_2]$ (R = alkyl or aryl; $Me_3[9]aneN_3$ = 1,4,7-rimethyltriazacyclononane). (Reproduced by permission of The Royal Society of Chemistry.)

botics, custom designed software, and rapid polymer characterization techniques. Most recently, the Mountford group, in collaboration with industrial coworkers, used high-throughput screening techniques to discover seven highly active ethylene polymerization catalysts [31]. A library of 47 imidotitanium compounds of the formula $[Ti(NR)(Me_3[9]aneN_3)Cl_2]$ (R = alkyl or aryl; $Me_3[9]aneN_3$ = 1,4,7-trimethyltriazacyclononane) were first prepared in two steps using semi-automated procedures. The compounds were then screened at 100 °C using MAO activation to reveal the first highly active, high temperature group IV imido-based ethylene polymerization catalysts (Fig. 12.7). This is a nice illustration of the utility of high-yielding coordination chemistry to access diversity within an interesting compound family.

12.7 Conclusions and Future Outlook

Clever chemical approaches, sophisticated parallel reactors, and thoughtful rapid screens represent advances that are highly significant to the polyolefin industry. The reports highlighted here suggest that high-throughput approaches to new olefin polymerization catalysts are now being implemented broadly in academic and industrial laboratories, and with continued development we can anticipate a systematic change in the way in which catalyst research is conducted. Successes from high-throughput programs will inevitably reach the commercial marketplace.

Further developments are certainly required, particularly in the area of polymer properties screening. Polymer properties drive catalyst research, and many conventional techniques designed to probe the physical properties of polymers require difficult sample preparations at scales that currently preclude high-throughput approaches. Given that small-scale polymerizations have been demonstrated to be effective for the discovery and optimization of new catalyst classes, we can anticipate a desire to acquire meaningful polymer property information at smaller and smaller scales. The integration of high-throughput approaches to catalyst discovery and rapid polymer property screening represents a new paradigm for the development of new catalysts and new polymer products.

12.8 References

1 For selected reviews on metallocenes, see: R. F. Jordan, *Adv. Organomet. Chem.* 1991, *32*, 325–387, N. Kashiwa, J.-I. Imuta, *Catal. Surveys Jpn.* 1997, *1*, 125–142; A. L. McKnight, R. M. Waymouth, *Chem. Rev.* 1998, *98*, 2587–2598; H. H. Brintzinger, D. Fischer, R. Mülhaupt, B. Rieger, R. M. Waymouth, *Angew. Chem., Int. Ed.* 1995, *34*, 1143–1170; M. Bochmann, *J. Chem. Soc., Dalton Trans.* 1996, 255–270; H. G. Alt, A. Köppl, *Chem. Rev.* 2000, *100*, 1205–1221.

2 W. Kaminsky, *Adv. Catal.*, 2001, *46*, 89–159.

3 (a) V. C. Gibson, S. K. Spitzmesser, *Chem. Rev.* 2003, *103*, 283–316; (b) G. J. P. Britovsek, V. C. Gibson, D. F. Wass, *Angew. Chem., Int. Ed.* 1999, *38*, 428–447.

4 *Handbook of Combinatorial Chemistry Drugs, Catalysts, Materials*, K. C. Nicolaou, R. Hanko, W. Hartwig (eds.), Wiley-VCH, Weinheim, 2002.

5 G. S. Constable, R. A. Gonzalez-Ruiz, R. M. Kasi, E. B. Coughlin, *Macromolecules* 2002, *35*, 9613–9616.

6 C. A. Nielsen, R. W. Chrisman, R. E. LaPointe, T. E. Miller, Jr., *Anal. Chem.* 2002, *74*, 3112–3117.

7 H. W. Turner, G. C. Dales, L. van Erden, J. A. M. van Beek, US 6306658, 2000; see also, US 6455316; 6582116; 6548116; 6489168; EP 1069942; and CA 2381014.

8 G. C. Dales, J. R. Troth, K. S. Higashihara, G. Diamond, V. Murphy, W. H. Chandler, T. G. Frank, C. J. Freitag, D. Huffman, EP 226867, 2002; see also, US 6485692 and 6566461.

9 R. B. Nielsen, A. Safir, R. Tiede, T. H. McWaid, L. van Erden, WO 01/93998, 2001.

10 P. Jähn, G. Wiessmeier, B. Krumbach, R. Rose, T. Seibert, R. Krautkrämer, EP 1256378, 2002.

11 A. Tuchbreiter, J. Marquardt, B. Kappler, J. Honerkamp, M.O. Kristen, R. Mülhaupt, *Macromol. Rapid. Commun.* 2003, *24*, 47–62.

12 C. Brändli, P. Maiwald, J. Schröer, *Chimia* 2003, *57*, 284–289.

13 A.R. Connolly, J.D. Sutherland, *Angew. Chem., Int. Ed.* 2000, *39*, 4268–4271.

14 V. Murphy, X. Bei, T.R. Boussie, O. Brümmer, G.M. Diamond, C. Goh, K. A. Hall, A.M. LaPointe, M. Leclerc, J.M. Longmire, J.A.W. Shoemaker, H. Turner, W.H. Weinberg, *Chem. Record* 2002, *2*, 278–289; R.C. Willson III, US 6514764, 2003, see also, 6333196, 6063633, and EP 883806.

15 P. Chen, *Angew. Chem., Int. Ed.* 2003, *42*, 2832–2847.

16 G.-J.M. Gruter, A. Graham, B. McKay, F. Gilardoni, *Macromol. Rapid Commun.* 2003, *24*, 73–80.

17 A. Safir, M. Petro, R.B. Nielson, T.S. Lee, J.M.J. Fréchet, US 6,475,391, 2002; see also US 6406632, 6416663 and 6260407.

18 T. R. Boussie, M. Devenney, EP 1160262, 2001.

19 P. Mansky, J. Bennett, US 6438497, 2002.

20 D.A. Hajduk, E. Carlson, R. Srinivasan, US 6484567, 2002.

21 (a) M.B. Kossuth, D.A. Hajduk, C. Freitag, J. Varni, submitted to *Macromol. Rapid Commun.* 2004; (b) D. Hajduk, E. Carlson, J.C. Freitag, O. Kolosov, US 2002/0023507.

22 T.R. Boussie, C. Coutard, H. Turner, V. Murphy, T.S. Powers, *Angew. Chem., Int. Ed.* 1998, *37*, 3272–3275.

23 T.R. Boussie, V. Murphy, K.A. Hall, C. Dales, M. Petro, E. Carlson, H. Turner, T.S. Powers, *Tetrahedron* 1999, *55*, 11699–11710.

24 M. Stork, A. Herrmann, T. Nemnich, M. Klapper, K. Müllen, *Angew. Chem., Int. Ed.* 2000, *39*, 4367–4369.

25 J. Kiesewetter, W. Kaminsky, *Chem. Eur. J.* 2003, *9*, 1750–1758.

26 Z.J.A. Komon, G.M. Diamond, M.K. Leclerc, V. Murphy, M. Okazaki, G.C. Bazan, *J. Am. Chem. Soc.* 2002, *124*, 15280–15285.

27 D.J. Jones, V.C. Gibson, S.M. Green, P.J. Maddox, *Chem. Commun.* 2002, 1038–1039.

28 J. Tian, G.W. Coates, *Angew. Chem., Int. Ed.* 2000, *39*, 3626–3629.

29 T.R. Boussie, G.M. Diamond, C. Goh, K.A. Hall, A.M. LaPointe, M. Leclerc, C. Lund, V. Murphy, J.A.W. Shoemaker, U. Tracht, H. Turner, J. Zhang, T. Uno, R.K. Rosen, J.C. Stevens, *J. Am. Chem. Soc.* 2003, *125*, 4306–4317.

30 G.M. Diamond, C. Goh, M.K. Leclerc, V. Murphy, H.W. Turner, WO 01/98371, 2001; see also US 6508894, 6420179, 6004617, 60030917, 6,248,540;, 6419881, EP 923590, 985678, 978499, and 983983.

31 N. Adams, H.J. Arts, P.D. Bolton, D. Cowell, S.R. Dubberley, N. Friederichs, C.M. Grant, M. Kranenburg, A.J. Sealey, B. Wang, P.J. Wilson, A.R. Cowley, P. Mountford, M. Schröder, *Chem. Commun.*, 2004, Advanced Article, Web Release Date January 14.

Subject Index

High-Throughput Screening in Chemical Catalysis
Edited by A. Hagemeyer, P. Strasser, A. F. Volpe, Jr.

ISBN: 3-527-30814-8